Introduction to PALEONTOLOGY

Stuart Sutherland, Ph.D.

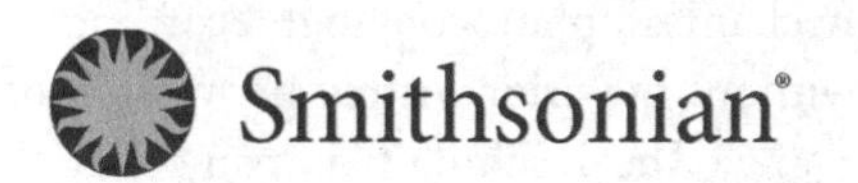

PUBLISHED BY:

THE GREAT COURSES
Corporate Headquarters
4840 Westfields Boulevard, Suite 500
Chantilly, Virginia 20151-2299
Phone: 1-800-832-2412
Fax: 703-378-3819
www.thegreatcourses.com

Printed in the United States of America

Stuart Sutherland, Ph.D.

Professor of Teaching
Department of Earth, Ocean and Atmospheric Sciences
The University of British Columbia

Stuart Sutherland is a Professor of Teaching in the Department of Earth, Ocean and Atmospheric Sciences at the University of British Columbia (UBC). He attended the University of Plymouth in the southwest of England, where he received a degree in Geology in 1987. In 1992, Professor Sutherland was awarded a Ph.D. in Geological Sciences from the University of Leicester for his studies on Silurian microfossils called Chitinozoa. His thesis examined the distribution and taxonomy of these fossils and considered the enigmatic biological affinities of the group and their usefulness in paleoceanographic studies.

After receiving his Ph.D., Professor Sutherland took a temporary teaching position at Brunel University in west London, where he first realized that he had a passion for teaching geology and paleontology. In 1994, he started postdoctoral research at the Natural History Museum in London, working with other paleontologists in an attempt to understand the Devonian organic-walled microfossils of the Cantabrian Mountains of northern Spain. With his earlier teaching experience still in mind, he acquired a teaching degree from Sheffield Hallam University in 1995 while still working for the museum.

In 1998, Professor Sutherland emigrated to Canada and eventually secured a faculty position at UBC's Vancouver campus. His interests at UBC are diverse but in general center on Earth history and paleontology with a particular focus on teaching. At UBC, Professor Sutherland has received the Killam Teaching Prize and the Faculty of Science Teaching Award, as well as the Earth and Ocean Sciences Teaching Award on three separate occasions. He has been mentioned as a "popular professor" among students in two editions of *Maclean's Guide to Canadian Universities*.

Professor Sutherland developed his lifelong fascination with rocks and fossils on family hikes in Derbyshire and the English Lake District. He now enjoys studying geology and paleontology in the beautiful environment of Vancouver and British Columbia.

Professor Sutherland's other Great Course is *A New History of Life*. ■

About Our Partner

Founded in 1846, the Smithsonian Institution is the world's largest museum and research complex, consisting of 19 museums and galleries, the National Zoological Park, and 9 research facilities. The total number of artifacts, works of art, and specimens in the Smithsonian's collections is estimated at 138 million. These collections represent America's rich heritage, art from across the globe, and the immense diversity of the natural and cultural world.

In support of its mission—the increase and diffusion of knowledge—the Smithsonian has embarked on four Grand Challenges that describe its areas of study, collaboration, and exhibition: Unlocking the Mysteries of the Universe, Understanding and Sustaining a Biodiverse Planet, Valuing World Cultures, and Understanding the American Experience. The Smithsonian's partnership with The Great Courses is an engaging opportunity to encourage continuous exploration by learners of all ages across these diverse areas of study.

This course, *Introduction to Paleontology*, offers a glimpse of our planet's extraordinary history through the fascinating science of paleontology. The course focuses on the flora and fauna that are featured in the collections at the Smithsonian's National Museum of Natural History, where scientists in the Department of Paleobiology have conducted cutting-edge research that helps to piece together Earth's ancient story. The foundation of that story is provided by fossils—the vital words on Earth's history pages—and the museum is full of fossil clues that make possible the exploration of the history of life, from Earth's earliest days to more recent times in our planet's history. ■

Table of Contents

SUPPLEMENTAL MATERIAL

Introduction to Paleontology

Scope

Of all the sciences, paleontology is probably one of the most narrative. It combines elements of geology, biology, ecology, and many other disciplines to peer back through time into vanished worlds. The pages of the story of Earth are written in rocks and fossils that require careful collection and interpretation, though.

In this course, we review the tools and techniques paleontologists use to breathe life into fossils and recreate ancient landscapes and oceans. We also discover how paleontological investigation can unpack complex events in Earth's history and how our understanding of these events is continuing to evolve as new fossils and new technologies present themselves.

We give special attention to the Smithsonian Institution, which has a rich history of paleontological research. For example, we meet the fourth Secretary of the Smithsonian, Charles Walcott, who discovered the now-famous Burgess Shale that changed our understanding of life just after the Cambrian explosion. And we learn how the Smithsonian's important research role has continued, with many new paleontological insights coming from work undertaken in the Department of Paleobiology in the Smithsonian's National Museum of National History.

The first part of the course examines some of the fundamentals of the science. As paleontology is a discipline rooted in time, we begin to come to grips with the immense extent of Earth's time by reviewing the deep history of the United States' capital, Washington DC. Following this, our attention turns to the fossils themselves, their diversity, and the variety of ways in which they can form. Finding, extracting, and preparing fossils will be covered, but also some of the techniques and technologies paleontologists have in their tool kit today.

Not all fossils represent the remains of large creatures, and we investigate how the behavior of organisms can be preserved and how a vast store of paleontological information exists in a plethora of tiny microfossils. Once found, extracted, and preserved, fossils have to be classified, and we consider some of the challenges paleontologists face when bringing taxonomic order to their finds. Fossils tell of the passage of time across eons, but we also discover how more familiar cycles of days, months, and years might be recorded in the fossilized remains of organisms.

Fossils can also help us unravel the geological dance of continents, and we investigate how paleontology is instrumental in the study of paleogeography. In addition, we also examine the evolution of the beautiful mineral heritage of Earth, a turn to geology that highlights the interplay between Earth's evolving mineral heritage and the development of the biosphere.

The rest of the course focuses on some important fossil groups and events in Earth's history, starting with the birth of paleontology at the dawn of life, potentially in the deep, dark ocean. The bewildering diversity of the biosphere today is generally traced to an explosion of life early in the Cambrian period, but we consider the potential roots of this explosion in even earlier times. The most successful group of organisms during and following the Cambrian explosion was arguably the arthropods, so we look to their origins and their fascinating evolution. Life has not had an untroubled journey, though, with 5 major extinctions recorded through time. We investigate the Devonian extinction, which occurred around 360 million years ago and may have had a series of potential triggers, and also the greatest culling event that the planet has witnessed, which occurred at the end of the Permian about 252 million years ago.

We also examine fossils that are still highly debated in the scientific community, such as the enigmatic but wonderfully bizarre *Spinosaurus*, claimed to be one of the largest carnivorous dinosaurs that ever lived, from around 30 million years before the iconic *Tyrannosaurus rex*. Next, we turn to the fossil record for a series of mammals that would progressively throw away their limbs for the sea to become the beautiful and diverse whales, dolphins, and porpoises that thrill us in today's oceans. We also consider the critical role that flowering plants (including the grasses) have had in the biosphere, from their evolution during the Mesozoic era to the profound influence that they have had in driving the evolution of other creatures.

As we move toward the end of the course, the wonderful Komodo dragons from Indonesia—and the tiny humans, *Homo floresiensis*, that lived with them on the island of Flores—take us to times much closer to our own. Another member of our family, *Homo neanderthalensis*, is covered, as are some of the mammoths and mastodons that wandered the Earth at that time. We will discover what fantastic new insights ancient DNA is providing in our understanding of both mammoths and Neanderthals. We conclude with a survey of the possible challenges the biosphere will face as it marches into the future and the role that the science of paleontology may have in charting that future.

Paleontology is a powerful tool we can use to wander through Earth's ancient past. With every fossil found and new technique developed, our picture of that past comes a little more into focus and the journey becomes ever more fascinating. By the end of this course, you should have a clearer understanding of the science and practice of paleontology—and how it can bring our planet's rich history back to life. ■

Lecture 1 History on a Geological Scale

All fossil creatures, and the vanished worlds they lived in, help us understand our place in space and time. They can also act as a vital benchmark for our appreciation of the Earth as it is today and perhaps provide clues to its future. The National Museum of Natural History is full of fossil clues to Earth's past, and aided by the collections—and some of the cutting-edge research in the Department of Paleobiology—this course will explore the history of life from Earth's earliest days to more recent times in our planet's history.

A Walk through Geological Time

- The Earth is 4.54 billion years old, a fantastically long time when compared to the lifetime of a human. As such, in an attempt to comprehend the age of the Earth, an analogy is often used. A common tool is to condense all of Earth history into one calendar year.

- On this scale, the Earth forms in the first second of January 1. This is a time when our solar system was a crowded place with a variety of rocky planets zipping around close to the Sun, and there would likely have been collisions on a colossal scale. Some of these encounters would add mass to growing young worlds while others probably obliterated each other in cataclysmic events.

- We start to find the first abundant fossils, those that possessed shells, on the calendar around November 18, fairly late in the year, and animals with 4 legs, or tetrapods, don't stride onto land until around December 1. The dinosaurs went into extinction on December 26, and Stonehenge was built just 30 seconds before midnight on December 31.

- However, given that the National Museum of Natural History is located along the front yard of the United States, better known as the National Mall, let's use that as our timeline. On that timeline, the origin of Earth, 4.54 billion years ago, can be placed at the Washington Monument, with today represented by the United States Capitol building.

- The distance between these 2 iconic Washington DC buildings can cover 4.54 billion years in just 2.87 kilometers, or 1.35 miles. On this scale, depending on your stride length, each step you take will be between 1 million and 2 million years.

- Let's start outside of our timeline—before the formation of Earth. The point that is just 29 meters in front of the Washington Monument makes it around 4.6 billion years ago, or 60 million years before Earth's first day. If you could transport yourself back in time, you would be in open space. There would be no Earth—just a nebula of dust and gas.

- We have places at a similar stage in their evolution in our galaxy today, such as the Orion Nebula, in which gas and dust are collapsing under gravity to form new stars and planets. Our solar system grew in the same way, perhaps initiated by the gravitational nudge from the death of an old star when it went supernova.

- Let's move forward in time to the first day of Earth and the Washington Monument. On our timeline, this is day 1, with the formation of a rocky planetary body that will evolve over the next 4.54 billion years into the Earth we know today. The Earth was heated after it formed, a combination of kinetic energy released from the impacts of the remaining debris in the solar system and from the concentration of radioactive elements in the young planet's interior.

- Washington DC at this time would have been a magma ocean, just like the rest of the planet, and it would take time for the Earth to cool and for its first solid skin, the crust, to form.

- If we go about 19 meters from the Washington Monument, we come to a very significant event in our story: the formation of Earth's Moon, probably the result of a cataclysmic collision with a Mars-sized object, sometimes called Theia, about 4.5 billion years ago. This would have serious consequences for the development of life on our planet, including tides, various lunar cycles, and day length. The Moon has also helped stabilize the Earth's "wobble," allowing for the relatively benign seasons we enjoy.

- Some of the oldest evidence of Earth's solid crust comes from Australia in the form of fragments of an older rock contained in a younger rock. This is called a conglomerate. Isotopic analysis of those fragments, contained in that conglomerate and specifically from crystals called zircons, indicate that Earth had a solid crust at about 4.4 billion years. On our timeline, that places us just 67 meters away from the Washington Monument.

- It would appear that the Earth had a "surface" of sorts very early in its history. In addition, isotopic analysis of those zircons, using different ratios of various stable elemental isotopes, hints at liquid water, too. The presence of liquid water in these distant times opens up the possibility that life may have a much older history than we initially thought.

- At 356 meters from the Washington Monument, on the north side of the mall is the National Museum of American History and on the south side is the United States Department of Agriculture. On our timeline, we are at 3.8 billion years ago—we have completed a little more than 16% of our walk—and at this point in history, Earth, and the rest of the inner solar system, was in a meteor and comet shooting gallery. This would last about 300 million years and is called the late heavy bombardment period.

- If life had evolved around 4 billion years ago, it probably had to survive deep in Earth's crust, because some have suggested that these early impact events may have in effect sterilized the Earth's surface. It is not until around 400 million years later, at 3.4 billion years, that we find our first fossils. On our timeline, we are at the eastern edge of the National Museum of American History.

- By the time we walk past the National Museum of Natural History, at 2.4 billion years, life would be enduring another crisis: a super glaciation called a snowball Earth event that would encase our planet in ice for millions of years. On our timeline, we have covered more than 47% of Earth's history.

- Associated with the end of the snowball event would be a rise in oxygen levels and the deposition of rocks in banded-iron formations, rich in iron oxides, demonstrating that our atmosphere was evolving. This change was caused by photosynthetic bacteria releasing oxygen as a waste product and, in the process, changing our planet forever.

- Following this rise in oxygen, life on Earth would go through a series of significant events. About 2.1 billion years ago, or 1011 meters from the Capitol building, we see the emergence of eukaryotic life. Eukaryotes are essentially all life that is not a bacteria or a virus.

- By 1.2 billion years ago, or 578 meters from the Capitol building, we have evidence of the first multicellular life-form—*Bangiomorpha*, probably a simple red algae—and by 720 million years ago, or 346 meters from the Capitol, the snowballs had returned and then end 650 million years ago, in the middle of the Capitol reflecting pool. We have now completed more than 85% of our walk through time.

- Larger creatures emerge at 541 million years—in front of the Ulysses S. Grant Memorial—and with them, evolution kicks in to overdrive. We can see the world just after that explosion of life at 505 million years. This is the time of deposition of the Burgess Shale, a rock unit from western Canada that was discovered in 1909 by Charles Walcott, former Secretary of the Smithsonian. The fossils preserved in the Burgess Shale provide a unique window into the explosion of complex.

- DC would have looked very different during this explosion of life. In Rock Creek Park, about 8.5 kilometers north of the National Mall, we find sediments deposited at about the same time. They tell us that DC was then on the edge of a deep ocean called Iapetus next to the continent of Laurentia.

- These rocks are called turbidites and formed as sediments tumbled over the continental edge and into deeper water. These rocks often show bands representing individual flows of sediment. Coarser or heavier components would settle out of the flow, first producing a sedimentary structure that is called graded bedding, representing the settling of material out of that sediment avalanche.

- But now things are going to change quickly as DC witnesses a series of tectonic pileups. The first one occurs by First Street Southwest at 460 million years ago on our timeline, with just about 10% of our timeline remaining. A series of volcanic islands that existed in that ocean, into which the Rock Creek Park sediments were deposited, would collide with North America. This raised the ocean floor and started the building of mountains in a north-south direction on the eastern continental edge of this ancestral North American "paleocontinent."

- But this was just the beginning. Around 100 million years later, microcontinents that include parts of what is now western Europe would slam into this part of the world, raising the mountains even higher and causing magmas to be intruded into the deformed rocks.

- Then, at 320 million years ago, Africa collides with this growing continental landmass. This is moving us toward the formation of a supercontinent called Pangaea and in the process raises the mountains even more. The remnants of those mountains are the Appalachians, which were as high as the Alps or the Himalayas when they were young.

- After this collision, and just 32.8 meters closer to the Capitol building, life in DC and around the world goes into crisis at 252 million years ago. Life on Earth would be laid to waste, with more than 90% of all species going into extinction—the greatest of the 5 major mass extinctions our planet has faced.

- This is a time of runaway global warming, probably triggered by titanic volcanic activity centered in what is today Siberia—global warming that may have also led to the production of toxic hydrogen sulphide in the Earth's oceans and the release of even more greenhouse gases as methane stored in ocean sediments destabilized and escaped into the atmosphere.

- It is the erosion of the mountains that formed due to the continental collisions that give us the next rocks we find in DC, and on our timeline, we have reached the foot of the Capitol building. This is when dinosaurs during the Cretaceous period, dating to 110 million years, fit into our timeline.

- From this point, the dinosaurs would have another 44.5 million years to rule the planet, but then on the steps of the Capitol building on our timeline, at about 66 million years ago, around 2300 kilometers to the southwest of DC, a 10-kilometer object comes screaming into the atmosphere and slams into Yucatán, ending the reign of those magnificent beasts.

- Paleoanthropologists estimate that our species, *Homo sapiens*, evolved around 200,000 years ago. On our timeline, that places us just under 5 centimeters from the front of the Capitol building. The maximum advance of ice, in the last glacial period of the current ice age, was about 22,000 years ago, which is just about 1 centimeter from the end of our timeline.

- The date of arrival, and origins, of the first people in North America is currently somewhat in flux, but an early North American culture known for their stone tools, called the Clovis culture, is generally agreed to be found

from about 13,000 years ago, just more than half of a centimeter on our scale. About 0.1 centimeters later, the last glacial period in the current ice age ends.

- Around 0.3 centimeters before the end of our timeline, at 6000 years ago, many miles away from DC, we have evidence of the founding of one of Earth's first cities, Uruk, in what is now modern-day Iraq. In this last 0.3 centimeters of our timeline is effectively all of what we could call recorded human history. Everything that we consider ancient on a human timescale is dwarfed by the immensity of the age of the Earth.

Questions to consider:

1. What should we consider to be Earth's day 1?
2. Why is there no complete record of Earth's history on our planet?

Suggested Reading:

Fortey, *Earth*.

Levin and King Jr, *The Earth Through Time*.

PALEOMAP Project, http://www.scotese.com.

Lecture 1 Transcript

History on a Geological Scale

How well do you know your local history? I mean, really know it? How far can you go? For example, consider this magnificent building, the National Museum of Natural History in Washington, DC. The current building was opened for business in 1910, but the Smithsonian Institution of which it is a part was founded in 1846 from a legacy provided by Englishman James Smithson, who incidentally had never visited the USA nor had any relatives or contacts here. His reasons for leaving all this money to the people of the USA is, well, kind of hazy.

The name for the city in which the Smithsonian is based was signed into law by its namesake President George Washington. You can consider DC's origins, therefore, to go back to 1791, when Washington commissioned Monsieur Pierre L'Enfant to draw up plans for a new city on the northeast shore of the Potomac River. But we can take the history of DC back even further. We know European explorers were here by about the early 17th century, but by then the area was already inhabited by a number of Native American peoples, including the Nacotchtank, who lived in a village of the same name along the Anacostia River. Archeological evidence indicates that Native Americans had settled this area by at least 4000 years before the present era.

This ancient history, though, is barely scratching the surface of the past of the area we now call Washington, D.C There are creatures far more fierce and scary than the political animals of today lurking in this city's past. For example, there are certain sedimentary rocks around Washington DC—sandstones, conglomerates, and mudstones—that have a story to tell, a story that takes us back 110 million years; a story that involves a rather scary-looking tooth.

These are the rocks that the fossils are found in, and as a geologist and paleontologist, I've been trained to kind of reverse engineer sedimentary rock sequences like this. We take clues from the type of sediments we find; the structures they preserve, like ripple marks produced by an ancient river, or cracks left in mud when a pond dries out in the sun; and, of course, any fossils those sediments contain. And from them, we can recreate lost environments and landscapes. The picture we get of DC back then, 110 million years ago, is of a swampy area a bit like you see in this image of the Mississippi river delta today. If you had hiked though the ancient landscape of DC, you'd have had to work your way across numerous meandering streams and rivers. There would have been ponds accumulating clays and silts, plant debris, and also fossils like that tooth I mentioned.

We also know that DC was close to the edge of a new ocean that had split the supercontinent of Pangaea from north to south. This was the young Atlantic Ocean, much narrower than it is today, of course, but widening with every year that passed. And if you were able to walk along the National Mall at 110 million years ago, this is the landscape and animals you would have seen. The forests would have been dominated by conifers and particular forms that are related to modern cypress trees. There was also a relative of the modern sequoia. We find fossils not only of its foliage but some of its cones, too.

Although we have no body fossils yet, probably due to the delicate nature of their bodies, there were also likely frogs, early birds, insects, primitive mammals, and, soaring overhead, majestic pterosaurs. And, of course, there were dinosaurs, some of them small and agile like little dromaeosaurus, a bipedal carnivore about 1.5 meters long and relative to velociraptor, one of those dinosaurs that were made famous in the *Jurassic Park* movie. And some giants too: the bones of pleurocoelus, a large sauropod plant-eating dinosaur, have been found with adults probably reaching up to around 9.1 meters high and around 15.2–18.2 meters long.

We also find the remains of a yet unidentified terror of this ecosystem. Remember that tooth? The tooth and associated fragments of bone found in the sediments around DC belong to a large predatory dinosaur, probably similar to acrocanthosaurus. As of yet we don't have enough material to make a definite classification but there was certainly something big and scary lurking near this swamp during the Cretaceous of DC.

Now, we humans are a relatively new species, so this world of dinosaurs 110 million years ago feels unimaginably ancient. Paleoanthropologists estimate our species, *Homo sapiens*, to be about 200,000 years old. As such, we have a very limited perspective on what a changeable place our home planet is. This is the power and the wonder of paleontology and its subdiscipline, paleobiology. It provides us with an ability to paint pictures of ancient worlds, but much more than that, fossils also help us address important questions, questions like how and when did large portions of the biosphere collapse, and how did some species manage to adapt to extremes in climate or cataclysmic natural disasters? Fossils provide insights into the pace of evolution over time and insights into how pervasive life can be. But they also serve as a warning that nothing lasts forever, and that entire groups of creatures can disappear overnight in the blink of a geological eye.

Fossils are vital words on the pages of Earth's history, and through this series I will address some of those questions, questions that the Department of Paleobiology at the National Museum of Natural History has been addressing for over 50 years. The museum is full of fossil clues to Earth's past, and aided by some of the collections and some of the cutting edge research in the Department of Paleobiology, we can explore the history of life from Earth's earliest days through to more recent times in our planet's history.

Paleontology is a broad church composed of various subdisciplines that draws on many other areas of science including biochemistry, mathematics, engineering, and many others, but something I should note here. I've been referring to the Department of Paleobiology at the National Museum of Natural History, and not the Department of Paleontology. Now, the difference might sound subtle but it's an important one, as it underlines the fact that, although research here is obviously paleontological in nature, the scientists within the department are engaged in much more than an exercise of classifying and cataloging fossils.

But don't get me wrong. As we shall see, the curation of fossil collections is a vital underpinning element of any museum or research department. The research carried out in the National Museum of Natural History, though, is quantitative and process-oriented, an approach that is essential if we are to fully understand ancient creatures and their ecology and make valid and reasoned comparisons to the biosphere today. In many of the lectures in this series we will examine some of the exciting research that is currently being undertaken at the museum, research that is being used to open up windows into lost histories and lost ecosystems buried all around us. Just now, though, I'd be grateful if you'd allow me to take you on just a little diversion.

When my mom—or "me mum," as you say if you come from the north of England—was born in Stockport just south of Manchester in 1927, the National Museum of Natural History had been open for just 17 years. It was brand new. This is a picture I have of my mom back in 1935. She'd been selected to be the May Queen that year and been told that May Queens don't grin, which explains the practiced scowl in this photograph.

And this is my grandmother Jane Murphy taking tea with a friend in 1911. Her mother, my great-grandmother, was Sarah Hannah Eyre, and still further back you can trace my mom's family all the way back to the 14th century of Derbyshire, and perhaps even earlier, right back to the Norman invasion of Britain back in 1066.

I know a lot of this from the stories my mom would tell us when we visited Derbyshire, and as a child, the length of time she was describing sounded absolutely incredible—almost 1000 years of history. And why am I taking you on this brief tour of my family history in a series about fossils and ancient life? Well, as human beings, we filter time through the lens of a human lifetime, which at present, depending on where you live, ranges from about 50 to 100 years. This is why, when considering the vast depths of Earth's time, just like our description of an ancient Washington, DC, the enormous scale of geological time can be difficult to grasp. It is a vital concept for us to come to terms with, though, if we are to gain a proper understanding of the history of life on our planet. So, I'd like to place our human perception of the passage of time into the context of Earth's history. For that, we need to understand the concept of deep geological time. Now, this is not be confused with depth in a physical miles or kilometers sense, but in a temporal sense of years and eons.

The Earth is 4.54 billion years old, a fantastically long time when compared to the lifetime of a human. That is 4 billion 540 million years, or, written out as numbers, 4,540,000,000 years. For me, that is just a big number with a lot of zeros. As such, in an attempt to wrap my head around the age of the Earth, I often fall back on analogy.

A common tool is to condense all of Earth's history into 1 calendar year. On this scale, the Earth forms in the first second of January 1. This is a time when our solar system was a crowded place full of a variety of rocky planets zipping around close to the Sun and, just like a busy freeway at rush hour, there would have likely have been collisions, but on a colossal scale. Some of those encounters would add mass to growing young worlds, while others probably obliterated each other in cataclysmic events the stuff of disaster movies.

But let's move forward on our calendar to when we start to find the first abundant fossils, those that possess shells. On our calendar, that puts us around November 18, fairly late in the year. And animals with 4 legs—tetrapods—don't stride onto land until about December 1. Continue with this scale, the dinosaurs just went into extinction on Boxing Day, December 26, and Stonehenge was built just 30 seconds before midnight on December 31.

However, given that the National Museum of Natural History is located along the front yard of the United States—better known as the National Mall—I thought we might use that as our timeline. On that timeline, I'm going to place the origin of Earth at 4.54 billion years ago at the Washington Monument, with today represented by the Capitol Building. Join me as I take a walk between these two iconic DC buildings covering 4.54 billion years in just 2.87 kilometers, or 1.35 miles. On this scale, depending on your stride length, each step you take will be between 1 million to 2 million years.

But let's start outside of our timeline, before the formation of Earth. This point is just 29 meters in front of the Washington Monument, making it around 4. 6 billion years ago—60 million years before Earth's day 1. If you could transport yourself back in time, you would be in open space: no Earth, just a nebula of dust and gas. We have places at a similar stage in their evolution in our own galaxy today. This is the Orion Nebula, and it's in a place like this that our solar system was born. In the Orion Nebula, gas and dust are collapsing under gravity to form new stars and planets. Our solar system would grow in the same way, perhaps initiated by the gravitational nudge from the death of an old star when it went supernova.

But let's move forward in time to the first day of planet Earth and the Washington Monument. For us, on our timescale, this is day 1, with the formation of a rocky planetary body that will evolve over the next 4.54 billion years into the Earth we know today. The Earth was heated after it formed, a combination of kinetic energy released from impacts of the remaining debris in the solar system, and from the concentration of radioactive elements in the young planet's interior. Washington, DC at this time would have been a magma ocean, just like the rest of the planet, and it would take time for the Earth to cool and for its first solid skin, the crust, to form. But let's keep on moving.

If we go just a little over 19 meters from the Washington Monument, we come to a very significant event in our story, the formation of Earth's moon, probably the result of a cataclysmic collision with a Mars-sized object—sometimes called Theia—about 4.5 billion years ago. This would have serious consequences for the development of life on our planet, including tides, various lunar cycles, and day length. The Moon has also helped stabilize the Earth's wobble, allowing for the relatively benign seasons we enjoy today.

Some of the oldest evidence of Earth's solid crust comes from Australia in the form of fragments of an older rock contained in a younger rock. It's a rock that we call a conglomerate. Isotopic analysis of those fragments contained in that conglomerate, and specifically from crystals called zircons, indicates that the Earth had a solid crust at about 4.4 billion years. On our timeline, that places us just 67 meters away from the Washington Monument.

So it appears that the Earth has a surface of sorts very early on in its history. In addition, isotopic analysis of those zircons using different ratios of various stable elemental isotopes hints at liquid water, too. The presence of liquid water in these distant times opens up the possibility that life might have a much older history than we initially thought. But we have to move on.

One of the oldest complete rocks we have on Earth is the Acasta gneiss from the Northwest Territories of Canada, dated to around 4 billion years. That would be way up here on 14th Street on our timeline, but I would like to jump forward to a troubled time in the early Earth's history at 356 meters from the Washington Monument. On the north side of the National Mall is the National Museum of American History, and on the south the United States Department of Agriculture. On our timeline, we are at 3.8 billion years ago. We have completed just a little over 16% of our walk, and at this point in history Earth was getting a hammering.

This would last about 300 million years and is called the Late Heavy Bombardment period. This is the time that Earth and the rest of the inner solar system was in a meteor and comet shooting gallery. If life had evolved about 4 billion years ago, as I hinted at earlier, it probably would've had to survive deep in Earth's crust, as some have suggested that these early impact events may have in effect sterilized the Earth's surface.

It is not till around 400 million years later, at 3.4 billion years ago, that we find our first fossils. On our timeline, we are at the eastern edge of the National Museum of American History. These precious fossil bacteria come from a fossilized beach in Australia. They are the first fossil evidence we have that our planet was alive, heralds of all the spectacular life that is to follow.

By the time we walk past the National Museum of Natural History at 2.4 billion years, life would be enduring another crisis: a super glaciation called a snowball Earth event that would encase our planet in ice for millions of years. On our timeline, we have covered over 47% of our Earth's history. Associated with the end of the snowball event would be a rise in oxygen levels and the deposition of these beautiful rocks, banded iron formations rich in iron oxides—effectively, rust—demonstrating that our atmosphere was evolving. This change was caused by photosynthetic bacteria releasing oxygen as a waste product, and, in the process, changing our planet forever.

Following this rise in oxygen, life on Earth would go through a series of significant events. About 2.1 billion years ago, 1011 meters from the Capitol Building, we see the emergence of eukaryotic life. Eukaryotes are essentially all life that is not a bacteria or a virus. This is also a stage of life represented by the Gunflint chert, a rock around 1.88 billion years old found along the shores of Lake Superior. And on our timeline, it would be only 867 meters from the Capitol Building.

By 1.2 billion years ago, 578 meters from the Capitol Building, we have evidence of the first multicellular life form *Bangiomorpha*, probably a simple red algae. And by 720 million years ago, just 346 meters from the Capitol, the snowballs had returned, and then end here at 650 million years ago right in the middle of the Capitol Reflecting Pool. We have now completed over 85% of our walk through time.

It would be true to say that, through most of our walk through time so far, life has not been exactly, well, inspiring. In fact, for the past 1926 meters—around 88% of the National Mall walk—life has been mostly microbial slime. But following that last snowball, things start to get interesting. Larger creatures emerge at 541 million years, just in front of the Ulysses S. Grant Memorial, and with them evolution kicks into overdrive. We can see that world, just after that explosion of life, here at 505 million years. This is the time of deposition of the Burgess Shale, a rock unit from western Canada that was discovered in 1909 by Charles Walcott, former Secretary of the Smithsonian Institution.

The fossils preserved in the Burgess Shale provide a unique window into the explosion of complex life, like this trilobite Olenoides, but also a unique set of weird and wonderful creatures that evolved rapidly into the Cambrian oceans. DC would have looked very different during this explosion of life. In Rock Creek Park, about 8.5 kilometers north of the National Mall, we find sediments deposited at about this time. They tell us that DC was then on the edge of a deep ocean called Iapetus next to the continent of Laurentia.

But now, hold onto your hats, as things are going to change fast as DC witnesses a series of tectonic pileups. The first one occurs here, by 1st Street Southwest, at 460 million years ago on our timeline, with just about 10% of our timeline remaining. A series of volcanic islands that existed in that ocean into which the Rock Creek Park sediments were deposited would collide with North America. This raised up the ocean floor and started the building of mountains in a north-south direction on the eastern continental edge of this ancestral North American paleocontinent.

But this was just the beginning. Around 100 million years later microcontinents, that include parts of what is now western Europe, would slam into this part of the world, raising the mountains even higher and causing magmas to be intruded into the deformed rocks. And then at 320 million years ago—bam!—Africa collides with this growing continental landmass.

This is moving us toward the formation of a supercontinent called Pangaea, and in the process raises the mountains even further. The remnants of those mountains are the Appalachians, as high as the Alps or the Himalayas when they were young. After this collision, and just 32.8 meters closer to the Capitol Building in DC, all round the world life goes into crisis at 252 million years ago. Life on Earth would be laid to waste with over 90% of species going into extinction, the greatest of the Big Five mass extinctions our planet has faced. This is a time of runaway global warming probably triggered by titanic volcanic activity centered in what is today Siberia, global warming that may have also led to the production of toxic hydrogen sulfide in the Earth's oceans and the release of even more greenhouse gases as methane stored in the ocean sediments destabilized and escaped to the atmosphere.

It is the erosion of the mountains that formed due to the continental collisions we described back there that give us the next rocks we find in DC, and on our timeline we have reached the foot of the Capitol Building. This is where those dinosaurs we mentioned earlier, dating to 110 million years ago during the Cretaceous period, fit into our timeline. From this point, the dinosaurs would have another 44.5 million years to rule the planet. But then, here on our timeline on the steps of the Capitol Building about 66 million years ago around 2300 kilometers to the southwest of DC, a 10-kilometer object comes screaming into the atmosphere and slams into the Yucatán, ending the reign of those magnificent beasts.

DC would witness an impact from space much closer to home 35 million years ago. An object smashed into North America just 200 kilometers to the southeast of DC in what is now Chesapeake Bay, and this with just 0.7% of our timeline remaining. Although not as large an impact as the dinosaur-exterminating event 31 million years earlier, it would still have been a spectacular and disturbing sight. The 40-kilometer crater formed a depression that diverted the course of rivers and still causes sinking of the land in this area as the material in the crater continues to settle.

As you can see, we are now really starting to run out of space on our timeline, so I'm going to employ a vital piece of scientific instrumentation to help us—this ruler. We think our species *Homo sapiens* evolved around 200,000 years ago. On our timeline, that places us just over this circular fountain and just under 10 centimeters from the front of the Capitol Building.

The maximum advance of ice in the last glacial period of the current ice age was about 22,000 years ago, and that's just about 1 centimeter from the end of our timescale. Although ice did not reach DC, it had a dramatic effect on the Washington area by lowering sea levels and the production of outwash plains. It diverted rivers and streams and had major effects on vegetation and climate.

The date of arrival and origins of the first people in North America is somewhat in flux at the moment, but an early North American culture known for their stone tools called the Clovis culture is generally agreed to be found at around 13,000 years ago, just over half a centimeter on our scale. Point-one centimeters later, the last glacial period in the current ice age ends, and around 0.3 centimeters before the end of our timeline—at 6000 years ago, many miles away from DC—we have evidence of the founding of one of Earth's first cities, Uruk, in what is now modern-day Iraq. The transition from B.C. to A.D. at 2016 years ago is 0.097 centimeters and just too small to be shown, even on our ruler scale.

So here, in this last 0.3 centimeters of our ruler, is effectively all of what you could call recorded human history. Everything that we consider ancient on a human timescale is dwarfed by the immensity of the age of the Earth. This reminds me of something that Carl Sagan, a childhood hero of mine, did back in 1990. Carl was one of the team of scientists involved with the Voyager spacecraft missions exploring the outer solar system. In 1990, when Voyager 1 was heading out of the solar system, he had the spacecraft turn to face the Earth and take this picture of the pale blue dot.

He asked us to picture this dot and appreciate that everyone who ever lived, lived there, on a mote of dust suspended in a sunbeam. In the same way, we can look at our timeline and consider that, as Carl put it, "every superstar, every supreme leader, every saint and sinner in the history of our species" lived there in the final few millimeters at the end of the National Mall.

So take a look again at the specimens we looked at earlier. The Acasta gneiss, some of the earliest Earth-born material that exists on the planet today, over 4 billion years old and 1927 meters away from the end of our timeline—our today. The trilobite Olenoides, a creature that lived in the echoes of life's big explosion, 243 meters away just in front of the Ulysses S. Grant Memorial. And a tooth from a DC long since vanished, barely the distance of an Olympic swimming pool away from the Capitol Building. And, once again, all our own history squeezed into the end of our 12-inch ruler.

All fossil creatures and the vanished worlds they lived in help us understand our place, not just in space but also in time. They can also act as a vital benchmark for our appreciation of the Earth as it is today, its health, diversity, and perhaps provide clues to its future in a world that is always on the move and ever-changing no matter what our perception of time may tell us. To put it another way, fossils help us reassess the phrase "as unchanging as the mountains." To a paleontologist, mountains are here today, gone tomorrow.

Of all the wonderful museums at the Smithsonian, it is the National Museum of Natural History in particular that speaks to this deep geological time. In this series, we will visit some of the players in Earth's story that are featured in the collections and in the research at the museum. We will examine how the scientists in the Department of Paleobiology have helped pieced together Earth's ancient story. It's going to be an interesting walk.

But before we start on our hike through the ancient worlds represented by Earth's deep history, we should consider the raw materials of the science of paleontology—the fossils themselves. In our next lecture, let's get up close and personal with fossils.

Lecture 2

Life Cast in Ancient Stone

In this lecture, you will learn about paleontology, including how paleontology developed as a science, what the chances are of becoming a fossil, what the common modes of fossilization are, and what exceptional preservation is. Although only a tiny portion of life on Earth has become fossilized, that portion still represents an enormous cache of material for future paleontologists to examine. New discoveries of fossil bonanzas and new approaches and techniques in paleontology and paleobiology will likely continue to surprise and delight generations of scientists to come.

The Rise of Paleontology

- Fossils have been a part of human culture for a long time. A very early reference to fossils comes from the 6th century B.C. Greek philosopher Xenophanes, who concluded from his examination of fossil fish and shells that water must have covered much of the Earth's surface at one time in the past, meaning that he understood that there was a deeper historical narrative to our planet that could be told by the use of fossils.

- Similar ideas were proposed in 1088 by the Chinese naturalist Shen Kuo, who found fossils in the Taihang Mountains and decided that they indicated that shorelines had shifted over time. He also found bamboo fossils in Shaanxi province, a part of China that is currently too dry for bamboo to grow, and concluded that climate change must have occurred at some time in the past.

- Thoughts on how creatures could become fossils were proposed in 1027 by the Persian polymath Avicenna, who speculated that fossils may have formed when a carcass was bathed in "petrifying fluids."

- These are all really modern concepts—concepts of sea level and climate change, the passage of vast amounts of time, and the processes that operate in the Earth to form fossils—and in the West, they would be largely ignored or forgotten.

- Things would change, though, as our understanding and appreciation of fossils accelerates when we move into the 17th century, the age of reason. There are many important figures who have contributed to the development of the science of paleontology in this period.

- Among them is Robert Hooke, an English natural philosopher who would make a significant contribution in his famous book *Micrographia*, published in 1665. The book was the first examination of the very small, as revealed by Hooke's microscope. From his study of fossil wood, Hooke would conclude that fossils were once-living organisms that have been transformed into rock by the petrifying action of mineral-rich water.

- Another important figure is George Cuvier, a French naturalist and philosopher who would make several significant contributions, including providing us with the concept of extinction in 1796. When comparing the jaws of living elephants with fossil jaws of something similar but distinct—we know it as the mammoth—Cuvier concluded that some species that used to exist on Earth were no longer around.

- It was one of Cuvier's students, Henri de Blainville, who would give the study of fossils a name. At first, he chose the term "paleozoologie" in 1817, but by 1822, after a number of iterations, he settled on the more inclusive "palaeontologie," which would cover both fossil plants and animals.

- William Smith would demonstrate the usefulness of fossils as a tool for correlating strata. By 1815, he published a groundbreaking geological map of England, Wales, and southern Scotland. Importantly, Smith demonstrated that fossils provided the context for the development of a geological timescale. Once constructed, this scale would allow scientists to correlate across regions, ensuring that they were on the same page of Earth's history and, in doing so, start to tell the story of Earth's deep evolution.

- This is an idea that would be taken up by Smith's brother-in-law, John Phillips, who in 1841 published the first geological timescale. He divided geological time into 3 of the eras we use today: the Paleozoic, Mesozoic, and Cenozoic. Although the names and dates on the timescale would change considerably over time, paleontologists and geologists now had a yardstick that they could use to delve into the past.

- Another important development was the discovery of fossils in a quarry in the Neander Valley of Germany. Described by schoolmaster Johann Carl Fuhlrott and anatomist Hermann Schaaffhausen in 1856, the fossils were identified as belonging to a group of humans that were quite different from any modern people. They would eventually be named *Homo neanderthalensis*. With these fossils, paleontology became part of our story, too. This was just 5 years after anatomist and founder of the Natural History Museum in London Richard Owen gave us the word "dinosaur."

- Charles Darwin, with the publication of *On the Origin of Species* in 1859, would eventually provide the first hints of the mechanism behind the changing suites of fossils paleontologists had been finding. From here, the field of paleontology would explode into numerous disciplines and subdisciplines, becoming an extremely important part of academic studies at universities and museums all over the world.

- North America has its share of famous paleontologists, too. For example, Othniel Marsh and Edward Cope would expand our understanding of dinosaurs during the "great dinosaur rush" in Colorado, Nebraska, and Wyoming in the late 1800s. In 1909, one of the most famous secretaries of the Smithsonian, Charles Walcott, would stumble across the Burgess Shale in British Columbia in the Canadian Rockies.

Becoming a Fossil

- The chances of anything becoming a fossil are pretty slim. The fact we have fossils at all speaks to the sheer numbers of individuals and species that have existed through time. With all their countless billions, it would

Charles Darwin

only take a tiny fraction to fossilize to leave a substantial fossil record in the rocks.

- But let's consider those that do make it. What factors did they have in their favor? How do you maximize your chance of becoming a fossil? First, being in the right place increases your chance of becoming a fossil. To form a fossil, you need to get your body buried as quickly as possible—out of the way of scavengers and preferably sealed from oxygen, or in reduced oxygen conditions. That isn't going to happen on an open plain or in a high mountainous region, for example.

- Because being buried in sediments is probably your best bet for becoming a fossil, organisms that live in aquatic environments, such as lakes or rivers, will have a greater chance of becoming a fossil. On the whole, aquatic creatures that live in the oceans and other water bodies that are receiving vast quantities of sediment via rivers and streams will have a greater preservation potential than terrestrial, land-based organisms.

- In addition to where an organism lives, another important factor is what it is made of. Any creature that has a significant development of hard parts has a greater chance of preservation and a greater representation in the fossil record than soft-bodied organisms.

- This is a persistent bias that paleontologists have to be aware of when reconstructing ancient ecosystems from fossil sites, especially when, in some settings, soft-bodied creatures lacking any skeletal components may have made up a large part of the animal assemblage.

- But considering hard parts, life has used a wide variety of materials for protection and structural support. Calcium carbonate, or calcite, is a very common mineral used by many organisms, including bryozoans, corals, brachiopods, mollusks, and many arthropods and echinoderms.

- Examples of silica-secreting organisms include sponges and the radiolarians, tiny marine protists that secrete exquisite ornament-like structures out of biological glass.

- Calcium phosphate, usually in the form of the mineral apatite, is used for the skeletal elements (bones and teeth) of vertebrates and the feeding apparatus of extinct chordates called conodonts.

- The varied skeletal components of all these creatures will behave differently under different environmental and rock-forming, called diagenetic, conditions. Even slightly different forms of the same mineral can react very differently to the processes of fossilization.

- For example, both ammonites and brachiopod use calcium carbonate in their shells, but not all calcium carbonate is the same. Ammonites are composed of a mineralic form of calcite called aragonite, while some brachiopods use the more typical calcite. Organisms composed of aragonitic shells are more likely to be altered to calcite during fossilization, a transformation that very often removes any fine internal details of the shell. If an organism is already composed of calcite, then there is likely a greater chance of detail being preserved.

- Another consideration is the proportion of organic material present in the mineralized material. For example, trilobites would have formed extremely robust cuticles, impregnated by the organic molecule chitin, but with a very high proportion of calcium carbonate.

- Fellow arthropods the Malacostraca (includes shrimps, crabs, and lobsters) and the Diplopoda (the millipedes) have a much higher proportion of organic material in their exoskeletons. As a result, they have a reduced preservation potential and a poorer fossil record.

- Insects also have a modest preservation potential, but because of their sheer abundance, they have a better fossil record than would be predicted from their exoskeletal durability alone.

- But it's not just the durability of the materials that we have to consider. Another factor is how that material is organized. For example, sponges are composed of discrete structural elements called spicules. The various scaffolding units of sponges are more common in the fossil record than fossils of the original complete organism. Corals, however, secrete a single robust skeletal element, and as a result, the preservation potential of the entire organism is better than that of the sponges.

Modes of Fossilization

- A common mode of fossilization is the production of molds and casts. Molds are negative impressions of an organism that preserve information about the surface of a creature. A mold will commonly be produced when circulating pore waters moving through a sediment dissolve away the original skeletal material. Paleontologists will occasionally inject epoxy resin into the mold and then dissolve the surrounding rock to free the cast. Casts can form the same way in nature when mineral-rich waters deposit various minerals into fossil molds.

- Fossils can also form by a process called carbonization, in which a process of distillation, caused by the heat and pressure of burial, preferentially removes the hydrogen and oxygen of soft tissue, leaving the carbon behind. This a common mode of preservation of many land-plant fossils.

- Some of the most spectacular preservation occurs when mineralizing fluids percolate through sedimentary units. Minerals are precipitated in spaces between the skeletal material of shells and other original structural materials, hardening and stabilizing the fossil. This mode of preservation, called permineralization, can preserve wonderful detail.

- The same process can occur in some circumstances when organic material becomes completely replaced by mineralizing fluids, a process called petrifaction. This can occur in both plant and animal fossils.

- There are rarer modes of fossilization that can produce spectacular material. Perhaps the most beautiful are those fossils trapped in amber. This is an important mode of fossilization for insects and spiders, but other life-forms, such as small vertebrates and plants, have also been preserved.

- Amber forms when resin, produced by a number of types of trees but particularly coniferous trees, is secreted to heal an injury or act as a defense. This oozes down a tree trunk, sticking and trapping creatures as it goes. Once the resin is buried, pressure and temperature will increase due to the overburden of sediments. This causes the organic chemicals in the resin to oxidize and polymerize, eventually hardening into amber and preserving the creatures it trapped in fantastic detail.

- Another example of exceptional preservation, housed at the Smithsonian, comes from northwest Montana along the edge of Glacier National Park. The particular rocks in question come from a unit called the Kishenehn Formation from the Middle Eocene about 46 million years ago.

- The deposit is called an oil shale due to the high amount of organic material it contains, and it formed in the calm shallow regions of a lake. The sediments are finely laminated and represent seasonal changes in deposition. Such laminations are called varves. During warmer periods, probably during spring and summer, when organic production in the lake was high, dark organic-rich layers are deposited. These alternate with more windblown mineral material from the cooler part of the year.

- Perhaps the most famous example of exceptional preservation is the Burgess Shale, more than 65,000 specimens from which are housed in the Smithsonian's National Museum of Natural History's Department of Paleobiology. The creatures were buried in an underwater avalanche of fine mud in low-oxygen conditions, preserving exceptionally fine details of the structure of their soft parts.

Questions to consider:

1. In which environments should we expect the greatest potential preservation?

2. How much of the Earth's biosphere was never preserved in the fossil record?

Suggested Reading:

Benton and Harper, *Introduction to Paleobiology and the Fossil Record*.

Bryson, *A Short History of Nearly Everything*.

Lecture 2 Transcript

Life Cast in Ancient Stone

Kids love fossils. Of all the exhibits in a museum that really gets a school group enthusiastic, it's the fossil hall. I guess I'm just a kid that never grew up, for I get the same feeling when I wander through a collection like this at the Smithsonian. I think the wonder comes in part from the imagination—as magnificent as lions, elephants, and whales are, we know what they look like, know how they fit into the rest of the ecosystem, the reality we currently live in.

For fossils, it's different though. Until the invention of a time machine that can travel to the past—which Professor Hawking informs us is not going to happen—there will always be that element of the uncertain. That is not say we can't get very close to understanding how these creatures once lived and interacted, but there will always be room for the imagination to play.

And that is where the wonder comes from. This fascination is wonderfully demonstrated in the National Museum of Natural History and the research that is conducted in the Department of Paleobiology. So, in this lecture I'd like to ask: What is paleontology? Kind of a fundamental question, given the name of this series, so let's look at how paleontology developed as a science, what the chances are of becoming a fossil, what the common modes of fossilization are, and what is exceptional preservation?

Fossils have been a part of human culture for a long time. You can imagine the legends that must have developed from dinosaur bones weathering out of a hillside, or a trackway of ancient footprints preserved in solid rock. I wonder how many stories of dragons' and giants' fossils like these have generated around a campfire at night.

A very early reference to fossils comes from the 6th century B.C. and the Greek philosopher Xenophanes. Like most Greek philosophers, he wandered around the Mediterranean having an opinion on, well, pretty much everything really, from the nature of religion and the gods to a sort of critical rationalism. Xenophanes would also bemoan the rewards and reverence given to popular athletes, while philosophers, like Xenophanes, were underappreciated and ignored. Perhaps he just really wanted his picture on a piece of pottery. Apart from being somewhat miffed by the superstar athletes of the time, Xenophanes would also conclude from his examination of fossil fish and shells that water must have covered much of the Earth's surface at one time in the past—an important concept, showing that he understood that there was a deeper historical narrative to our planet that could be told by the use of fossils.

Similar ideas were proposed by the Chinese naturalist Shen Kuo in 1088, another wonderful overachiever who, finding fossils in the Taihang Mountains, decided that they indicated shorelines had shifted over time. He also found bamboo fossils in Shaanxi province, a part of China that is currently far too dry for bamboo to grow, and concluded that climate change must have occurred at sometime in the past—all this 900 years before Al Gore. Thoughts on how creatures could become fossils were proposed in 1027 by Persian polymath Avicenna, who speculated that fossils must have formed when a carcass was bathed in petrifying fluids.

These are all really modern concepts, concepts of sea-level change and climate change, the passage of vast amounts of time, and the processes that operate in the earth to form fossils, and in the West they would largely be ignored or forgotten.

Things would change, though, as our understanding and appreciation of fossils accelerates when we move into the 17th century, the age of reason. There are so many important figures who have contributed to the development of the science of paleontology in this period it would be difficult to cover them all, but here's just a selection.

This is Robert Hooke, an English natural philosopher who would make a significant contribution in his famous book *Micrographia* published in 1665.

The book was the first examination of the very small as revealed by Hooke's microscope. From his study of fossil wood, Hooke would conclude that these and other fossils were once living organisms that had been transformed into rock by the petrifying action of mineral-rich water.

Incidentally, the image we have here of Hooke here is somewhat conjectural, as no contemporary portrait of him exists. Hooke was a somewhat, shall we say, prickly character who managed to get onto the wrong side of, well, pretty much everybody, but also Isaac Newton, who it is rumored, when he became president of the Royal Society, made sure no images of Hooke remained. Sometimes the kids just don't play nicely in the scientific playground.

Another extremely important figure is Georges Cuvier, a French naturalist and philosopher. Cuvier would make several significant contributions, including providing us with the concept of extinction back in 1796. When comparing the jaws of living elephants with fossil jaws of something similar but at the same time distinct—we know it as the mammoth—Cuvier concluded that some species that used to exist on Earth were no longer around, a somewhat troubling concept to some who though that this idea challenged the notion of God's perfect and complete creation, a bit like a smile missing a number of important teeth.

Indeed, this concept of extinction preyed heavily upon the mind of the third president of the United States, Thomas Jefferson. Jefferson was an avid naturalist and intrigued by the bones of ancient giants like the Columbian mammoth. Jefferson himself proposed the name Megalonyx, meaning Giant Claw, in 1797 for the bones of a giant ground sloth from western Virginia, although at the time he believed them to be the remains of some sort of large lion. Jefferson firmly believed in the completeness of nature, and expected that these giants would be found alive and well west of the Mississippi River. He even asked the explorers Lewis and Clark to keep a watch out for Megalonyx and other fantastic fossil creatures on their expedition of 1804–1806.

It would be one of Cuvier's students, though, Henri Marie Ducrotay de Blainville—I love saying that—who would finally give the study of fossils a name. At first he chose the term paleozoologie in 1817, but by 1822, after a number of iterations, settled on the more inclusive palaeontologie, which would cover both fossil plants and animals. As such, thanks to Henri, I and Dr. Ross Geller from *Friends* are both paleontologists.

The son of a blacksmith, William Smith, would demonstrate the usefulness of fossils as a tool for correlating strata. By 1815, he published a groundbreaking geological map of England, Wales, and Southern Scotland, earning him the name Strata-Smith. Importantly, Smith demonstrated that fossils provided the context for the development of a geological timescale. Once constructed, this scale would allow scientists to correlate across regions, ensuring they were on the same page of Earth's history, and in doing so start to tell the story of Earth's deep evolution.

This is an idea that would be taken up by Smith's brother-in-law John Phillips, who in 1841 published the first geological timescale. He divided geological time into 3 eras we use today: the Paleozoic, Mesozoic, and Cenozoic. Although the names and the dates on the timescale would change considerably over time, paleontologists and geologists now had a yardstick which they could use to delve into the past.

Another important development was the discovery of fossils in a quarry in the Neander Valley of Germany. They were described by schoolmaster Johann Carl Fuhlrott and anatomist Hermann Schaaffhausen in 1856. They were identified as belonging to a group of humans that were quite different to any modern people. They would eventually be named *Homo neanderthalensis*. And so, with these fossils, paleontology became part of our story, too. And this is just 5 years after anatomist and founder of the Natural History Museum in London, Richard Owen, gave us the word dinosaur, probably the first scientific term any child has known ever since.

And, of course, who could forget this man, Charles Darwin, who, with the publication of *The Origin of Species by Means of Natural Selection* in 1859, would eventually provide the first hints of the mechanism behind the changing suites of fossils paleontologists had been finding. From here, the field of paleontology would explode into numerous disciplines and subdisciplines, becoming an extremely important part of academic studies in universities and museums all over the world. North America has had its share of famous paleontologists, too.

For example, Othniel Marsh and Edward Cope, who would expand our understanding of dinosaurs during the Great Dinosaur Rush in Colorado, Nebraska, and Wyoming in the late 1800s. Marsh and Cope, of course, are also the paleontological poster boys for grudge bearing in paleontology as they tried to financially and socially ruin each other's attempts to wreck the scientific career of the other. And coming closer to home, we should also include one of the most famous secretaries of the Smithsonian Institution, Charles Doolittle Walcott. In 1909, he would stumble across the Burgess Shale in British Columbia in the Canadian Rockies—but more of Charles Walcott later.

What about the actual fossils, without which of course there would be no science of paleontology? How can you turn a soft, organic, squishy organism into a fossil that can survive earthquakes, rising mountains, closing oceans, and colliding continents for millions, perhaps even billions, of years? Just what are the chances of some organism becoming a fossil?

To be honest, the chances of anything becoming a fossil are pretty slim. The fact that we have fossils at all speaks to the sheer numbers of individuals and species that have existed through time. With their countless billions, it would only take a tiny fraction of them, though, to fossilize to leave us with a substantial record in the rocks.

But let's consider those that actually do make it. What factors do they have in their favor? How do you maximize your chance of becoming a fossil? As my real estate agent would say: location, location, location. Just as being in the right place might maximize your chances of making a killing on the housing market, so being in the right place increases your chance of becoming a fossil.

An example of a worst place to be could be something like a high mountainous region or an eroding plain. Now, this might appear odd when we consider an environment like the African savanna, so full of life—shouldn't that be a great place for a fossil to form? The problem is, once, as John Cleese would have it, "you kick the bucket, shuffle off this mortal coil, or go to join the choir invisible," your body just tends to just lie out there in the open. In the case of the African savanna, you probably wouldn't lie out there for very long. Scavengers like hyena and vultures would very quickly dismember the carcass, with whatever remained—which is probably not very much by then—rotting to nothing under the influence of numerous microbes in our oxygen-rich atmosphere.

To form a fossil, you need to get your body buried as quickly as possible, out of the way of the scavengers and preferably sealed from oxygen, or in at least reduced oxygen conditions. Now, this just isn't going to happen on an open plain, but if the subject in question happened to live close to a body of water—a river or a lake—you now have a chance to being in an environment where you might be able to bury your corpse with sediment.

As being buried in sediments is probably your best bet for becoming a fossil, it follows that organisms that live in aquatic environments like lakes and rivers will have a greater chance of becoming a fossil. It also follows that, on the whole, aquatic creatures that live in the oceans and other water bodies that are receiving vast quantities of sediment via rivers and streams will have a greater preservation potential than terrestrial, land-based organisms.

So where you live is important, but there's another important factor to consider as well, and that is what you're made of. Soft-bodied, squishy organisms like this octopus will have a much reduced preservation potential compared to, for example, one of its close relatives, the nautilus. Of course the body parts of the nautilus will decay just like the octopus, but a significant portion of the original creature, its shell, will have a greater chance of preservation. This is why the fossil record of octopi is considerably poorer than that of their shelled relatives.

Any creature that has a significant development of hard parts, therefore, will have a greater chance of preservation and a greater representation in the fossil record. This is a persistent bias paleontologists have to be aware of when reconstructing ancient ecosystems from fossil sites, especially when, in some settings, soft-bodied creatures lacking any skeletal components may have made up a large part of the animal assemblage.

But, considering hard parts for a while, life has used a variety of materials for production of structural support. Calcium carbonate or calcite is a very common mineral used by many organisms, including bryozoans, corals, brachiopods, mollusks, and many arthropods and echinoderms like sea urchins. Examples of silica-secreting organisms include sponges and the radiolarians—tiny marine protists about 0.1 to 0.2 millimeters in size that secrete these exquisite ornament-like structures out of biological glass. Calcium phosphate, usually in the form of the mineral apatite, is used for the skeletal elements—that's bones and teeth—of vertebrates and the feeding apparatus of the extinct chordates called conodonts.

Now, the varied skeletal components of these creatures will behave differently under different environmental, rock-forming—what we call diagenetic—conditions. Even slightly different forms of the same mineral can react very differently to the processes of fossilization. Consider these 2 fossils: an ammonite cephalopod, relative of the nautilus; and a brachiopod, a creature that might look like a clam but is actually more related to the bryozoan moss animals. Both creatures use calcium carbonate in their shells, but not all calcium carbonate is the same. The ammonite is composed of a mineralic form called aragonite while some brachiopods use the more typical calcite.

The main difference in the 2 forms of calcium carbonate is in the arrangement of the atoms that make up the crystal lattice. This might sound like an esoteric difference, given that chemically they're very similar, but when the 2 substances are buried they behave quite differently. Aragonite is only really stable at surface temperatures and pressures while calcite is stable over a much broader range.

The practical result is that organisms composed of aragonitic shells are more likely to be altered to calcite during fossilization, a transformation that might not result in the destruction of the fossil but very often removes any of the fine internal details of the shell. If an organism is already composed of calcite, like the brachiopod in our example here, then there is likely a greater chance of detail being preserved, like you can see in the laminations in this fossilized shell of a Carboniferous brachiopod. The laminations here represent the individual layers that the brachiopod secreted over its lifetime, like rings on a tree.

Another consideration is the proportion of organic material present in the mineralized material. For example, trilobites, like these from the Lower Devonian of Oklahoma, would have formed extremely robust cuticles impregnated by the organic molecule chitin but with a very high proportion of calcium carbonate. Fellow arthropods the Malacostraca that include shrimps, crabs, and lobsters have a much higher proportion of organic material in their exoskeletons, and as a result they have a reduced preservation potential and a poorer fossil record.

Insects also have a modest preservation potential, but because of their sheer abundance they have a better fossil record than would be predicted from their exoskeletal durability alone. But it's not just the durability of the materials—calcium carbonate, silica, calcium phosphate, cellulose, and chitin—that we have to consider. Another factor is how the material is organized. For example, arthropods consist of numerous exoskeletal body parts called sclerites. These tend to separate after death, giving us vast quantities of components of specimens in the fossil record.

The same is true for sponges, which are composed of discrete structural elements called spicules, some as large as 2 millimeters but many less than 60 microns in size. As a result, the various scaffolding units of sponges, like you can see here, are also common in the fossil record—more common than fossils of the original complete organism. It's a bit like an archeologist finding the bricks and rocks that make up a large building rather than the building itself. Corals, however, like these from the Ordovician period of southern Ohio, secrete a single robust skeletal element, and as a result the preservation potential of the entire organism is better than that of the sponges.

So you see how a whole number of factors come into play when considering the preservation potential of a fossil: location and mode of life of the original organism, the presence or absence of hard parts, the composition of those hard parts, and whether the organism contributes several cast molts during its lifetime. And of course there is pure blind luck. So, fossils like these beautiful silicified marine organisms from a Permian reef in Texas, although quite abundant, are still a marvel when you consider the vast numbers of individuals that never made it into the fossil record.

We have to remember, though, that each of these preservational factors we have described creates a bias, with certain environments and/or organisms having a greater representation in the fossil record. An understanding of preservational mechanisms can also help us with miscomprehensions about the fossil record. For example, a common question I've been asked about the meteor impact that killed off the dinosaurs and a whole bunch of other creatures at the end of the Cretaceous 66 million years ago is: "If a meteor wiped out the dinosaurs, why don't we find their fossils all packed around the horizon that marks the impact event?"

That is a good question and is answered nicely by something called the Signor-Lipps effect. This was proposed by Philip Signor of the University of California, Davis, and Jere Lipps of the University of California Museum of Paleontology, and is all about the factors we have been discussing. Basically the Signor-Lipps effect states that it is very unlikely that the very first, or the very last, member of a particular fossil species in time will be found. This is due to the points that we are able to sample at any given time in the geological record.

Consider the African savanna again. This is how we imagine it, isn't it? To quote Mr. Cleese once more: "Herds of wildebeest sweeping majestically across the plain." But let's be honest: even on a safari vacation, you still have to go looking for these creatures to photograph them. Most of the time the African savanna looks like this, with apparently not much happening at all. Now add to that the fact there is only a minute proportion of those animals that live in that environment actually become fossils, and you can see why we may have a lack of fossils at any one geological horizon.

There is more, though. When you are looking at a geological horizon in time, you are mostly looking at a discontinuous 2-dimensional slice through what was once a vast 3-dimensional surface upon which creatures may or may not have been fossilized. As such, the chance of actually locating a dinosaur fossil at a very specific horizon in time, even after such a catastrophic event, is very small indeed.

So you can see it is probably only a very small proportion of any iteration of the biosphere that is going to get preserved, and there will likely be biases that may control the number and types of fossils that we do find. I don't want to sound too disheartening here, though—the fossil record is still very powerful and an enlightening tool, but should be treated carefully.

Given that, let us look at some of the ways in which creatures become fossils. A common mode of fossilization is the production of molds and casts. Molds are negative impressions of an organism that preserve information about the surface of a creature. A mold will commonly be produced when circulating pore waters moving through a sediment dissolve away the original skeletal material. Paleontologists will occasionally inject epoxy resin into the mold and then dissolve the surrounding rock to free the cast. Casts can form in the same way in nature when the mineral-rich waters deposit various minerals into fossil molds.

Fossils can also form by a process called carbonization. Here, a process of distillation caused by the heat and pressure of burial preferentially removes the hydrogen and oxygen of the soft tissue, leaving the carbon behind. This is a common mode of preservation of many land plant fossils but also of these beautifully preserved graptolites that used to float through the Silurian oceans over 423 million years ago.

Some of the most spectacular preservation, though, occurs when mineralizing fluids percolate through sedimentary units. Minerals are precipitated in spaces within skeletal materials of shells and other original structural materials, hardening and stabilizing the fossil. This mode of preservation is called permineralization and can preserve wonderful detail. For example, this image was taken on a scanning electron microscope and shows the cancellous—that's spongy—tissue from the femur of a dinosaur magnified many times. You can even make out the honeycomb structure that added support and strength to the bone faithfully preserved.

The same processes can occur in some circumstances when organic material becomes completely replaced by the mineralizing fluids, a process called petrifaction. This can occur in both plant and animal fossils, but possibly one of the most spectacular examples comes from the petrified conifer forests of Arizona. An example of one of these trees stands outside the museum today.

The fossils come from the late Triassic period, about 225 million years ago, when Arizona was humid and subtropical. The trees were living on the edges of streams and rivers, some growing up to 200 feet tall. Occasionally, they would fall into the water and get buried in sediment that contained high proportions of volcanic ash that was rich in silica. It's this volcanic ash that is the key to their preservation. Fluids carrying this dissolved mineral percolated through the sedimentary pile, seeping into the waterlogged trees and replacing their organic material with silica.

The wonderful colors you see in many of these fossils are due to secondary minerals such as iron oxides like hematite that gives us the reds and purples, and limonite, which gives us the yellow. So numerous are these fossil trees that in the past they've been used as building materials. Although internal detail is not always preserved in them, some of the logs and some of the animal bones found in these deposits show detail even down to the cellular level.

There are, of course, other rarer modes of fossilization that can produce spectacular material. Perhaps the most beautiful are those fossils trapped in amber. This is an important mode of fossilization for insects and spiders, but other life-forms like small vertebrates and plants have been preserved, as well. Amber forms when resin produced by a number of trees, but particularly coniferous trees, secrete resin to heal an injury or act as a defense. This oozes down the tree trunk, sticking and trapping creatures as it goes. Once the resin is buried, pressure and temperature will increase due to the overburdened pressure of the sediments, and this causes the organic chemicals in the resin to oxidize and polymerize, eventually hardening into amber, and preserving the creatures it trapped in fantastic detail.

Between 1981 and 1985, through the efforts of entomologist Don R. Davis, the Smithsonian acquired a wonderful collection from the Dominican Republic between 21 and 15 million years old. This collection is housed in the Department of Paleobiology and is a valuable resource today to fossil insect researchers throughout the world. It was collected by retired army major Jacob Brodzinsky and his wife Marianella Lopez-Peña.

The Brodzinsky/Lopez-Peña collection comprises just over 5000 specimens, providing a wonderful window into the early Miocene insects, mites, centipedes, and spiders, but also leaves, flowers, and even a bird feather. It is particularly useful for paleontologists as the amber is very pale, which allows a better view of the fossils. The preservation is just spectacular. In this sample a small mite was clinging to the leg of this winged ant when it was trapped in the resin. And here, a new species of wasp: just look at the level of detail down to the fine hairs on the insect's body.

Another—as we might say in my hometown of Manchester—gob-smacking example of exceptional preservation housed at the Smithsonian comes from northwest Montana along the edge of the Glacier National Park. This particular rock in question come from a unit from the Middle Eocene about 46 million years ago, a bit more than twice the age of that Dominican amber we just described. The deposit is called an oil shale due to the high amounts of organic material it contains, and it formed in the calm, shallow regions of a lake. The sediments are finely laminated and represent seasonal changes in deposition.

Such laminations are called varves. During warmer periods, probably during spring and summer when organic production was high, we see dark, organic-rich layers. These alternate with more windblown material, mineral material from cooler parts of the year. Dale Greenwalt, a volunteer and researcher at the Smithsonian, has been collecting beautifully preserved insects from this formation for a number of years. Most are small, but the level of preservation is remarkable, even down to the microscopic veins in wings, and, most unusually for fossils, remnants of pigment, too.

Such a collection of exceptional fossil finds is often called a Lagerstatten, from the German "storage-place." The preservation here was likely due to a number of factors. One would have been reduced oxygen conditions caused by the blooming algae in the lake during the summer and the spring. As the algae died and sank, they would rapidly use up oxygen as they decayed. Such low oxygen conditions arrested decay and restricted scavengers. It's also been proposed that microbial mats in the lake may have also helped by sticking the insects and so stabilizing them on the floor of the lake.

But the preservation of these fossils is even more significant than we originally thought. While examining a fossil of a mosquito, Dale noticed that one of the specimens had an abdomen that looked black and swollen. Using a scanning electron microscope and a mass spectrometer, a team led by Dale investigated the molecules preserved in the sample and found large amounts of iron, specifically in the insect's abdomen, and traces of a molecule called porphyrin, a molecule that binds iron and oxygen in blood. This suggests that this was the mosquito's last meal. This is 46-million-year-old fossilized blood.

Exceptional preservation is not restricted to fossils from geologically recent times, though. The museum holds some beautiful fossils dating to 286–245 million years of the Middle Permian from what is today the Glass Mountains of western Texas. The Glass Mountains are part of an ancient reef system that formed in a shallow inland sea called the Delaware Basin, now exposed in the Apache Mountains and Guadalupe Mountains. The reefs around the margin of the Delaware Basin record shallow warm-water marine life of incredible diversity—a complete reef ecosystem.

Some of these animals made their skeletons from calcium carbonate, the same composition as the limestone rocks in which they're now formed. During fossilization, though, silica replaced the calcium carbonate of the fossils. This is an example of permineralization. It also presented paleontologists with a wonderful opportunity, as silica is resistant to acids that will dissolve the limestones. As a result, the limestone can be etched with acid, releasing beautifully preserved silicified fossils. In 1939, Smithsonian scientist G. Arthur Cooper, at that time a member of the National Museum of Natural History Department of Geology, began collecting blocks of limestone and building a fantastic collection of these Permian reef creatures delicately spun in glass.

And, of course, there is perhaps the most famous example of exceptional preservation, the Burgess Shale, which provides a wonderful glimpse of early life on Earth over half a billion years ago. As we noted earlier, they were first discovered in 1909 by Charles D. Walcott, with over 65,000 specimens housed in the Department of Paleobiology. The creatures were buried in an underwater avalanche of fine mud in low oxygen conditions, preserving exceptionally fine details of the structure of their soft parts. But, again, more of the Burgess Shale in a later lecture.

This is just a brief selection of some of the fantastic fossils that are available to paleontologists. We haven't even considered examples of more recent fossils that may contain remnants of DNA preserved by freezing or drying processes. Again, we'll get to those in a later lecture. Hopefully, though, it's whetted your appetite to investigate what else the geological record holds, and perhaps for a visit to the National Museum of Natural History, where you can see some examples of these beautiful fossils on display, or examine some of these fossils at the museum's Q?rius Center that houses fossils for public examination and study.

Although, as we have seen, only a tiny proportion of life on Earth has become fossilized, that portion still represents an enormous cache of material for future paleontologists to examine. New discoveries of fossil bonanzas, the Lagerstatten, and new approaches and techniques in paleontology and paleobiology will likely continue to surprise and delight generations of scientists to come.

Lecture 3

Tools of the Paleontological Trade

In this lecture, you will consider some of the tools and techniques used by paleontologists, and you will discover how new technologies are opening up windows into the past in a way that would have astounded the founding fathers and mothers of geology and paleontology in the 18th century. You will learn how fossils are found in the field, how fossils are collected, how fossils are prepared, what the new tools of the trade are, and how life is given to fossils through scientific illustration and reconstruction.

Finding Fossils

- Many paleontologists were first trained as geologists before specializing in the study of fossils. There are many very good reasons for this. In part, an appreciation of geology helps place fossils in the context of a dynamic Earth system, which in turn has implications for the way in which we interpret the fossils we find. In addition, paleontologists have to rely on a number of basic geological principles and skills to track down and accurately record the fossils they find.

- In particular, there are principles regarding the manner in which rocks—sedimentary mostly—are deposited one on top of another over time. These ideas basically state that the sequential deposition of sediments means that the oldest layers, or strata, will be at the bottom of the pile and the youngest at the top.

- Geologists know that our dynamic planet rarely allows the thin crust we live on to stay still for long. Horizontal strata more often than not will become tilted or folded over time as the continents wander, collide, and raise mountains, twisting and distorting the geological pages of Earth's history book.

- This can complicate matters when we are trying to read Earth's story, which is why a vital skill for any paleontologist out in the field is the ability to create and read geological maps.

- Geology is very rarely beautifully exposed. The story we want to tell is often covered by a soil profile, vegetation, asphalt, or an inconvenient shopping mall. A paleontologist is often only presented with a fragmentary glimpse of the geology at the surface in the form of limited "outcrops," with little evidence of what the geology is doing in the subsurface. It is from these limited views that geologists create a map—a hypothesis—of both the seen and unseen geology below the surface.

- Even though aerial and satellite photography and gravity and magnetic surveys can help with mapping today, the geoscientist still has to rely mostly on getting down on the ground and hiking along outcrops of rocks. Like the field kits of the first geological mapmakers, basic field kits include a geological hammer, a hand lens, a compass (with a clinometer for measuring the dip of strata), and a notebook (to record findings). GPS and electronic data storage devices are also used.

- When complete, the map is tested by continued mapping or in some cases by drilling boreholes to see if your subsurface predictions are actually matched by the rocks you recover in core, predictions that might be confirmed by characteristic rock types and/or fossils. In this way, the geology of an area, especially when that area might be geologically complex, is revisited and refined—tweaked so that the model we produce comes closer and closer to the reality of the rocks in the Earth.

- A good geological map can help a paleontologist predict the location of strata of particular interest across the landscape with fossils themselves tying those strata into a temporal framework. A map therefore can help paleontologists zero in on the pages of Earth's history that they are interested in and also helps them understand the wider temporal context of the fossils that they find.

- Even though a map may help you focus in on the area you should be looking for fossils, there may still be a lot of hunting around to find the fossils once you're in the field. A good start is to eyeball the ground for fragments of fossils in what is called float, or loose pieces of rock that have been eroded from an outcrop that actually contains the fossils.

Collecting Fossils

- But what about collecting the fossils once they have been located? This will vary depending on what fossils you are finding, but most fossils, such as the shells of various marine creatures, can often be collected by the application of hammer, chisel, crowbars, and a little muscle, making sure that eyes are protected by safetly glasses because many rocks have a high silica content and splinter into dangerous shards when hit. Once recorded, the fossils are wrapped to protect them and placed in a bag with an identification number.

- This becomes trickier when dealing with large fossils in rock. Sometimes a small pick and a hammer just aren't going to be enough. That's when you might see a field paleontologist employing a jackhammer or a backhoe.

- In addition, it is generally impractical to extract large fossils from their rock matrix in the field, so once the specimen has been exposed, it is extracted with the adjacent rock matrix still attached.

- Plaster and burlap straps are applied to the specimen, forming a jacket, and once hardened, the fossils can be removed and transported back to the lab. Sometimes, given the remotness of sections being studied, this could require a helicopter.

- The experience of collecting fossils is a little different for micropaleontologists, who don't have the luxury of seeing their fossils in the field. Most microfossils are fractions of a millimeter in size and often

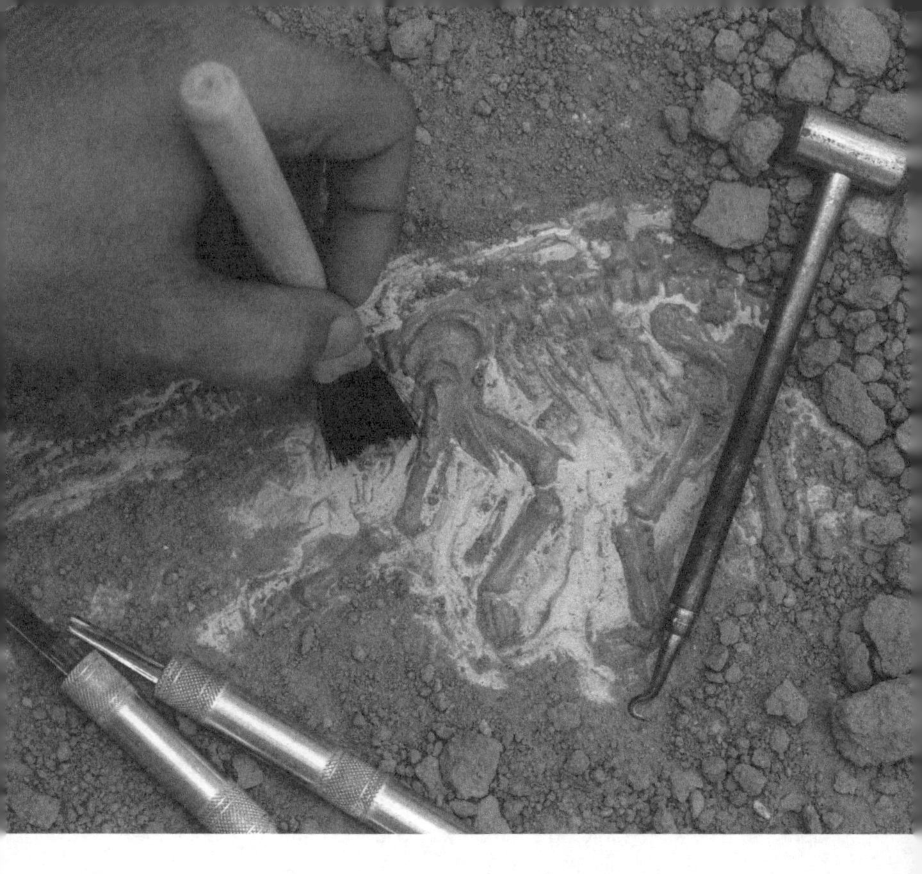

impossible to see, even with a 20x hand lens. The best micropaleontologists can do is find the right kind of rocks that might contain the fossils and hope that they will find the fossils when they get back to the lab.

- Common to all fossil collection is recording as much detail as possible regarding where the fossils were found—not just spatially on a map, but stratigraphically so that their vertical (time) and lateral (geographic) relationship to other specimens can be assessed. To preserve both their original geographic and stratigraphic location, fossils are often recorded on a stratigraphic log, which is a vertical representation of the strata that are being studied.

Preparing Fossils

- Back at the lab, in the case of larger specimens, the fossils have their plaster jacket removed, and the long and careful process of removing the fossil from the rock matrix begins. A number of tools are used for this, including the air scribe, which acts like a miniature jackhammer, chipping away at the rock matrix. When getting close to removing the majority of the matrix, the air scribe's impacts may "pop off" the last bits of rock, leaving the fossils clean and exposed.

- If a fossil is too fragile, or the matrix is too hard, fossil preparators may use gentle grinding tools to help separate the fossil from the rock. When getting too close to the fossil, tiny picks and needles are used to clean up the specimen.

- Various adhesives are an essential part of a fossil preparator's toolkit, too. Thick solutions are useful for rejoining large broken fossils. For fragile specimens, a thin solution can be applied that penetrates into cracks and pores, strengthening the fossil from within. After a fossil is rejoined, it is often placed in sand that holds the pieces in the correct positions until the adhesive sets. When complete, spectacular detail can be revealed.

- The preparation of fossils at the other end of the scale, with microfossils, is somewhat different. The most common method of preparing microfossils involves the use of various often-nasty acids, such as hydrofluoric acid, to dissolve away the rocky matrix.

Studying Fossils

- In Robert Hooke's famous publication of *Micrographia* in 1665, a whole new world was revealed—the world of the very small. Hooke's beautiful drawings, such as those of the flea and the compound eye of a dragonfly, were instrumental in promoting the early use of the microscope in understanding the natural world.

- Since then, microscopes have been used in many branches of paleontology for studying various aspects of fossils, commonly by making a thin section of the rock that reveals the anatomy of well-preserved fossils as light passes through them. Optical microscopy has its limitations, though.

- Practically, you have a maximum magnification of about 1500x due to the wavelength of light that limits the resolution of the microscope. In addition, there is a problem with depth of field in viewing specimens that have much relief. As such, optical microscopes essentially provide a flat image.

- Fortunately, we can use something other than photons of light to make images of the very small. A scanning electron microscope (SEM) uses electrons rather than photons. Because electrons have a much shorter wavelength than light, SEMs have a much greater resolution than optical microscopes. The resolution of an SEM can range up to around 300,000x.

- Some of the electrons fired at the object from the SEM travel deeper in the specimen, get absorbed, and cause a release of x-rays. These x-rays can then be used to determine the composition on the object being studied; all you need is an SEM fitted with an x-ray detector. This technique, called energy-dispersive x-ray spectroscopy, has been used by a research team headed by Dr. Conrad Labandeira at the Smithsonian's National Museum of National History to look at exceptionally preserved material in Jurassic lake sediments in northeastern China.

- There are other tools for determining the composition of materials, and in some cases, determining the relative proportions of very specific isotopic components of a material can provide vital environmental information about the past. An isotope is a variant of an element that differs only in the number of neutrons it contains in its nucleus.

- Some isotopes are unstable and decay into more stable elements over various time periods; others are stable and hang around in the environment. Isotopes of carbon, oxygen, sulphur, nitrogen, and a whole bunch of others react in very specific ways to different environmental factors that speak to various events in Earth's past.

- It is usually igneous rocks that are used in radiometric dating. Igneous rocks form as magma or lava cools, forming crystals that trap small amounts of radiometric material. This can then be used to date the rock. This technique has permitted the dating of materials from many periods of Earth's history, including some of the most ancient.

- A technique currently being used at the Smithsonian Institution involves capturing precious fossils in 3 dimensions on a computer. The digitization program at the Smithsonian can capture incredible detail from a specimen, using millions or billions of points of measurement on its surface.

Illustrating and Reconstructing Fossils

- Just because we have new technologies available to us does not mean that we abandon more traditions tools. This is nowhere better seen than in the power of paleontological art and illustration. Science meets art when we need to reconstruct ancient environments and the organisms that lived in them. The Smithsonian has a rich history of paleontological art.

- Even with all the new advances in imagining and data manipulation, the role of the scientific illustrator is still vital both in research and in public display of materials. Very often, an illustration can highlight features that might be too subtle to be picked out on a photograph and also correct for problems, such as poor depth of field or distortion, that can occur with a camera lens.

- Paleontologists require a wide variety of visual material to illustrate their work, including the reconstruction of fossil specimens, restorations of ancient animals and plants, and various diagrams, graphics, and maps to help illustrate research. Often, drawings reveal structure, anatomy, and features that are not readily grasped in photographic images.

- The collaboration between artist and scientist in the reconstruction of past environments is inspiring. The reconstruction of paleoenvironments begins with consultations between artist and paleontologist, perhaps with the scientist making a rough initial sketch.

- From materials such as specimens, photographs, and a range of other background material, the scientific illustrator begins to bring life to lost landscapes and the animals and plants that populated them. The result is the point at which art and science meet to produce wonderful images that breathe life into worlds long since vanished. These images are the products of all the fieldwork, preparation, analysis, and interpretation that is part of the science of paleontology.

Questions to consider:

1. What critical information is lost when a fossil cannot be tied to where it was originally recovered?
2. What are the advantages of more traditional artistic representations in paleontology when compared to modern visualization techniques?

Suggested Reading:

Taylor, *DK Eyewitness Books*.

Thompson, *The Audubon Society Field Guide to North American Fossils*.

U.S. Department of the Interior: Bureau of Land Management, "Hobby Collection," http://www.blm.gov/wo/st/en/prog/more/CRM/paleontology/fossil_collecting.html.

Lecture 3 Transcript

Tools of the Paleontological Trade

We've been aware of fossils and their significance as messengers from the past for quite a while now. At times, their interpretation may have differed quite radically from our understanding today, as you can see in this wonderful rendering of the ancient oceans of England by Henry De la Beche, but paleontology is a dynamic, ever-changing science. It's one of the reasons I love it.

Paleontology as a modern science is a relatively new discipline with its roots in the European enlightenment of the 18th century. At this time, reductionism—an approach to understanding nature by reducing it into smaller, simpler parts—and scientific rigor were hot intellectual movements of the day. It was in this environment that paleontology and geology would grow and become sciences in their own right. It's from early pioneers, like James Hutton here, that we get much of the common core of the way we practice the science today. However, as a discipline, paleontology has not stood still, and new technologies and techniques have pushed the science forward, particularly in the last 30 years, with ever-increasing speed.

So, in this lecture, I'd like to consider some of the tools and techniques used by paleontologists and how new technologies are opening up windows into the past in a way that would've just astounded the founding fathers and mothers of geology and paleontology back in the 18th century. So let's ask: How do you find fossils in the field? How do we collect fossils? How are fossils prepared? What are the new tools of the trade that we can use? And look at how we can breathe life into fossils through scientific illustration and reconstruction.

Many, but not all, paleontologists were first trained as geologists before specializing in the study of fossils—that was my background. There are many very good reasons for this. In part, an appreciation of geology helps place fossils in the context of a dynamic Earth system, which in turn has implications for the way in which we interpret the fossils we find. But also, a paleontologist has to rely on a number of basic geological principles and skills in order to track down and accurately record the fossils they find. In particular, there are principles regarding the manner in which rocks—sedimentary, mostly—are deposited, one on top of another, over time.

These ideas basically state that the sequential deposition of sediments means that the oldest layers, or strata, will be at the bottom of a pile and the youngest at the top. Now, this is not rocket science I'm sure you'll agree, but it is a fundamental concept if we are to read these strata and the fossils they contain, like peeling the pages back of a book. As geologists, we know that our dynamic planet rarely allows for the thin crust we live on to stay still for very long. Horizontal strata, more often or not, will become tilted or folded over time as continents wander, collide, and raise mountains, twisting and distorting the geological pages of Earth's history book. This can complicate matters when we are trying to read Earth's story, which is why a vital skill for any paleontologist or geologist out in the field is the ability to create and read geological maps.

Geology is very rarely beautifully exposed like you see here in the Grand Canyon in Arizona. The story we want to tell is often covered by a soil profile, vegetation of some sort, asphalt, or an inconvenient shopping mall. A paleontologist is often only presented with a fragmentary glimpse of the geology at the surface in the form of limited outcrops with little evidence of what the geology is doing in the subsurface. It is from these limited views that geologists create a map, a hypothesis, of both the seen and the unseen below the surface. Even though aerial and satellite photography, and gravity and magnetic surveys, can help with mapping today, the geoscientist still has to rely mostly on getting down on the ground and hiking around along the outcrops of rock.

Like the first geological mapmakers, such as William Smith who produced the really first large-scale geological map, the basic equipment of geological hammer, hand lens, compass—with a clinometer for measuring the dip of strata—and a notebook to record your findings is still the basic kit in the field. GPS and an electronic data storage device are certainly used, but any geologist worth their salt should only need the basics and just a little bit of hard work.

When complete, the map is tested. After all, a geological map is basically a hypothesis based on what is often incomplete information. It is tested by continued mapping or, in some cases, by drilling boreholes to see if your subsurface predictions are actually matched by the rocks you recover in core, predictions that might be confirmed by characteristic rock types and/or fossils, for example. In this way, the geology of an area, especially when those areas might be geologically complex, is revisited and refined, tweaked so that the model we produce comes closer and closer to the reality of the rocks in the Earth.

A good geological map can help a paleontologist predict the location of strata of particular interest across the landscape with fossils themselves tying those strata into a temporal framework. A map, therefore, can help a paleontologist zero in on the pages of Earth's history that they are interested in, and also helps that paleontologist understand the wider temporal context of the fossils that they find.

Even though a map may help you focus in on an area you should be looking for fossils, there may still be a lot of hunting around to find the fossils once you're in the field. A good start is to eyeball the ground for fragments of fossils in what we call float. These are loose pieces of rock that have been eroded from an outcrop that actually contains the fossils you're looking for. This was how Smithsonian Secretary Charles Walcott zeroed in on the fantastic fossils of the Burgess Shale by first finding loose specimens in a scree slope—a scree slope is a loose rock debris flow—and carefully tracing them upslope till he and his family found the now famous fossil-bearing strata in Walcott's quarry, a beautiful area of the world if ever you get to British Columbia.

In the same way, during July 2000, Neil Shubin and his colleagues investigating strata on Ellesmere Island in the Canadian Arctic would track down fragments of bone to a bluff, which, in Shubin's words, were kicking out bones. Some of these bones would be one of the most significant fossils of recent times: Tiktaalik, a fish that was on its way to becoming a land-based tetrapod, a transitional form—that is, a fossil that is partway between one type of creature and another. It's basically a fossil that is showing that evolution happens.

But what about collecting the fossils once they've been located? Well, this will vary depending on what fossils you are finding, but for most, like the shells of various marine creatures, fossils can often be collected by the application of hammer, chisel, crowbars, and just a little muscle, making sure your eyes are protected by safety glasses, as many of the rocks will have a high silica content and splinter into dangerous shards when hit.

Once recorded, the fossils are wrapped to protect them and placed in a bag with an identification number. This, of course, becomes a little more tricky when dealing with something like this. Wrapping this in a copy of the *National Enquirer* and stuffing it in your backpack isn't really an option, and sometimes a small pick and a hammer just aren't going to be enough. And that's when you might see the field paleontologist employing a jackhammer or a backhoe—probably grinning with absolute delight.

In addition, it is generally impractical to extract large fossils from their rock matrix in the field, so once the specimen has been exposed, it is extracted with the adjacent rock material still attached to the fossil that you're interested in. Plaster and burlap straps are then applied to the specimen, forming a kind of a jacket, and once hardened, the fossils can be removed and transported back to the lab. Sometimes, given the remoteness of the sections being studied, this could require a helicopter and another opportunity for a happy paleontologist to indulge their inner 6-year-old.

My own experience of collecting fossils has been a little different, though. As a micropaleontologist, I don't have the luxury of seeing my fossils in the field, as much as I really would've liked to at some times. Most microfossils are fractions of a millimeter in size and often impossible to see, even with a powerful hand lens when you're out in the field. The best we can do is find the right kind of rocks that might, just might, contain the fossils we're looking for. For example, the microfossils I study, the chitinozoa, and the planktonic forams of Dr. Brian Huber of the Department of Paleobiology, are exclusively marine.

Of course, the right age of rocks is vital, too. For my chitinozoa, that ranges from the Lower Cambrian period, about 510 million years ago, through to the end of the Devonian period, about 359 million years ago. That spans some 151 million years of the geological record. After that, we just have to sample the strata and hope that we find the fossils when we get back to the lab—oh please. Of course, it does have the advantage—if conducting fieldwork, particularly in the United Kingdom—of allowing micropaleontologists to be the first back at the pub after a day in the field and not worrying about spending days of labor excavating a large and troublesome dinosaur.

But common to all fossil collection is recording as much detail as possible regarding where the fossils were found, not just spatially but also stratigraphically. So, we're not just plotting them on the map; we also need to know the vertical time extent, as well as their lateral geographic relationships. This is why fossils collected illegally by fossil hunters have much less value scientifically, especially as context is so important when considering the specimens that are recovered. If removed from their contextual setting, a fossil loses all of the associations that might be able to be drawn from all the other information that might be included in the rocks from which they originally were found.

To this end, and to preserve both their original geographical and stratigraphical location, fossils are often recorded on a stratigraphic log. A strat log, as we often call it, is a vertical representation of strata that are being studied. They are often depicted like this: a vertical set of rocks that you might find exposed, for example, in a quarry face or in cliffs along the coastline, even if the rocks themselves in outcrop are tilted or folded. Such a vertical column just makes time-based arrangements of fossils just easier to understand and interpret.

Back at the lab, in the case of the larger specimens, the fossils have their plaster jacket removed and the long and very careful process of removing the fossils from the rock matrix will begin. A number of tools are used for this, including the air scribe, which acts a bit like a miniature jackhammer, chipping away at the rock matrix. When getting close to removing the majority of this matrix, the air scribe's impacts may actually pop off the last bits of rock, leaving fossils beautifully clean and exposed. If that doesn't happen, it's time to break out picks and needles, removing remaining matrix grain by painstaking grain.

If a fossil is too fragile, or the matrix is too hard, fossil preparators may use gentle grinding tools to help separate them from the rock. Once again, though, when getting to close to the fossil, it is only tiny picks and needles that are used to clean up the specimen. As you can see, a steady hand and a lot of patience is required for this job.

Various adhesives are an essential part of a fossil preparator's tool kit as well. Thick solutions can be used to rejoin large broken fragments of fossil. For fragile specimens, a thin solution can be applied that penetrates into the cracks and pores, strengthening the fossil from within. After a fossil is rejoined they are often placed in sand that holds the pieces in the correct position until the adhesive sets. When complete, spectacular detail can be revealed, detail like you can see on fantastic reconstructions of dinosaurs, or in the beautiful and delicate spines and processes that have been uncovered from these exquisitely prepared trilobites released from their rocky matrix after 100s of millions of years.

As we noted with regards to collecting specimens, the preparation of fossils at the other end of the scale, with microfossils, is somewhat different. The most common method of preparing microfossils involves the use of various, often very nasty, acids such as the hydrofluoric that I had to use in my research to extract my chitinozoa. In this way, you're able to dissolve away the rocky matrix and reveal the fossils.

Once you have them, you have to concentrate the residue of material up, and then, once you have that concentrated residue, spread them onto a microscope slide, or pick through them using fine brushes, bristles, or fine pipettes. After all the exciting collection and preparation of your specimens, there is another stage that needs to be completed, whatever type of paleontologist you are: the fossils need—in fact, must—be curated.

This is often one aspect of paleontology that is overlooked by the media, which is unfortunate, as curation of material—safely storing, recording, and documenting—is vital to the science. If we did not have dedicated curators at museums and universities, all we would have are rooms full of undocumented fossils slowly falling apart in dusty draws and cabinets of little to no use to science.

The National Museum of Natural History currently holds over 126 million items, and counting. Samples ranging from fossilized pollen to bones of *Tyrannosaurus rex*; samples of algae or slabs of a giant sequoia tree; tiny crustaceans to giant squids; insects trapped in amber and samples of DNA; the skulls of whales and the dung of ground sloths; the Hope Diamond and rocks from the moon. What a wonderful palace of treasures this is.

But let's consider now some of the tools we use to study our fossils. In Robert Hooke's famous publication of January 1665, *Micrographia*, a whole new world was revealed, the word of the very, very small. Hooke's beautiful drawings, such as these of the flea and the compound eye of a dragonfly, were instrumental in promoting the early use of the microscope in understanding the natural world. Hooke would also turn his attention to fossils through his study of petrified wood. In observing that structures he found in petrified wood very closely resembled those he could see in rotten oak, he concluded that wood would be tuned to stone if water percolated though its tissues, precipitating minerals.

Since then, microscopes have been used in many branches of paleontology for studying various aspects of fossils, commonly by making a thin section of the rock that reveal the internal anatomy of well-preserved fossils as light passes through them. In this magnified thin section of an Ordovician limestone—the scale bar here is about 2 millimeters—you can see a fragment of a trilobite and spine from an echinoderm, a sea urchin.

Optical microscopy has its limitations, though. Practically, you have a maximum magnification of about 1500× due to the wavelength of light that limits the resolution of the microscope. In addition, there is a problem with depth of field in viewing specimens that have much relief, such as, for example, an entire trilobite exoskeleton. As such, optical microscopes essentially provide a flat image.

Fortunately, we can use something other than photons of light to make images of the very small. In 1924, Louis de Broglie postulated that beams of electrons behaved in a manner similar to beams of light. As such, it would be possible to produce a microscope that used electrons rather than photons. In addition, as electrons have a much shorter wavelength than light, they would also have much greater resolution.

This idea would be taken forward by Max Knoll and Ernst Ruska during the 1930s, who, by 1935, had constructed the first crude scanning electron microscope, or SEM as we call them. By 1965, the Cambridge Instrument Company produced the first commercial machine, and SEMs have been an important part of many paleontologists' tool kits ever since.

Basically, an SEM is composed of an electron gun, which fires a beam of electrons into a vacuum chamber. A vacuum is necessary as electrons don't travel very far in air. The electrons are often generated by heating a metal filament; commonly tungsten, as it has a very high melting point. These electrons, being negatively charged, are then attracted to, and accelerated by, a positively charged electrode, or anode.

Rather than glass lenses, the electron beam in an SEM is focused by electromagnetic coils, and then the beam moved from side to side by another magnetic coil such that it scans over the surface of the object being studied. Some of the electrons get absorbed by the object, which then emits electrons back to the SEM, while others get reflected or backscattered. The electrons are collected by detectors and the specific pattern of electrons received used to help resolve a detailed image that's displayed on a monitor.

The resolution of an SEM is far greater than an optical microscope and can range up to around 300,000×. This is why they are often placed in the basements of buildings due to the sensitivity of instruments to vibrations, such as, for example, heavy truck traffic passing by a museum or a university. To increase magnification, the microscope just scans over a much smaller area of the object. An image is built through repeated scans, row by row, producing a more 3-dimensional image when compared to the flatter images of optical microscopes.

SEMs have another trick up their sleeves, though. Some of the electrons fired at the object travel deeper into the specimen, get absorbed, and cause a release of X-rays. These X-rays can then be used to determine the composition of an object being studied. All you need is an SEM fitted with an X-ray detector. This technique, called energy-dispersive X-ray spectroscopy, or EDS for short, has been used recently in a research team headed by Conrad Labandeira at the National Museum of National History. Conrad and his team were looking at exceptionally preserved material in Middle Jurassic and Early Cretaceous lake sediments in northeast China.

The fossils they found include insects in the order Neuroptera, the group that that modern lacewings belong to, that appeared to be taking on very butterfly-like habits by visiting plants for nectar about 40 million years before butterflies evolved. One of the fossils, *Oregramma illecebrosa*, even possessed eyespots very similar to modern owl butterflies that use them to trick predators that a large animal is nearby. The eyespots of this insect were analyzed using EDS.

EDS relies on the fact that each element emits characteristic X-rays when excited by a beam of electrons in an SEM. As a result, EDS allows for the composition of a specimen to be determined. From this analysis, it was proposed that these Jurassic lacewings were using the chemical melanin to create their eyespots, just like modern butterflies do today.

We have other tools for determining the composition of materials, though, and in some cases determining the relative proportions of very specific isotopic components of a material can provide vital environmental information about the past. An isotope is a variant of an element that differs only in the number of neutrons it contains in its nucleus. Some of these isotopes are unstable and decay into more stable elements over various time periods; others are stable and hang around in the environment.

For example, consider 2 isotopes of the element carbon: carbon-12 and carbon-13. Carbon-12 is by far the most abundant, comprising around 98.93% of all elemental carbon. Carbon-12 has 6 electrons and, in the nucleus, 6 protons and 6 neutrons, thus giving us an atomic mass number of 12. Carbon-13, however, has 7 neutrons in its nucleus, giving it the higher mass number of 13.

So what? Well, importantly in nature, different stable isotopes of elements get preferentially fractionated by differing processes over time. For example, in the case of carbon-12 and its heavier sister isotope carbon-13, carbon-12 is preferentially used by living things. This means that, proportionally, there will be a higher concentration of carbon-13 in the general environment, as a lot of carbon-12, the lighter isotope, is being concentrated by living things within the biosphere. During events such as a mass extinction, the uptake of carbon-12 gets severely reduced, flooding the environment with more carbon-12 and diluting the carbon-13 isotope, an important isotopic signal of past crises in the biosphere.

It's not just carbon, though. Isotopes of oxygen, sulfur, nitrogen, and a whole bunch of others react in very specific ways to different environmental factors that speak to various events in Earth's past, some of which we will examine in this series. To obtain these isotopes, though, we need a very specific piece of equipment called a mass spectrometer.

The substance you want to study gets placed in a vacuum chamber and then is bombarded with electrons. This knocks ions off the sample and transforms them into positive ions. These are then accelerated toward a negatively charged plate and the ion beam passed through a magnetic field. The lighter ions, such as carbon-12, are bent further than the heavier ones, such as carbon-13, creating an ion spectrum that can be read on a detector.

Mass spectrometers are used in another regard too, but this time using unstable radioactive isotopes. When a radioactive element decays, it emits radiation and is transformed into a different, more stable element. This is called radioactive decay. Each radioactive isotope has a specific and constant rate of decay from the original parent material to the stable daughter product. The rate of decay is described by the half-life of the isotope. That is the time it takes for half of the parent element to decay into the daughter product.

For example, potassium-40 decays into argon-40 with a half-life of about 1.25 billion years. That means that after 1.25 billion years, half the potassium-40 in a rock will have become argon-40. If a sample contained equal amounts of potassium-40 and argon-40, therefore it would be 1.25 billion years old. If the sample contained 3 atoms of potassium-40 for every 1 atom of argon-40, it would be 625 million years old, and so on.

It is usually igneous rocks that are used in radiometric dating. Igneous rocks form as magma, or lava, cools, forming crystals that trap small amounts of radioactive material. This can then be used to calculate the age of the rock. This technique has permitted the dating of materials from many periods of

Earth's history, including some of the most ancient. For example, Earth's oldest rocks from northern Canada, the Acasta gneiss, dated to about 4 billion years old; Moon rocks dated at around about 4.4–4.5 billion years; and meteorites, basically debris left over from the formation of the solar system, dated to 4.4–4.6 billion years. We will look more at how we date rocks and fossils, including using radiometric dating, in another lecture of this series.

Although I've included SEMs and mass spectrometers under new tools, they've been around for quite a while. There are now many new techniques, though, that would've appeared incredible even when I started my PhD back in the 1980s, much of it related to advances in computing power and innovations in instrumentation. An example currently being used at the Smithsonian Institution involves capturing precious fossils in 3 dimensions on a computer. The digitization program at the Smithsonian can capture incredible detail from a specimen using millions, sometimes billions, of points of measurements on its surface.

These wonderful pictures were provided by Conrad Labandeira at the museum, and are X-ray computer tomography images of a mite and a pseudoscorpion trapped in Baltic amber. The power of digitization has also been demonstrated recently by Smithsonian paleontologist Nick Pyenson, who uncovered a fantastic fossil site in the remote Atacama Desert of Chile. In 2010, a construction crew widening the Pan-American Highway exposed a remarkable whale fossil site. Nick had to document and move these fossils in a hurry before the area was paved over. The whole site was captured digitally by technicians from the Smithsonian Institution. To find out how they did this, catch the lecture on whale evolution later in this series.

Just because we have new technologies available to us, though, doesn't mean that we abandon more traditions tools. This is nowhere better seen than in the power of paleontological art and illustration. Science meets art when we need to reconstruct ancient environments and the organisms that lived in them. The Smithsonian has a rich history of paleontological art.

In 1995, dinosaur collections manager Michael Brett-Surman discovered a set of approximately 1200 illustrations of dinosaur skeletal material on top of a specimen storage cabinet. What he found were beautiful ink wash drawings that had been prepared for the famous paleontologist Othniel Charles Marsh in the late 19th century.

Even with all the wonderful new advances in imagining and data manipulation, the role of the scientific illustrator, like Mary Parrish in the Department of Paleobiology, is still vital both in research and in public display of materials. Very often an illustration can highlight features that might be too subtle to be picked out on a photograph, and also correct for problems such as poor depth of field or distortion that can occur with a camera lens.

Paleontologists require a wide variety of visual materials to illustrate their work, including the reconstruction of fossil specimens, restorations of ancient animals and plants, and various diagrams, graphics, and maps to help illustrate research. Often, drawings reveal structure, anatomy, and features that are not really grasped in photographic images.

For me, though, it is the collaboration between artist and scientist in the reconstruction of past environments that I find the most inspiring. The reconstruction of paleoenvironments begins with consultations between artist and paleontologist, perhaps with the scientist making a rough initial sketch.

From materials such as specimens, photographs, and a range of other background material, the scientific illustrator begins to bring to life lost landscapes and the animals and plants that populated them. The result is the point where art and science meet to produce wonderful images such as this, breathing life into worlds long since vanished. This image and others like it are the product of all that fieldwork, preparation, analysis, and interpretation that is part of the wonderful science of paleontology.

Lecture 4

How Do You Fossilize Behavior?

With the tools of logic and deduction, paleontologists can act as detectives to piece together the lives of long-dead creatures. In this lecture, you will discover a powerful class of fossils that essentially is the fossilized behavior of organisms: trace fossils. You will learn what trace fossils are and how they form; what they tell us about the evolution of life; what traces creatures leave about how they moved, fed, and built a home; and how fossilized behavior can track changes in an environment.

Trace Fossils

- Trace fossils are found in both marine and terrestrial environments and can be made by a variety of creatures. In sediments, they can be tracks and trails and burrows and borings. In the world of plants and insects, they record plant damage produced by feeding, egg depositing, pollinating, and a whole host of other activities.

- Ever since life became big, it has been interacting with the environment in a very physical manner. Creatures have been disturbing the physical structures that occur in sediments, such as fine laminations or ripple marks, and basically giving things a good mix, what is called bioturbation.

- The study of trace fossils is called ichnology and can essentially be regarded as the study of fossilized behavior. Unlike body fossils, which consist of the actual parts or impressions of an organism, trace fossils have limited use in biostratigraphy—the dividing up and correlation of rocks in a time sense—but trace fossils have several advantages in other areas of paleontology.

- First, trace fossils often develop under specific environmental conditions, making them great for paleoenvironmental interpretation. Another distinct advantage is that you can be certain that a trail or footprint has not been moved. This is a problem with body fossils: If you are using them to interpret an ancient environment, you have to make sure that the creature has not been moved from one environment to another post-mortem. But for a trace fossil, where you find it is where it formed.

- Second, they can also give us an appreciation of the activity of soft-bodied creatures that rarely fossilize. Trace fossils have been studied for a long time, although initially many of these lines and squiggles in rocks were often misidentified as seaweeds or worms. Dinosaur footprints are a little easier to interpret but were often regarded as footprints of huge flocks of birds.

- This misidentification of some of these tracks and trails as worms and plants may explain why trace fossil are named like fossil plants and animals, using the Latin binomial system with a genus and species name.

- The fact that we are not dealing with an individual species but a type of fossil behavior can cause some confusion, though. A trace fossil is given a generic Latin name—for example, *Rusophycus*. However, because this is a type of behavior—in this case, where an organism rested—*Rusophycus* can be produced by a whole range of different organisms.

- Another difference we have to consider relates to the different things organisms may do on a day-to-day basis. Just as a person potentially could leave multiple different traces in wet sand by making sand castles, digging for clams, walking, and running, any individual fossil organism could be responsible for an entire range of traces, depending on what it was doing.

- This potential confusion when studying and naming trace fossils is why we tend to classify them by the behavior they represent and not by the creatures that produced them.

- There are several types of traces that can be found. *Repichnia* are traces that an animal makes as it moves. *Fodichnia* describe various feeding

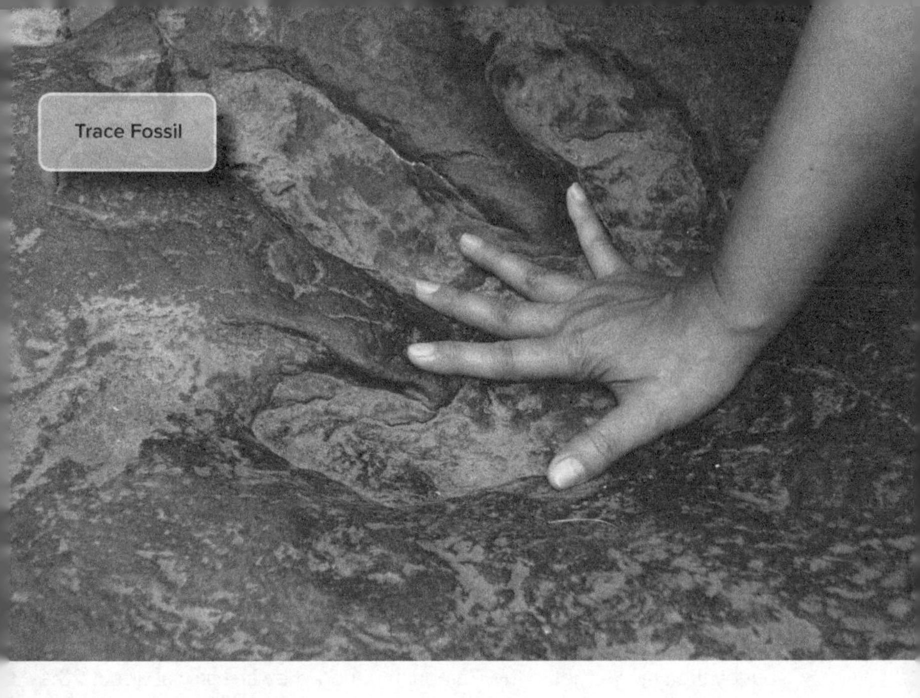

burrow structures. *Domichnia* are interpreted as places where an organism actually lived. *Cubichnia* covers all types of resting traces, such as *Rusophycus*.

- The divisions between some of these classifications might not be hard and fast. Perhaps a creature's living burrow could also act as a feeding burrow. However, these classifications provide a framework on which we can start to hang various types of fossil behavior.

Evolution and Diversification of Life

- Trace fossils are an important part of a paleontologist's arsenal for interpreting past behavior and environment, but there is another important component to the study of trace fossils: understanding the early evolution of the biosphere and the diversification of animal life.

- The first fossil evidence of life is found around 3.4 billion years ago in the fossil microbes found in the Strelley Pool sandstone of western Australia. In the same section, though, laminated structures have been found that have been interpreted as stromatolites, which are commonly produced by photosynthetic cyanobacteria. These bacteria trap grains of sediment in their sticky, mucilaginous sheath and then move upward through the sediment, creating a new layer. In this way, they commonly "dome upward," producing column-like sedimentary structures.

- Stromatolites can be classified as trace fossils because it is not the cyanobacteria that are being preserved—just the laminations and the structures they produce. For billions of years—although we find individual, single-celled microfossils—stromatolites are really the only large-scale evidence of life we have.

- Their diversity over time also has a story to tell. Throughout the Precambrian period, they are a common component of many shallow marine settings, but by the time we get to the Cambrian period, they are only at about 20% of their former abundance.

- This change in stromatolite abundance tells us that grazing organisms are now an important part of the ocean system. Grazers are now feeding on the bacteria on top of the sediments before they get a chance to form the beautiful columns and domes we see in the Precambrian.

- This change is called the Cambrian substrate revolution, where not only grazing animals were staring to have an impact on the planet, but also burrowing animals, who would churn the sediment still further. After the evolution of various worms and arthropods during the proliferation of life called the Cambrian explosion, sediments in shallow marine environments went from fairly firm, stabilized by microbial mats, to mushier.

- Since the Cambrian, stromatolites have been marginalized, restricted to extreme environments, places where the water may be too salty or too oxygen deficient for grazing creatures. Areas like these provide a rare window into the way the world must have looked billions of years ago, before the evolution of grazers.

- Another type of trace fossil, coprolites (fecal pellets), may have also made a significant contribution to the Earth system at this time. By producing large fecal structures that sink and deliver carbon rapidly to the ocean floor and sediments, they lessened the oxygen demand of surface waters. This may have promoted oxygen enrichment in the oceans and allowed for the evolution of larger and even more complex creatures.

- Generally, geological boundaries are marked by the incoming of a distinctive fossil species that can be correlated widely. There is one point in geological time, though—probably one of the of the most important points in the history of the planet—where it is trace fossils and not body fossils that are used to define the base of a geological period.

- The trace fossil assemblage *Treptichnus pedum* not only defines the base of the Cambrian period but also the whole Phanerozoic eon, marking a transition from the Precambrian into a new world full of complex life.

- Geological periods are defined at sections called international stratotypes, places where rocks are used to define a particular boundary in time within geological history. For the base of the Cambrian, the transition from the Precambrian to the Cambrian, and the entire Phanerozoic eon, is taken at a place called Fortune Head in Newfoundland.

- It is here that the appearance of *Treptichnus*, and an association of other trace fossils, defines the base of the Cambrian. It is thought that the creature producing this trace fossil was a creature called a priapulid worm, probing and searching in the sediment, either preying on or scavenging for small invertebrates just in or on the surface.

- The significance of this assemblage is profound. It marks a critical event in the biosphere in which organisms are now starting to dynamically interact with the physical environment, an evolutionary change in the behavior of life and the biosphere.

Advanced Behaviors

- Trace fossils can provide insights into how life leaves a record of some pretty fundamental behavior—specifically, how creatures moved, how they fed, and how some of them built a home.

- A mobile fauna is a big leap forward for the biosphere. By the time we get to the Cambrian, various creatures, such as trilobites, are engaged in a variety of different activities, from furrowing through sediments to skipping across the sediment surface.

- But not all marine creatures simply "meander" around on the surface of the ocean floor. Sometimes, movement is needed in another direction—when a catastrophe occurs. There is a class of trace fossils called *Fugichnia*, or escape traces.

- We can't talk about movement traces without mentioning dinosaur tracks. Although we have to be careful when interpreting such tracks, we can glean some very important information from them. For example, if you have a set of prints, you can calculate stride length. Also, when we shift from a walk and into a run, stride length will increase.

- Trace fossils have also helped answer questions about how marine vertebrates moved. A long paleontological debate has centered around how the Triassic semiaquatic marine reptile *Nothosaurus* swam. Did they sweep their limbs in a figure-8 motion (the mode of locomotion penguins use today), or did they employ a rowing motion with their limbs?

- Paleontologists from the University of Bristol and the China Geological Survey found a series of pairs of slot-like tracks preserved in mudstones from Yunnan in southern China. Analysis of the orientation and size of the traces indicate that they were made by the animal's forelimbs as they moved over the seafloor, using a rowing action in unison—not a figure-8 motion.

- Ever since the biosphere has started purposely moving around, it has been generating trace fossils. But complex life doesn't just move; it also has to eat and, in doing so, produces a whole additional type of trace fossils.

- One of strangest types is *Paleodictyon*, a honeycombed-shaped structure found in quiet deepwater environments from the Cambrian to the present day. A type of trace called *Agrichnia*, a farming trace, has been interpreted as organisms deliberately cultivating bacteria on the ocean floor. But probably the most common feeding traces are those made by deposit feeders, which are animals that process sediment for organic material and nutrients.

- *Domichnia* are dwelling traces. Some can be simple vertical burrows, such as *Skolithos*, or more complex ones, such as *Ophimorpha*, a trace decorated with "sediment balls" that the trace maker stuck on the outside of the burrow. Horizontal-branching dwelling traces called *Thalassinoides* can be generated by a number of creatures, including acorn worms, fish, and crustaceans. Recently, some unexpected dwelling traces have come to light in both western Montana and Victoria, Australia, that show that some dinosaurs dug burrows.

Environmental Analysis

- As useful as trace fossils are in paleobiology, they are also extremely useful in environmental analysis. Trace fossil assemblages, or ichnofacies, respond rapidly to the environmental conditions in which they form.

- During the 1960s, the utility of trace fossils in paleoenvironmental interpretation really took off. Pioneering work by researchers such as German paleontologist Adolf Seilacher and later Robert Frey have expanded our appreciation of trace fossils in this regard.

- The general character of these trace fossil ichnofacies has remained fairly consistent from the Cambrian to the present day. The implication of this is that although the producers of the traces have changed through time, the manner in which they were responding to the environment—their behavior—has not changed very much.

- The ichnofacies are named by reference to a particular characteristic trace fossil of each assemblage. *Nereites* ichnofacies are characterized by horizontal, meandering or spiral-feeding surface traces and sometimes the bacterial farming traces. *Cruziana* ichnofacies are very busy trace fossil assemblages full of feeding, moving, living, and resting traces. *Skolithos* ichnofacies are dominated by vertical burrows and are usually associated with coarser sandy material.

- Trace fossil assemblages not only tie down specific environments, but they can also help chart changes in environmental conditions over time, whether that is long-term change, such as with variations in sea level, or short-term events, such as storm surges.

- Sediment-based trace fossils, as well as evidence for the interactions between plants and arthropods, are an extremely valuable tool that we can use to understand the behavior of extinct organisms. They also chart critical environmental changes in a way that body fossils sometimes cannot.

Questions to consider:

1. What trace fossils might we humans be leaving for future paleontologists to discover?
2. Why can trace fossils rarely be tied to any one particular organism?

Suggested Reading:

Lockley, *Tracking Dinosaurs*.

Seilacher, *Trace Fossil Analysis*.

Lecture 4 Transcript

How Do You Fossilize Behavior?

Unusual circumstances can preserve exceptional fossils, from insects preserved in exquisite detail in amber to the intricate detail of tissues and cells revealed in coal ball peels to the soft-bodied creatures of the Burgess Shale over half a billion years old preserved on the side of a mountain. Such fossils tell us much about the paleontology and paleoanatomy of past life, details of how bodies were organized, and even details of soft parts that don't usually fossilize.

But how about how those animals lived? How they moved? How they ate? Did they live as loners, or did they associate in vast herds or shoals? Can we tell anything about their day-to-day life in the same way that a field biologist can about creatures living today? Can we add greater depth to the picture of a fossilized creature beyond skin, bone, and muscle?

This might seem like an impossible task for the paleontologist. The material we are looking at is, after all, as dead as a doornail—or as dead as my career as an underwear model, if you like—but the tools of logic and deduction can help paleontologists act as detectives to piece together the lives of long dead creatures. In previous lectures, I've demonstrated how the fossils themselves can provide clues that can be used to bring dead bones and shells back to life, but there is something else, too, a powerful class of fossils that I'd like to bring to your attention today, a class of fossils that essentially is fossilized behavior of organisms.

So in this lecture let's ask, what is a trace fossil and how do they form? What do trace fossils tell us about the evolution of life? What traces do creatures leave about how they moved, fed, and built a home? And how can fossilized behavior track changes in environments?

These are trace fossils, very recent ones. They show that an animal passed this way not long after this cement was laid. From the prints, you can tell many things: the gait or stride of the animal; potentially how heavy and how large it was. It also ties this animal to this specific environment, and something else about its behavior, too—in this case, a dog's unswerving interest in trotting across cement that you've just spent several hours getting just right.

This is also a trace fossil. It shows an insect larva, how it fed on a leaf, its host plant. Not only do we know the name of the host plant from a perceptive botanist, but we can also know the group of insects that produced this mine on the leaf from an equally perceptive entomologist. And if we are truly interested in this interaction, we can determine what tissues this larva preferred to feed on, and in this way, something about the evolutionary history of insects that mine leaves for a living.

Trace fossils are found in both marine and terrestrial environments and can be made by a variety of creatures. In sediments, they can be tracks and trails and burrows and borings; in the world of plants and insects, they record plant damage produced by feeding, egg depositing, pollinating, and a whole host of other activities. Ever since life got big, it has been interacting with the environment in a very physical manner. Creatures have been disturbing the physical structures that occur in sediments like fine laminations or ripple marks and basically giving things a good mix, what we call bioturbation.

The study of trace fossils is called ichnology and can essentially be regarded as the study of fossilized behavior. Unlike body fossils that consist of the actual parts or impressions of an organism, trace fossils have limited use in biostratigraphy—that's the dividing up and correlation of rocks in a time sense. But trace fossils have several advantages in other areas of paleontology.

Firstly, trace fossils often develop under specific environmental conditions, making them great for paleoenvironmental interpretation. Another distinct advantage is that you can be certain that a trail or footprint has not been moved. This is a problem with body fossils if you're trying to use them to interpret an ancient environment, as you have to make sure that the creature has not been moved from the environment to another environment postmortem. But for a trace fossil, where you find it is where it formed.

Secondly, they can also give us an appreciation of the activity of soft-bodied creatures that rarely fossilize. Trace fossils have been studied for a long time, although, initially, many of these lines and squiggles in rocks were often misidentified as seaweeds or worms. Dinosaur footprints are a little easier to interpret, of course, but were often regarded as footprints of huge flocks of birds.

This misidentification of some of these track and trails as worms and plants may explain why trace fossil are named like fossil plants and animals, using the Latin binomial system with a genus and a species name. The fact that we are not dealing with an individual species but a type of fossil behavior, though, can cause some confusion; you just ask some of my long-suffering students. For example, consider this trace fossil. It is given the generic Latin name of *Rusophycus*. However, as this is a type of behavior—in this case, where an organism rested—*Rusophycus* can be produced by a whole range of different organisms: by a polychaete worm, a snail, a shrimp, and a trilobite, just to name a selection.

Another difference we have to consider relates to the different things organism may do on a day-to-day basis. Just as I could potentially leave multiple different traces in wet sand by making sand castles, digging for clams, walking, running—or, more commonly in my case, tripping and falling flat on my face—any individual fossil organism could be responsible for an entire range of traces just depending on what it was doing at the time.

Consider this visualization. A single trilobite is shown moving from a resting position, where it generates one of those *Rusophycus* resting traces, to starting to a furrowing behavior called *Cruziana*, and then picking up its skirts—well, actually, its pleural flanges—and doing what we call in Britain a runner and scarpering on out of there. In this trace fossil, we can tell based on the orientation of the marks in the *Cruziana* trail that this individual approached from the right side, chewing through the sediment as it moved, and eventually, after a busy time plowing through the muck, decided to take a break and rest for a while, leaving another *Rusophycus*, those resting traces.

There are, of course, more than just resting, feeding, and moving traces. Just see how many traces the common fiddler crab can leave behind, from a living burrow, various types of walking traces, a radiating trace of processed sand that the animal leaves when it's feeding, and poop. Yep, poop is a trace fossil, too. This potential confusion when it comes to studying and naming trace fossils is why we tend to classify them by behavior they represent and not by the creatures that produced them. I would like to give you a brief description of the types of trace—by behavior—that can be found. We will look at some specific examples in more detail a little later on.

These are *Repichnia*, traces that an animal makes as it moves. The *Cruziana* left behind by a trilobite trail or the footprints of a dinosaur are *Repichnia*. *Pascichnia* are traces left by animals not only moving but grazing at the same time. There is also evidence of what can only be described as paleontological farming, the so-called *Agrichnia*—for example, the honeycombed *Paleodictyon* that we'll meet later.

Fodinichnia describe various feeding burrow structures. *Domichnia* are interpreted as traces where an organism actually lived; for example, a prairie dog tunnel system. And *Cubichnia* covers all types of resting traces, like the *Rusophycus* we described earlier and this impression of a resting starfish called Asteriacites from the Devonian of northeast Ohio.

As you may have realized, the divisions between some of these classifications may not be hard and fast—perhaps a creature's living burrow could also act as a feeding burrow. It does however provide a framework upon which we can start to hang various types of fossil behavior. As you can see, trace fossils are an important part of a paleontologist's arsenal for interpreting past behavior and environment, but there is another important component to the study of trace fossils, that is in understanding the early evolution of the biosphere and the diversification of animal life.

The first fossil evidence of life is found around 3.4 billion years ago in the fossil microbes found in the Strelley Pool sandstone of western Australia. In the same section, though, laminated structures have been found that have been interpreted as stromatolites. Stromatolites are laminated structures commonly produced by photosynthetic cyanobacteria. These bacteria trap grains of sediment in their sticky mucilaginous sheath and then move upward through the sediment, creating a new layer. In this way, they commonly dome upward, producing column-like sedimentary structures. You can see modern examples exposed above water in the Hamelin Pool Marine Nature Reserve in western Australia.

Stromatolites can be classified as trace fossils as it is not the cyanobacteria that are being preserved, just the laminations and the structures they produce. For billions of years, although we find individual, single-celled microfossils, stromatolites are really the only large-scale evidence of life that we have. Their diversity over time also has a story to tell. Throughout the Precambrian, they are a common component of many shallow marine settings, but by the time we get to the Cambrian period, they are only at about 20% of their former abundance.

This change in stromatolite abundance marks a critical part of the story of life on Earth. It tells us that grazing organisms are now an important part of the ocean system. Grazers are now feeding on the bacteria on top of the sediments before they get a chance to form the beautiful columns and domes we see in the Precambrian.

This change is called the Cambrian substrate revolution, where not only grazing animals were starting to have an impact on the planet, but also burrowing animals too, who would churn the sediment still further. After the evolution of various worms and arthropods during the proliferation of life we call the Cambrian explosion, sediments in shallow marine environments went from fairly firm—stabilized by all those microbial mats—to more, well, mushy.

Since the Cambrian, stromatolites have been marginalized, restricted to extreme environments, places where the water may be too salty or too oxygen poor for grazing creatures. Areas like these in Shark Bay, Australia provide a rare window into the way the world must have looked billions of years ago before the evolution of grazers.

Another type of trace fossil, coprolites or fecal pellets—poop—may have also made a significant contribution to the Earth's system at this time. By producing large fecal structures that sink and deliver carbon rapidly to the ocean floor and sediments, they lessened the oxygen demand on surface waters. This may have promoted oxygen enrichment in the oceans and allowed for the evolution of larger and even more complex creatures. We should all be thankful for poop.

But there is one more interesting item regarding the evolution of trace fossils that I'd like to highlight. Remember I said that trace fossils are pretty useless when it comes to biostratigraphy—that is, timing events in geological history. Generally, geological boundaries are marked by the incoming of a distinctive new fossil species that can be correlated across a wide area. For example, the base of the Carboniferous period is defined by the appearance of a conodont microfossil, *Siphonodella sulcata*, in a geological section in Brittany in northwestern France.

There is one point in geological time, though, probably one of the most famous and important points in the history of the planet, where it is a trace fossil and not a body fossil that are used to define the base of a geological period. But not only a period, for the trace fossil assemblage *Treptichnus pedum* not only defines the base of the Cambrian period but also the whole Phanerozoic eon, marking a transition from the Precambrian into a world full of complex life.

Geological periods are defined at sections called international stratotype sections, places where rocks are used to define a particular boundary in time within geological history. For the base of the Cambrian, the transition from the Precambrian to the Cambrian, and the entire Phanerozoic eon, is taken at a place called Fortune Head, Newfoundland. It is here that the appearance of *Treptichnus* and an association of other trace fossils defines the base of the Cambrian. It is thought that the creatures producing this strange trace fossil was a creature called a priapulid worm, probing in the sediment, preying upon or scavenging for small invertebrates just in or on the surface of the sediment.

The significance of this assemblage is profound, though. It marks a critical event in the biosphere in which organisms are now starting to dynamically interact with the physical environment, an evolutionary change in the behavior of life and the biosphere.

Let's focus in on some of that advanced behavior now. Let's see how trace fossils provide insights into how life leaves a record of some pretty fundamental behavior, specifically how creatures have moved, how they fed, and how some of them built a home. A mobile fauna is a big leap forward, if you'll pardon the pun, for the biosphere. As we saw earlier, by the time we get into the Cambrian, various creatures, like trilobites, are engaged in a variety of different activities, from furrowing through sediments to picking up their skirts and scurrying across it.

But not all marine creatures simply meander around the surface of the ocean floor. Sometimes movement is needed in another direction when a catastrophe occurs—well, a catastrophe if you're a marine invertebrate. I'm thinking here of a class of trace fossils called *Fugichnia*, or escape traces. One such trace, *Diplocraterion*, is a U-shaped burrow that shows curved laminations or spreite on the inside and outside of its U-shaped burrow. The outer spreite form as the organism movies rapidly, adjusting its position upward in response to a pile of sediment being very rapidly dumped upon it.

It would be wrong to assume that all trace fossils are found in marine deposits, though, and to illustrate this, here is a movement trace probably created by an insect like a bristletail—insects closely related to silverfish—that were walking, and sometimes jumping, across a tidal flat during the Lower Permian about 270 million years ago in southern New Mexico. After the initial description of these unique traces and trails in 1997, the same types of traces are now being found in several other localities worldwide. Now that we know what to look for, these traces are turning up everywhere.

Of course, we can't talk about movement traces without mentioning dinosaur tracks. We have to be careful when interpreting these, though. Just from this example here, a bunch of questions present themselves. For example, were these tracks made all at once or at different times? If at different times, how soon after: seconds, minutes, hours, days? Also, how many individuals? We obviously have 2 species of print, but was it just 2 individuals or many? But we can glean some very important information about such tracks. For example, if you had a set of prints, you can determine stride length, the distance between one footfall and the next footfall. Also, when we shift from a walk and into a run, stride length will increase.

A British zoologist, Robert McNeill Alexander, saw an opportunity here when he noticed a relationship between stride length and an animal's speed, providing you know the distance from the hip to the ground. He also noted that this relationship was true whether the animal was a quadruped, like my friend Jacob here, or a biped like an emu, human, or, of course, a 2-legged dinosaur.

He developed a formula that expressed this relationship, where U is the velocity, g is the acceleration of free fall due to gravity, d is the stride length, and h the height of the hip from the ground's surface. The whole thing is simplified to this. There is an obvious problem here though. Because we're looking at a trace fossil, the individual who produced the trace is often unknown. In the case of dinosaurs, fortunately, we have an out, as there is a predictable relationship between hip height and foot length for each dinosaur group.

Hip height is found to be about 4 to 6 times foot length, depending on what group of dinosaurs you are dealing with. From this, we think that the walking speed for most dinosaurs is about 1 to 4 meters per second, that's similar to humans, with running speeds around 10 to 15 meters per second for some of the smaller, carnivorous, bipedal dinosaurs. The fastest dinosaur is clocked at 54 kilometers per hour. That's about 35 mph, similar to a fast racehorse today.

Trace fossils have also helped answer questions about how marine vertebrates moved, too. Consider the Triassic semiaquatic marine reptile Nothosaurus. A long paleontological debate has centered around how these animals swam. Did they swim with their limbs in a kind of a figure-of-eight motion? This is the mode of locomotion penguins use today. Or did they employ more of a rowing motion with their limbs?

Paleontologists from the University of Bristol and the geological survey in China found a series of pairs of slot-like tracks preserved in mudstones from Yunnan in southern China. Analysis of the orientation and size of these traces indicate that they were made by the animal's forelimbs as they moved over the seafloor using a rowing motion in unison, not a figure-of-eight motion.

But there's more, too. These fossils also provide insight into some of the hunting behavior of these animals. Some traces have been interpreted as foraging trails. Apparently Nothosaurus was using its forelimbs to disturb the mud on the ocean floor, and then snap at fish or crustaceans as they tried to swim away—all this biomechanical and behavioral information about a long-extinct animal provided by marks and gouges in sedimentary rock.

And, of course, our branch of the biosphere has been leaving footprints in paleontological wet cement for many millions of years as well. Some of the first, and probably some of the most famous, are the Laetoli footprints in Tanzania discovered by Mary Leakey in 3.6-million-year-old volcanic ash during her 1976 field season. The prints were probably made by *Australopithecus afarensis*, the same species as the famous Lucy fossil. These footfalls record humans walking through ash made wet by rainfall. As we have noted, many things can be deduced from a footprint.

These show that our early ancestors walked by the heel of the foot first, followed by the toes, pushing off at the end of the stride, the so-called heel-strike toe-off walking, very similar to the way that we walk today. Other footprints belonging to our ancestors are being studied at the Smithsonian. Footprints of *Homo erectus* in Ileret, Kenya dating to 1.51–1.53 million years ago have been found showing how humans at this time were really starting to walk very much like we do today. Researchers, such as National Museum of Natural History scientists Dr. Richard Potts and Kay Behrensmeyer, are hoping that footprints will be found at this site that will provide even more insight into the evolution of human walking and running, too.

So, ever since the biosphere has started purposely moving around, it has been generating trace fossils. But complex life doesn't just move, it also has to eat, and in doing so produces a whole additional type of trace fossil. We have already mentioned one of strangest types earlier, *Paleodictyon*, a honeycombed-shaped structure found in quiet deepwater environments from the Cambrian to the present day. This particular type of trace is called an *Agrichnia*, a farming trace, and has been interpreted as organisms deliberately cultivating bacteria on the ocean floor.

But what could possibly be producing such regular geometric patterns? The deepwater submersible DSV *Alvin* went searching for modern *Paleodictyon* near volcanic vents that lie about 3500 meters underwater along the Mid-Atlantic Ridge. They found and took samples of structures very similar to *Paleodictyon* but unfortunately could not find the creature that produced them. The mystery continues.

But probably the most common feeding traces are those made by deposit feeders, animals that process sediment for organic material and nutrients. There are many different types but 2 common ones are *Rhizocorallium* and *Zoophycos*. *Rhizocorallium* is generally found as burrows that are shallowly inclined to the sediment bedding planes. They can be can be over a meter long—most are smaller—but the animal that produced these structures is still a mystery, one of many that keeps paleobiology such a fascinating and exciting discipline. *Zoophycos*, by contrast, produces a more vertical spiral trace as it moves through the sediment searching for goodies.

Of course, not all feeding traces are vertical; some just wander over the surface of the sediment. These particular sediment gulpers will generally produce traces that have no overlapping parts, a good strategy as you want to avoid processing the same area twice.

But everyone needs a place to call home. Dwelling traces, or *Domichnia* in trace fossil terminology, can be simple vertical burrows, like *Skolithos* here, or more complex ones like *Ophiomorpha*, a trace decorated with sediment balls that the trace maker stuck on the outside of the burrow. These are horizontal branching dwelling traces called *Thalassinoides* found in Jurassic limestone in southern Israel. Such burrows can be generated by a number of creatures like fish and crustaceans, but specifically by Thalassinidea, after which this particular type of trace is named.

Recently, some unexpected dwelling traces have come to light in both western Montana and Victoria of Australia that show that some dinosaurs dug burrows, and usually we actually find the remains of the creature producing these structures in the bottom of the burrows themselves. This is Oryctodromeus, meaning "digging runner," a genus of fairly small—about 6.8 feet long—ornithopod dinosaur from the mid-Cretaceous. The burrows in Montana are about 2 meters long and 20 centimeters wide.

The possibility of vertebrate burrowing behavior can likely be pushed back even further than this, though. In the Triassic rocks of Argentina's Ischigualasto Basin, a series of large burrows dating to around 230 million years ago have been discovered by paleontologist Carina Colombi and colleagues. It is thought that at least one type of these burrows was produced by cynodonts, a type of reptile that were closely related to the early mammals.

As useful as trace fossils are in paleobiology, they are also extremely useful in paeloenvironmental analysis, too. Trace fossil assemblages, or ichnofacies as they're known, respond rapidly to environmental conditions in which they form. During the 1960s, the utility of trace fossils in paleoenvironmental interpretation really took off. Pioneering work by researchers such as German paleontologist Adolf Dolf Seilacher and, later, Robert Frey have really expanded our appreciation of trace fossils in this regard.

It's interesting that the general character of these trace fossil ichnofacies has remained fairly constant from the Cambrian to the present day. The implications of this is that, although the producers of the traces have changed through time, the manner in which they were responding to their environment—their behavior—has not changed very much at all. Or put it another way: it's a bit like a play that's been running for 100s of millions of years, but the script has largely stayed the same, but the actors occupying the roles has changed many, many times. The ichnofacies are named by reference to one particular characteristic trace fossil of each assemblage. Let's look at a very small selection from the deep to shallow water.

The *Nereites* ichnofacies is characterized by horizontal, meandering, or spiral feeding traces, and sometimes the bacterial farming traces that we mentioned earlier. Vertical burrows aren't very common. This assemblage is indicative of very quiet, low-energy water conditions typically found in deeper oceanic settings at a distance from shore.

The *Cruziana* ichnofacies is a very busy trace fossil assemblage, full of feeding, moving, living, and resting traces. This facies is usually characterized by sands and silts deposited in shallow marine settings with a high biodiversity.

The *Skolithos* ichnofacies is our shallowest facies and is dominated by vertical burrows, and is usually associated with coarser sandy material like you would find close to shore in a surf zone. Here, shifting sediment on the ocean makes living on the surface of the substrate difficult. This explains why creatures that burrow vertically tend to dominate.

You can see how tempting it would be to link different ichnofacies to water depth, and there is a bit of a relationship, but probably a more useful factor to consider would be water turbidity. Consider, for example, the case of the Cardium Formation, a rock formation of the late Cretaceous of Alberta in Canada. The sediments demonstrate an alternation in conditions between quiet and more turbid water.

During quiet times, the sediments are dominated by sands and silts and are characterized by typical *Cruziana* ichnofacies. The traces are mostly produced by mobile carnivores and deposit feeders chewing through or plowing over the fine-grained nutrient and organic-rich sediment. This relatively quiet environment is occasionally interrupted, though, by coarser clean wash sands that contain a typical vertical orientated *Skolithos* assemblage, demonstrating many of those escape *Fugichnia* burrows that we described earlier.

These clean sands have been interpreted as storm events, with sediment being washed back from nearshore environments into deeper waters by storm surge ebb currents. During these events, the quieter water *Cruziana* fauna would either be killed or displaced and replaced by the high-energy *Skolithos* fauna. When the storm surge currents waned and normal sedimentation resumed, the *Cruziana* fauna returned.

So, trace fossil assemblages not only tie down specific environments, they can also help chart changes in environmental conditions over time, whether that is long-term change, as you might expect with variations in sea level, or just short-term events, like the ones we've just examined from the Cardium Formation.

All of these sediment-based trace fossils, as well as evidence for the interactions between plants and arthropods, are an extremely valuable tool that we can use to understand the behavior of extinct organisms. They also chart critical environmental changes in a way that body fossils sometimes cannot provide. We've barely scratched the surface of the study of ichnology and plant-arthropod interactions. Hopefully, though, you can appreciate what a vital part fossil behavior in all of its variety has to play in the interpretation of our planet's history.

Lecture 5

Taxonomy: The Order of Life

In this lecture, you will consider one of the fundamental underpinning pillars of paleontology: the science of classifying and naming organisms—the science of taxonomy. To some, this may sound trivial, but without it, there would be no paleontology. In this lecture, you will learn who Carl Linnaeus was and what Linnaean classification is, how taxonomy is different for paleontology, and why classification is important in paleontology.

Linnaean Classification

- In 1735, Carl Linnaeus published the first edition of *Systema naturae*, which had a profound effect on biology and paleontology. In this book, all of creation is organized into 3 major kingdoms. Each of those kingdoms is divided into subgroupings of class, order, genus, and species—significantly fewer than the subdivisions we have today. Naturalists before Linnaeus often used a somewhat arbitrary grouping of creatures—for example, groupings that comprise all creatures that live in water or all domestic animals. Linnaeus was one of the first to group genera into higher taxa based on somewhat logical similarities.

- Linnaeus's 3 kingdoms are the animal kingdom, the plant kingdom, and the mineral kingdom.
  - The animal kingdom is comprised of Mammalia (mammals), Aves (birds), Amphibia (including retiles and non-bony fish), Insecta (all arthropods, not just insects), and the Vermes (basically all other invertebrates, including worms, mollusks, and echinoderms).
  - For the plant kingdom, Linnaeus creates a system of 24 classes of plants based on the number and organization of a plant's sexual organs,

Carl Linnaeus

the male stamens and female pistils and related reproductive features. This wasn't without controversy; the way Linnaeus would focus on the sexuality of his classification offended some.

  - In Linnaeus's time, many believed that minerals possessed a basic "life force," and as such, minerals form part of Linnaeus's system of classification. The mineral kingdom was divided into Petrae (rocks), Minerae (minerals and ores), and Fossilia (fossils and aggregates).

- Linnaeus published 12 editions in his lifetime, continually revising and updating his classifications. The manner in which we order life today is quite different than Linnaeus's original efforts, but much of the legacy of his efforts are still with us. We still have a hierarchical organization of life, and we still have the scientific binomial system.

- In the days before the *Systema naturae*, naming creatures could be quite messy. Take, for example, the tomato. Prior to Linnaeus, it went by the rather grand and long-winded name of *Solanum cauke inermi herbaceo, folis pinnatis incises*: "The solanum with the smooth stem which is herbaceous and has incised pinnate leaves." Under the Linnaean binomial system, it becomes *Lycopersicon esulentum*—much less of a mouthful.

- Because of the hierarchical system of *Systema naturae*, you don't need to list all the descriptive components of a species; all you need is the name of the genus followed by the name of the species—for example, *Passer domesticus* (the house sparrow) and *Acheta domesticus* (the house cricket).

- It is a simple but powerful system. The species, or specific, name of any member of a genus could be used for other, different genera, such as *domesticus*, which is also used as the species name of several other plants and mammals. However, genera will always be unique.

- Today, the system of classification used by many biologists and paleontologists is called cladistics, which considers the "shared and derived" characteristics of creatures when classifying them, rather than a superficial "appearance." For example, under cladistics, there is no grouping called "fish," or "class Pisces," as the group is traditionally understood.

- When you actually study the characteristics of certain fish—for example, a lungfish and a cod—you will find that a lungfish shares more features in common with a frog than it does with a cod, even though a lungfish under the Linnaean system would be classified under Pisces, "fish."

- In cladistics, by contrast, groupings that only contain all of their descendants can be considered as a legal classification, or what is defined as monophyletic.

- In classification using cladistics, relationships between organisms are illustrated using a cladogram, a branching diagram of relationships supported by derived character states. Cladograms are not evolutionary trees, and ancestors are not shown at branching points.

How Is Classification Different for Paleontology?

- Linnaeus placed fossils within his mineral kingdom under Fossilia. Unlike rocks and minerals, the binomial system for paleontology persisted—which is understandable, given that we are dealing with former life—and the zoological or botanical codes of taxonomy that apply to living animals and plants likewise apply to fossil forms, too.

- However, the problem that we have with fossils compared with living creatures is that, as fossils, a lot of the information that could be used to classify these creatures is simply gone. As such, drawing the lines between species can be difficult.

- Imagine how different human beings can look depending on their sex, age, historical background, and environment. Add to that the problem that paleontologists may be dealing with incomplete, fragmentary, or otherwise modified material and the problem is compounded.

- For a biologist, differences within the same species can be tested by simply observing that living species, watching how a species develops and changes over time and recording differences that might occur to the same species due to environmental factors. And if you observe 2 individuals mating—however different they may look superficially—and producing viable offspring, you can be sure that they are of the same, or at the least very closely related, species.

- Today, biological classification is further aided by studying the genetic similarity between creatures, allowing us now, more than any time before, to start to place life into real groupings based on real genetic similarity. This, with the exception of rare and fairly recent fossils, is not available to the paleontologist.

- Three of the most iconic dinosaurs can help illustrate some of the problems that paleontologists face. *Brontosaurus*, *Stegosaurus*, and *Triceratops* were discovered by famous paleontologist Othniel Marsh in the late 1800s.

- *Brontosaurus* was part of a treasure of dinosaurs recovered from the western United States by famous dinosaur hunters during the 1870s. Marsh discovered the skeletons of 2 partial sauropod dinosaurs—the group to which these 4-legged, long-necked dinosaurs belong—and sent them to the Peabody Museum at Yale. He named the first specimen *Apatosaurus ajax*, or "deceptive lizard."

- In 1903, he named the second skeleton and decided it was sufficiently different—not only to be considered a different species, but also a completely new genus. As such, *Brontosaurus excelsus*, or "noble thunder lizard," was born.

- After Marsh described these 2 specimens, skeletons belonging to similar sauropod dinosaurs were excavated, and upon analysis, it was determined that one species fell on a morphological spectrum somewhere between *Apatosaurus* and *Brontosaurus*. As such, the differences between the 2 end members of the group didn't appear so extreme after all, and the skeletons, including the new species, were all placed in the same genus.

- Taxonomically, the first named specimen has precedence, so all of these animals became apatosaurs, with *Brontosaurus excelsus* renamed *Apatosaurus excelsus*.

- When you make a taxonomic determination, you are effectively proposing a hypothesis regarding the position of a particular living, or fossil, organism within the 4-billion-year-old tree of life. Any new information, such as data from new analytic techniques or additional specimens, may help revise that hypothesis.

- For *Brontosaurus*, new information would be released in 2015 in a paper by British and Portuguese paleontologists Roger Benson, Octávio Mateus, and Emanuel Tschopp.

- Determining when some fossil is sufficiently different from another, to be placed in an entirely new genus, is not strictly governed by any clear taxonomic guidelines. A judgment call has to be made.

- Since the time that the genus *Brontosaurus* had been demoted, however, new sauropod specimens had been recovered, and these paleontologists took advantage of the new discoveries to apply an extensive statistical analysis of the differences between various features of these animals.

- In concentrating their analyses on the broad group to which the apatosaurs (the diplodocid dinosaurs) belong, they found that the difference between widely accepted genera within the diplodocids were at the very least the same as the differences between *Apatosaurus* and what Marsh had originally described as *Brontosaurus.* This shows how taxonomy is dynamic and potentially subject to change with every new discovery that is made.

- An exceptional discovery of a "*Stegosaurus* graveyard" in Montana has permitted the analysis of a large well-preserved population of stegosaurs. Like *Brontosaurus*, this dinosaur was also originally discovered and named by Marsh in 1877 from the Jurassic Morrison Formation in southwestern Wyoming. Initially, Marsh thought that the plates of the *Stegosaurus* lay flat on its back like shingles—hence the name *Stegosaurus*, or "roofed lizard."

Marsh would later rethink his interpretation, giving us the classic spiky-backed dinosaur we know today.

- When considering the overall morphology of the stegosaurs and the microscopic bone structure, graduate student Evan Saitta of the University of Bristol was able to determine that all the stegosaurs found in the deposit were adults and that they all belonged to the same species, *Stegosaurus mjosi*.

- Even so, there was a particular difference he found in the shape and arrangement of plates on their backs: Some specimens had plates that were pointed and tall while others possessed plates that were broader and rounded. Saitta suggests that this could represent sexual dimorphism in the dinosaurs. He proposed that the broad, round plates belong to the males and the tall, pointed plates belong to the females.

- Although this hypothesis is not accepted as evidence of sexual dimorphism by all, this does illustrate how we need to be careful when naming our dinosaurs or any other fossils. It is possible that if fewer, more poorly preserved specimens were recovered, these 2 forms could have been interpreted as 2 different species, rather than male and female of the same species.

- *Triceratops* was first discovered near Denver, Colorado, in 1887 and was originally described by Marsh as a bison, *Bison alticornis*. However, he eventually realized that they belonged to a horned dinosaur he named *Triceratops*. A controversy would erupt regarding this dinosaur in 2009, when paleontologist Jack Horner from the Museum of the Rockies and his graduate student John Scannella would propose a hypothesis that would significantly reduce the number of dinosaurs we have on the books.

- They suggested that *Triceratops* and *Torosaurus*, another horned dinosaur discovered by Marsh, were the same species, with *Triceratops* being the juvenile and *Torosaurus* being the adult. They even proposed an intermediate “teenager” in the genus *Nedoceratops*.

- These dinosaurs look quite different, though. *Torosaurus* has a much larger frill, which is perforated with large oval holes—perforations that are lacking in *Triceratops*. Although these animals overlap in time, they were regarded as being so different that they were not only different species but also different genera. But Horner and Scannella take a different view, claiming that these differences just reflect different developmental stages of the same dinosaur.

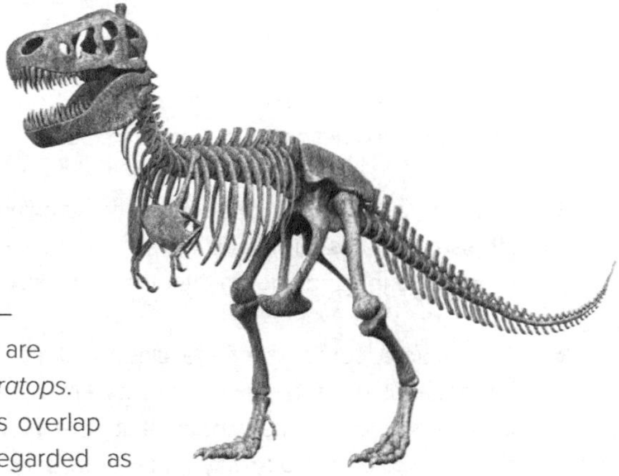

- If Horner and Scannella are proved to be correct, it would mean that we lose *Nedoceratops* and *Torosaurus* as valid Linnaean genera—all of them becoming different growth stages of *Triceratops*, which, as described first, takes taxonomic precedence.

- Horner thinks that this could be part of a wider problem. He estimates that perhaps more than 1/3 of all dinosaur species in the Late Cretaceous, where *Triceratops* is found, may never have existed. He believes that many may just represent different growth stages, misinterpreted as separate species.

Why Is Classification Important?

- It may appear that questions about classification are very academic, but just consider the debate sparked by the classification of *Triceratops*, *Nedoceratops*, and *Torosaurus*.

- If we accept the views of Horner and Scannella regarding the number of "real" dinosaur species at the end of the Cretaceous, we have a much more impoverished dinosaur population than was previously thought prior to the impact of the extraterrestrial body that marks the Cretaceous-Paleogene extinction 66 million years ago.

- This gives us a very different understanding of the paleoecology and stresses that this formerly successful and very biodiverse group may have been experiencing prior to their final extinction.

- The ongoing process of the classification of fossils helps deepen our understanding of biodiversity over time. Taxonomy refines our focus through the deep-time window that paleontology affords us. This is vital, as paleontology is our only long-term benchmark against which we can compare modern changes in the biodiversity and current health of our ecosystem. In fact, paleontological taxonomy and classification could prove vital in charting our planet's future.

Questions to consider:

1. What problems do palaeontologists face when attempting to classify fossils?
2. With a better understanding of the relationships between organisms, do groups like "fish" and "reptiles" make sense anymore?

Suggested Reading:

Blunt, *Linnaeus*.

Foote and Miller, *Principles of Paleontology*.

Lecture 5 Transcript

Taxonomy: The Order of Life

In 2007, the Smithsonian Institution was approached by the Swedish Embassy in DC and asked if they would like to borrow a very special item. This particular object would be displayed at the museum, under guard, for 2 days and would be the spark for a symposium at which international scientists would give lectures.

What was this object? What was it that could inspire such excitement? A fabulous diamond or ruby? An incredible dinosaur or mammoth fossil, perhaps? No, the object was a book, a first edition of a book that was actually the personal copy of the author. It was a book that was published in 1735, some 125 years before Darwin's *On the Origin of Species*, but a book, like Darwin's, that's had a profound effect on biology and paleontology. That book was the *Systema Naturae* and its author was a very colorful character called Carl Linnaeus.

In this lecture, we are going to be considering one of the fundamental, underpinning pillars of paleontology; the science of classifying and naming organisms—the science of taxonomy. Now to some this might sound somewhat, well, trivial, but trust me, without it there would be no paleontology. So, in this lecture, we will ask: Who was Carl Linnaeus, and what is Linnaean classification? How is taxonomy different for paleontology? And why is classification important in paleontology?

So who was this extraordinary man who could inspire an ad hoc international conference at the spur of a moment? Linnaeus was born in an exciting time, the time of the Enlightenment, also known as the Age of Reason. From the late 1600s to the early 1800s there was a great flowering of debate and questioning of traditional views and authorities, a time when there was much talk of societal reform and tolerance, a time when skepticism and science were the touchstone for debate.

It would see the emergence of philosophers like Descartes, Voltaire, and Hume, and scientists like James Hutton, the founder of modern Geology; Antoine Lavoisier, one of the founders of modern chemistry; and physicist Sir Isaac Newton. And there was Carl Linnaeus born 23 May, 1707, in the countryside of southern Sweden in Råshult. His father was a Lutheran minister but, probably more importantly for science, Nils Ingemarsson Linnaeus was also a keen amateur naturalist and avid gardener. It was probably from this that Linnaeus got his lifelong love of plants.

In 1727, he went to Lund University to study medicine, but transferred to Upssala a year later to continue his studies. Apparently the standard of medical training was not of the highest quality, so he spent most of his time just collecting plants. He got such a reputation for botany that he was asked to teach at the university as a lecturer.

By 1732, and only in his mid-20s, he was leading an expedition to Lapland where he'd become fascinated with the local Sami people and in particular their use of medicinal plants. He published details in *Flora Lapponica* in 1737, probably making him one of the world's first ethnobotanists. You can see Linnaeus pictured here, dressed in native clothing of the Sami. He is also holding a sprig of the twinflower, an emblem that features in many of the paintings of Linnaeus. The plant was named scientifically for him, *Linnaea borealis*, and described as shy and timid as the person after which it was named, which is somewhat amusing as Linnaeus, although never grand in stature—he was only about 5 feet tall—was certainly a grandmaster of self-promotion. He called himself the General of Flora's Army, and referred to his graduate students as his apostles. Not too surprising then that he wrote at least 5 autobiographies.

Linnaeus returned to his medical studies in 1735, but in the Netherlands rather than Sweden, where he would graduate as a medical doctor at your get-your-degree-quick University of Harderwijk. He would present a written thesis on the cause of ague—that's malaria—which was a common disease in Sweden during the 18th century and often called Uppsala Fever.

He identified that the ague was linked to areas with damp clay soils but assumed that the disease was caused by clay particles getting stuck in people's blood vessels. Although we know that this is wrong—Linnaeus was far too early for the high-powered microscopes that could identify the parasites in mosquitoes that actually cause this disease—you still have to admire his observational skills with linking ague to poorly drained land.

Following this, he was questioned by examiners and required to examine and diagnose a patient. After this process, he received his degree, and in the space of just 2 weeks he was graduated as a doctor at the age of 28. He would specialize in the treatment of syphilis, financially a very prudent thing to do, as one of the main treatments for syphilis was mercury. Mercury would treat some of the symptoms but not the disease itself. As a result, your patients were always repeat patients, hence the phrase: One night in the arms of Venus leads to a lifetime on Mercury.

For paleontologists and biologists, though, 1735 would see something even more significant, because this was the year of publication of the first edition of that most important book we mentioned earlier, the book that was guarded at the Smithsonian Institution's National Museum of Natural History for 2 days over 270 years later and about 4000 miles away from where it was first published.

The *Systema Naturae* represents Linnaeus's desire to bring order to creation. He saw it as his mission to bring back to nature the order that had existed at the beginning. He became known, probably much to his great delight, as the second Adam, naming creation just as the first man had in the Garden of Eden. As he said about himself, his favorite topic next to plants: "Deus creavit, Linnaeus disposuit." That means: God created, Linnaeus organized. Just as Newton had uncovered God's mathematics, Linnaeus wanted to uncover God's classification.

In the *Systema* all of creation is organized into three major kingdoms. Each of those kingdoms is divided into subgroupings of class, order, genus, and species, significantly fewer than the subdivisions we have today. Naturalists before Linnaeus often used a somewhat arbitrary grouping of creatures, though—for example, groupings that comprise all creatures that live in water, or all domestic animals. Linnaeus was one of the first to group genera into higher taxa based on somewhat logical similarities.

So what were Linnaeus's three kingdoms? Firstly, the animal kingdom, comprising: the mammalia—that's the mammals; then the aves, or the birds; amphibia—that included, at that point, also the reptiles and the non-bony fish; insecta—that's all arthropods, not just the insects; and the vermes, basically all other invertebrates including worms, mollusks, and echinoderms, like sea urchins.

In the first edition of *Systema Naturae*, Linnaeus still included whales with fish, but by the 12th edition they were placed in the mammals. He also correctly placed bats in the mammals and, horror of horrors, included humans along with the apes in the primates.

Secondly, the plant kingdom. Here, Linnaeus creates a system of 24 classes of plants based on the number and organization of a plant's sexual organs, the male stamens and the female pistils and related reproductive features. This wasn't without controversy, though. The way Linnaeus would focus on the sexuality of his classification offended some, and, to be honest, Linnaeus did sometimes kind of get carried away in his descriptions. For example:

> The flowers' leaves serve as bridal beds which the Creator has so gloriously arranged, adorned with such noble bed curtains, and perfumed with so many soft scents that the bridegroom with his bride might there celebrate their nuptials with so much the greater solemnity.

You've got to admit, it's a bit over the top. Contemporary botanist Johann Siegesbeck called it a loathsome harlotry, which may also explain why Linnaeus named the stinky weed, St. Paul's wort, *Sigesbeckia orientalis* in his honor.

Thirdly, the mineral kingdom. Now, this might sound strange to us today, but in Linnaeus's time many believed minerals possessed a basic kind of life force, and as such form a part of Linnaeus's system of classification. It was divided into: the petrae, which is all the rocks; the minerae—that's minerals and ores; and the fossilia—fossils and aggregates.

Linnaeus published 12 editions in his lifetime, continually revising and updating his classifications. The manner in which we order life today is quite different than much of Linnaeus's original efforts, but much of the legacy of his efforts is still with us. We still have a hierarchical organization of life and, of course, we still have the scientific binomial system.

In the days before the *Systema Naturae*, naming creatures could be quite messy. Take, for example, the tomato. Prior to Linnaeus, it went by the rather grand and long-winded name of *Solanum cauke inermi herbaceo, folis pinnatis incises*, which basically means "the solanum with the smooth stem which is herbaceous and has incised pinnate leaves." Under the Linnaean binomial system, it becomes *Lycopersicon esulentum*—much less of a mouthful.

Because of the hierarchical system of the *Systema Naturae*, you don't need to list all the descriptive components of a species, all you need is the name of the genus followed by the name of the species: for example, *Passer domesticus*, the house sparrow; or *Acheta domesticus*, the house cricket. It's a simple but powerful system. As you can see, the species or specific name of any member of a genus could be used for other different genera, like *domesticus* in our example, which is also used as the species name of several other plants and mammals. However, genera will always be unique.

Today, the system of classification used by many biologists and paleontologists is a system called cladistics. Cladistics considers the shared and derived characteristics of creatures when classifying them, rather than a superficial appearance. For example, under cladistics there is no grouping called fish, or class Pisces, as the group is traditionally understood.

As we covered in *A New History of Life*, when you actually study the characteristics of certain fish—for example, a lungfish and a cod—you will find that a lungfish shares more features in common with a frog than it does with a cod, even though a lungfish under the Linnaean system would still be classified under Pisces, the fish. In cladistics, by contrast, only groupings that contain all of their descendants can be considered as a legal classification, or what is defined as, among those in the know, monophyletic.

In classification using cladistics, relationships between organisms are illustrated using a cladogram, a branching diagram of relationships supported by derived character states. For example, on this cladogram the tiger, gorilla, and human can all be classified as mammals that share common features such as hair—somewhat vicariously in my case.

If you just define a group called reptiles—that includes our friend the lizard here—as a group, it does not include all of its descendants: tigers, gorillas, and us in this case. And so reptiles is an illegal grouping and therefore are deemed paraphyletic under cladistic terminology, and you'll have the cladistic police knocking on your door. Reptiles are, however, part of a larger, legal monophyletic group that contains animals called the amniotes, a group that contains all animals that lay eggs adapted to be laid on land, or retain the fertilized egg within the mother.

Let's now consider how the classification schemes might be different when used by a paleontologist. Remember that Linnaeus placed fossils within his mineral kingdom under fossilia? Unlike rocks and minerals, the binomial system for paleontology persisted, understandable given that we're dealing with former life, and the zoological or botanical codes of taxonomy that apply to living animals and plants likewise apply to the fossil forms, too.

The obvious problem we have with fossils compared with living creatures is, well, it's just that. As fossils, a lot of the information that could be used to classify these creatures is simply just gone. As such, drawing lines between species can be difficult. Just imagine how different human beings can look depending on their sex, their age, their historical background, and what environment they might be living in. Add to that the problem that paleontologists may be dealing with incomplete, fragmentary, or otherwise modified material, and the problem is just compounded.

For a biologist, differences within the same species can be tested by simply observing that living species, watching how a species develops and changes over time, recording differences that might occur to the same species due to maybe environmental factors. And if you observe two individuals mating, however different they may look superficially and still producing viable offspring, you can be sure that they're of the same or at least very closely related species.

Today, biological classification is further aided by studying the genetic similarity between creatures, allowing us now—more than any time before, really—to start to place life into real groupings based on real genetic similarity. Once again, this—with the exception of rare and fairly recent fossils—is not available to the paleontologist.

To illustrate some of the problems that paleontologists face, I'd like to introduce you to some of the most iconic of the dinosaurs, *Brontosaurus*, *Stegosaurus*, and *Triceratops*, 3 very familiar dinosaurs discovered by famous paleontologist Othniel Charles Marsh in the late 1800s. Of the three dinosaurs we are considering here, I must admit I think *Brontosaurus* is my favorite. Back in the early 70s when I was at elementary school, I remember drawing pictures of this long-necked behemoth wallowing in tropical Jurassic swamps to support its vast bulk, one of the pervasive, and actually inaccurate, views of the ecology of the creature at that time. The thing is though, since 1903, there was no *Brontosaurus*.

Brontosaurus was part of a treasure of dinosaurs recovered from the western USA by famous dinosaur hunters like Marsh and Cope during the 1870s. Marsh discovered the skeletons of 2 partial sauropod dinosaurs—that's the group to which these 4-legged, long-necked dinosaurs belonged—and sent them to the Peabody Museum at Yale. The first specimen he got round to describing he named *Apatosaurus ajax* or "deceptive lizard." We have a bone of an Apatosaur right here.

Two years later, in 1903, he named the second skeleton and decided it was sufficiently different not only to be considered to be a different species but also a completely new genus. As such, *Brontosaurus excelsus*, or "noble thunder lizard," was born.

After Marsh described these two specimens, skeletons belonging to similar sauropod dinosaurs were excavated and, upon analysis, it was determined that one species fell on a morphological spectrum somewhere between

Apatosaurus and *Brontosaurus*. As such, the differences between the two end members of the group didn't appear so extreme after all, and the skeletons, including the new species, were all placed in the same genus. Taxonomically, the first named specimen has precedence, so all of these animals became apatosaurs, with *Brontosaurus excelsus* named to *Apatosaurus excelsus*. They had all become "deceptive lizards."

When you make a taxonomic determination, you're effectively proposing a hypothesis regarding the position of a particular living or fossil organism within the 4-billion-year-old tree of life. Any new information such as data from new analytic techniques or additional specimens may help revise that hypothesis. For our friend *Brontosaurus*, new information would be released in a paper by British and Portuguese paleontologists in 2015 by Benson, Mateus, and Tschopp.

Now, determining when some fossil is sufficiently different from another, to be placed in an entirely new genus, is not strictly governed by any clear taxonomic guidelines, a kind of a judgment call has to be made. Since the time that the genus *Brontosaurus* had been demoted, however, new sauropod specimens had been recovered, and these paleontologists took advantage of the new discoveries to apply an extensive—the paper was 300 pages long—statistical analysis of the differences between various features of these animals.

In concentrating their analyses on the broad group to which the Apatosaurs belong—that's the diplodocid dinosaurs—they found that the differences between widely accepted genera within the diplodocids were at very least the same as the differences between *Apatosaurus* and what Marsh had originally described as *Brontosaurus*. Much to every 6-year-old's delight, and mine—at least according to the authors of this paper—*Brontosaurus* could be officially resurrected as one of the coolest dinosaurs of the Upper Jurassic. This shows how taxonomy, far from being the dusty, old, unchanging science of classification, is dynamic and potentially subject to change with every new discovery that's made.

Another example of how we have to be careful with taxonomy comes from another dinosaur favorite: *Stegosaurus*. An exceptional discovery of a

Stegosaurus graveyard in Montana has permitted the analysis of a large, well-preserved population of stegosaurs. We all know what *Stegosaurus* looked like, right? There's one right back there—the dinosaur with the tiny head and the spiky tail and rows of plates along its back. Like *Brontosaurus*, this dinosaur was originally discovered and named by Marsh in 1877 from the Jurassic Morrison Formation in southwestern Wyoming. Initially, Marsh thought that the *Stegosaurus* plates lay flat on its back like shingles, hence the name *Stegosaurus* or "roofed lizard." Marsh would later rethink his interpretation, giving us the classic spiky-backed dinosaur we all know and love today.

The discovery of a stegosaur graveyard in Montana presented a rare opportunity to study a larger population of well-preserved specimens, an opportunity taken up by graduate student Evan Saitta of the University of Bristol in the United Kingdom. When considering the overall morphology of the stegosaurs and the microscopic bone structure, Saitta was able to determine that all the stegosaurs found in this deposit were adults and that they all belonged to the same species, *Stegosaurus mjosi*. Even so, there was a particular difference he found in the shape and arrangement of plates on their backs. Some specimens had plates that were pointed and tall, while others possessed plates that were broader and more rounded.

So what are we looking at here? Saitta suggests that this could represent sexual dimorphism in the dinosaurs, or, in other words, differences between boy stegosaurs and girl stegosaurs. He proposed that the broad round plates belonged to the males and the tall pointed plates to the females. He bases his hypothesis on the size of the rounded plates that are some 45% larger in area and may therefore represent a display investment by the male, not unlike the large antler rack you find on deer or moose.

Although this hypothesis is not accepted as evidence of sexual dimorphism by all, this does illustrate how we need to be careful when naming our dinosaurs or any other fossils. It's possible that if fewer, more poorly preserved specimens were recovered, these two forms could have been misinterpreted as two different species rather than the male and the female of the same species.

And so to our third dinosaur. This one was discovered near Denver, Colorado in 1877 and originally described by Marsh as a bison: *Bison alticornis*. However, he eventually realized they belonged to a horned dinosaur he named *Triceratops*, another crowd-pleasing dinosaur.

A controversy would erupt regarding this particular dinosaur in 2009 when paleontologist Jack Horner, from the Museum of the Rockies, and his graduate student John Scannella would propose a hypothesis at a conference at the University of Bristol that would significantly reduce the number of dinosaurs we have on the books. They suggested that *Triceratops* and *Torosaurus*, another horned dinosaur discovered by Marsh, were one and the same species with *Triceratops* being the juvenile and *Torosaurus* the adult. They even proposed an intermediate teenager in the genus *Nedoceratops*.

These dinosaurs look quite different, though. *Torosaurus* has a much larger frill that is perforated with a large oval hole, perforations that are lacking in *Triceratops*. Although these animals overlap in time, they're regarded as being so different that they were not only different species, they were also different genera.

However, Horner and Scannella take a different view. They claim that these differences just reflect different developmental stages of the same dinosaur. They believe *Triceratops* to be the juvenile, *Nedoceratops* the intermediate teenager—probably somewhat sullen and always wanting to borrow your car—and *Torosaurus* the adult who really didn't understand *Nedoceratops* and thought his music was all noise.

Horner and Scannella claim that microscopic study of bone textures among these animals show how the bone was transforming as the animal matured in what is called a metaplastic manner. For example, according to Horner, bone in *Triceratops* was reabsorbed from the center of the head shield as the dinosaur matured and then redeposited around the margin of the structure, forming a distinctive rim in the shield of *Torosaurus*.

This is still a very controversial hypothesis. Many paleontologists claim that the *Tri-Nedo-Toro* sequence doesn't always appear to be consistent. In addition, analysis is hampered by the relative scarcity of postcranial—that's anything other than the head—skeletons. Other criticisms have been leveled due to the presence of fused bones, indicating that growth had stopped in the proposed younger forms of the sequence. Nevertheless, Horner and Scannella point out that details of skull bone fusion are very variable in the some 100+ specimens they hold at the Museum of the Rockies. And so the debate continues. If Horner and Scannella are proved to be correct it would mean we lose *Nedoceratops* and *Torosaurus* as valid Linnaean genera, all of them becoming different growth stages of *Triceratops*, which, as described first, takes taxonomic precedence.

Horner thinks that this could be part of a wider problem. He estimates that perhaps over ⅓ of all dinosaur species in the late Cretaceous, where *Triceratops* is found, may never have existed. He believes many just may represent different growth stages misinterpreted as separate species as a particular dinosaur matures from a juvenile, through the terrible teens—could you imagine a teenage dinosaur?—and into an adult.

Other examples of mistaken species could include *Anatotitan*, that some have suggested is just an adult *Edmontosaurus*. And the wonderfully named *Dracorex hogwartsia*—yep, it was named for the *Harry Potter* books—and *Stygimoloch* may just be growth stages in the development of the hardheaded *Pachycephalosaurus*.

But why should we care about taxonomy and classification? It may appear that these questions are very academic, nothing more than the grumblings of academics about how to arrange their baseball cards: by year, by team, by state—does it really matter? But just consider the debate sparked by the classification of *Triceratops*, *Nedoceratops*, and *Torosaurus*. If we accept the views of Horner and Scannella regarding the number of real dinosaur species at the end of the Cretaceous, we have a much more impoverished dinosaur population than we previously thought prior to the impact of the extraterrestrial body that marks the Cretaceous-Paleogene extinction at 66 million years ago. This would give us a very different understanding of the paleoecology and stresses that this formerly successful and very biodiverse group may have been experiencing prior to their final extinction.

The ongoing process of the classification of fossils helps deepen our understanding of biodiversity over time. Taxonomy refines our focus through the deep time window paleontology affords us. This is vital, as paleontology is our only long-term benchmark against which we can compare modern changes in the biodiversity and the current health of our ecosystem. Far from being a dusty old science of the past, paleontological taxonomy and classification could prove vital in charting our planet's future.

Lecture 6

Minerals and the Evolving Earth

This lecture will consider the evolution of our planet with a focus on the evolution of Earth's minerals—a perspective that considers how minerals have influenced all of Earth's systems, including the biosphere and its history as revealed by paleontology. In this lecture, you will learn what the first minerals are, which minerals develop after the Earth formed, how we reach the wonderful diversity of minerals we see today, and what role life would have in that story.

The First Minerals

- The idea of looking at our planet through the lens of minerals was developed by Robert Hazen of the Carnegie Institution of Washington. Hazen and his colleagues, from various research institutions, proposed that many of the 4400 minerals we know of today have "coevolved" with the biosphere through time.
- Obviously, minerals don't mutate and evolve in a biological sense, but they have changed over time, both reflecting and influencing our evolving planet. As such, considering the changing mineral makeup of our planet also helps paleontologists appreciate factors that might be influencing the fossils they find through Earth's history.
- Hazen and his colleagues proposed 3 eras and 10 stages of Earth's mineral evolution. Each stage sees dramatic changes in the diversity of Earth's near-surface recoverable minerals.
- In the beginning—about 13.7 billion years ago, at the time of the big bang—there were no minerals. It is estimated that by about 377,000 years after

the big bang, the first hydrogen and helium atoms started to form. It is likely that during these cataclysmic events, some of the first microscopic crystalline minerals, around a dozen, would form, including diamond, graphite, and various silicates.

- These few primordial minerals have been named ur-minerals by Hazen, a reference to the ancient Sumerian city of Ur that marks some of the earliest evidence of complex civilization. These early ur-minerals would combine, mix, and react over time to form much of the complex world we know of today.

- To understand the mineral story of Earth, though, we have to move forward in time to about 9 billion years after the big bang—that's 4.6 billion years ago. This is a time before our familiar planets had formed. In their place was a vast cloud of hydrogen, helium, and dust—the dust probably comprising some of the early ur-minerals.

- This cloud was contracting and spinning under its own gravity, forming a concentration of material at its center. This is known as the T Tauri phase of a star's development. The star is not yet able to fuse hydrogen and initiate nuclear fusion but is still bright and radiant as it collapses under gravity.

- Even at this stage, though, it is still energetic and hot, with the young protostar heating up a disk of material that surrounds it. This is the protoplanetary disk, and it from this that the planets will eventually form. It is thought that around 60 mineral species get cooked, and thus form, in this particular stage of the solar system's development.

- Some of those early mineral phases have been preserved and occasionally fall to Earth as a class of meteorites called primitive chondrites. These developed as the dust and Sun-bathed minerals started to accrete together, initially due to electrostatic attraction, a bit like dust bunnies, and later, as they became larger, under gravity.

- An interesting feature of these meteorites are the small (around 1 millimeter) spherical chondrules. These probably represent molten droplets that were formed by flash heating as the early Sun cooked the materials in its surrounding protoplanetary disk—fascinating echoes of conditions in that early cloud before the Earth was born.

Meteorites

- Over time, these small chondrites would accrete together to form larger bodies. If larger than about 200 kilometers in diameter, heat from the decay of radioactive isotopes trapped inside these rock piles and heat generated by collisions would cause the interiors of these larger bodies to melt, or at least partially melt, and produce new suites of minerals.

- These so-called planetesimals would also differentiate under gravity, with heavier components, such as nickel and iron, sinking to the center of the mass. This creates a protoplanet with a basaltic, relatively light, lavalike crust surrounding a dense metallic core.

- Meteorites called achondrites are thought to represent the shattered crustal fragments of some of these protoplanets. Iron-nickel meteorites are likely their shattered metallic cores.

- This also tells us that the early solar system was a busy shooting gallery with multiple mergers, titanic collisions, and destruction of some of these early planetary bodies. By the end of this stage, all this activity would see the cumulative count of minerals rise to about 60.

Mineral Development after Earth's Formation

- After the Earth had formed and differentiated, the light scum of less dense minerals that remained close to the surface would have cooled to form a blackened basaltic skin. This black crust would be repeatedly recycled though, melting and generating magma that would undergo an important processes called fractional crystallization.

- As magma cools, its composition changes. This is because different minerals crystallize out of the melt at different times, depending on their melting point. As the magma continues to cool, minerals crystallize in order of their melting points, continually changing the composition of the remaining magma and the composition of the minerals it generates.

- This stage of magma differentiation is probably the level of mineral evolution that the Moon and Mercury reached but went no further. It is probably the presence of liquid water on Earth that allows mineral evolution to progress further. Our planet may have been cool enough for liquid surface water as early as 4.4 billion years ago. The interaction of minerals with water would allow for the number to rise to about 500 in the Earth system. This is also possibly the stage that the once-wet Mars may also have reached.

- Another significant development would be the formation of granitic rocks, rocks that contain lots of quartz and feldspar, which started to form in Earth around 4 billion years ago. As magmas were continually injected into the early crust, they would partially melt the surrounding crustal rocks, but with only the relatively less dense minerals melting, as it is these that have the lowest melting points.

- As a result, these magmas had a very different composition of less dense minerals than the parent rocks that were melted to form them. It is this process that would produce the granitic magma that would rise into higher levels of the crust, cool, crystalize, and form granites.

- Because granites are significantly less dense that basaltic rocks, they are very buoyant and tend to float on the surface of the dense rocks in the mantle. These accumulations of buoyant granitic rocks would be the seeds of the first continents.

- Granitic melts would continue to differentiate, helping concentrate rare and mostly lighter elements into granitic rocks. Through these processes, our mineral count is now around the 1000 mark.

- Earth, and possibly Venus, reached this stage of granite production and mineral evolution, but our own planet—probably uniquely in the solar system—would have more stages to pass through before its current final inventory was reached.

- The next stage involves the initiation of a process that we think is only found on our planet, at least in our solar system: plate tectonics, which describes the large-scale motions of the fractured plates that make up the Earth's outer surface, the lithosphere.

- At plate boundaries, parts of the lithosphere can slide past each other, but the lithosphere can also spread apart, generating new oceanic lithosphere. Some boundaries are marked by the collision of plates, forming large mountains, or by plates being destroyed as one is forced under another in a process called subduction.

- As far as we know, Earth is the only planet to have initiated extensive and prolonged plate tectonics. Plate tectonics is a significant reason for the complexity and diversity of Earth's geology and biosphere.

- The temperature and pressure regimes caused by different types of plate movements generated new minerals. Plate tectonics would also elevate mountains, exposing these newly formed minerals to weathering process and generating even more minerals. This process is still going on today.

- In addition, oceanic water, seeping into the crust at ocean-crust-generating mid-ocean ridge systems and also taken down into the mantle on subducting slabs of oceanic lithosphere, would alter preexisting rocks, creating new minerals. This process is still occurring at hydrothermal vent systems located at mid-ocean ridges today.

- It is at these vent systems where metals, in combination with sulfur, generate massive sulfide ore deposits. These processes have concentrated large quantities of metal ores.

- All this plate tectonics–related activity brings our mineral count to 1500.

The Role of Life

- The presence of abundant and very evident life probably explains the overwhelming bulk of the 4400 minerals on our planet today. We have paleontological evidence of life at around 3.4 billion years ago—bacteria that were metabolizing sulfur-based compounds. It would appear, however, that life initially had very little effect on increasing the mineralogical diversity of our planet.

- That would change dramatically, though, about 2.5 billion years ago, when we start to see significant numbers of certain microbes spreading across the planet—microbes that had developed a photochemical trick called photosynthesis.

- The earliest form of photosynthesis used hydrogen sulfide as a hydrogen donor to power the reaction, but later forms of photosynthesis would use water. The consequence of this would be the release of oxygen. This period in history is known as the great oxidation event and is probably the most important event in the diversification of Earth's mineralogy.

- Of the approximately 4400 known mineral species we have today, more than half of them are oxidized and hydrated products of other minerals, a situation that can only develop on a planet rich in free oxygen. It is at this point where we see the diversity and complexity of minerals outstripping anything else in our solar system.

- Another consequence of this availability of oxygen would be a dramatic change in the chemistry of the oceans. Prior to the great oxidation event, the Earth's oceans had been largely anoxic—that is, they contained little to no dissolved oxygen. As a consequence, unoxidized iron was the common form found dissolved in seawater.

- With the introduction of oxygen into this system, unoxidized iron was oxidized into insoluble minerals, such as magnetite and hematite, which would effectively form rust in the oceans that would settle out in layers on the ocean floor of continent shelves, alternating with layers of less iron-rich chert. These so-called banded-iron formations are some of the most iron-rich ores on Earth today and are the result of this significant change in the Earth system around 2.3 billion years ago.

- At around 1.85 billion years ago, the deposition of banded-iron formations ceases abruptly. This change marks the transformation of the land as oxygen, now no longer captured to form rust in the oceans, is released to the atmosphere and would start to oxidize minerals on the continents, turning many parts of the surface red.

- What follows, from 1.8 to 1 billion years ago, is known as the boring billion, which sees no new major innovations in life or minerals.

- Between 1 billion to 542 million years ago, the Earth would suffer a series of super glaciations, or snowball Earth events. It is thought that the end of each snowball would be associated with extreme weather conditions, which would thoroughly mix the oceans, flooding them with nutrients and causing a bloom of oxygen-producing cyanobacteria.

- The resulting increase in the availability of oxygen provided opportunities for creatures to evolve bigger bodies. This would set the stage for our next leap in the Earth system: the explosion of multicellular life-forms. By the time we get to the base of the Cambrian period, 542 million years ago, biology would be the main driving force in the formation of new minerals.

- The colonization of the planet by organisms—and, in particular, the movement of plants onto land—would see a vast increase in the amount of clay minerals being produced by biological weathering. Particularly important would be the effect land plants would have on the development of new types of organic-rich soils and the opportunities for more mineral formation. This expansion of the biosphere and organic carbon production would see the formation of more carbon-rich deposits.

- The explosion of biologically driven mineralogy would increase the total number of mineral species to the current level of about 4400—a product of our planet's long and complicated evolution and the prolonged development of its biosphere.

Questions to consider:

1. Because a mineral is loosely defined as a naturally occurring crystalline solid, is ice a mineral?
2. Could a complex mineralogy be used in the search for life on other planets?

Suggested Reading:

Chesterman, *The Audubon Society Field Guide to North American Rocks and Minerals*.

Hazen, "The Evolution of Minerals."

Lecture 6 Transcript

Minerals and the Evolving Earth

We humans tend to like to construct convenient boxes to describe the universe we see around us. Consider the manner in which we divide the sciences: physics, chemistry, biology—neat, convenient, little compartments, like specimens in a case. In reality, of course, the universe we live in is a lot more messy. For example, to truly understand a complex system like the Earth, we can't rely on any one of the traditional sciences but have to call on them all collectively to describe the interconnecting, ever-evolving planet from differing perspectives.

So, in this lecture, let's look at another way of considering the evolution of our planet, a perspective that has a focus not on the evolution of life but on the evolution of Earth's minerals, a perspective that considers how minerals have influenced all Earth's systems, including the biosphere and its history as revealed by paleontology.

Today, there are around 4400 known minerals. The crust of our planet is composed of around 90% of minerals we call silicates. These are minerals that contain a silica molecule composed of 1 atom of silicon bonded to 4 atoms of oxygen in various configurations with other elements like aluminum and calcium. One of the most common minerals is feldspar. Another very common silicate mineral in the crust is quartz, a mineral composed of the silicate molecule bonded just to itself in a 3-dimensional structure.

It is important to note that the majority of minerals, unlike the images you see in books or in museum collections, do not occur as these large and beautiful showy crystals, but are usually found as a mosaic association with other minerals in rocks. Rocks are effectively just aggregates of various minerals.

Only around 8% of the Earth's crust is composed of non-silicates such as oxides, sulfides, and carbonates. Carbonates, and in particular the mineral calcite, have been vital in the evolution of the biosphere. The crust, though, only represents just 1% of the Earth's total volume. Below the crust, down in the mantle, are silicates rich in magnesium and iron—for example, olivine. Olivine is common in rocks like peridotite, a heavy rock reflecting its higher iron content. By the time we get to the Earth's core, the main constituents are iron and nickel, liquid till about 2890 kilometers below the Earth's surface and then present as an iron-nickel alloy to the center of our planet.

This picture I've just painted is the result of over 4.54 billion years of planetary evolution and has had profound implications for our understanding of Earth's biosphere, so in this lecture, I want to consider the mineralogical evolution of our planet. So let's consider: What are the very first minerals? Which minerals developed after the Earth had formed? And how do we reach that wonderful diversity of minerals we see today, and what role would life have to play in that story?

The idea of looking at our planet through the lens of minerals was developed by Robert Hazen of the Carnegie Institute. Hazen and his colleagues from various research institutions proposed that many of the 4400 minerals we know of today coevolved with the biosphere through time.

Obviously minerals don't mutate and evolve in a biological sense, but they have changed over time, both reflecting and influencing our evolving planet. As such, considering the changing mineral makeup of our planet also helps paleontologists appreciate factors that might be influencing the fossils they find through Earth's history. They proposed 3 eras and 10 stages of Earth's mineral evolution. Each stage sees dramatic changes in the diversity of Earth's near-surface—that's recoverable—minerals. That might explain why Hazen's ideas were also covered in *The Economist*. Incidentally, Bob Hazen is also a *Great Courses* alumnus who presented a show on the origin and evolution of the Earth in which he expands these ideas further. Why don't you check it out?

But let's start in our story where the book starts: right at the beginning. In the beginning there were no minerals. By this beginning, I mean the very beginning some 13.7 billion years ago at the big bang. It's estimated that about 377 thousand years after the big bang the first hydrogen and helium atoms started to form, and probably by about 560 million years after the big bang stars would form from the collapsing clouds of these gases. Ultimately, the death of these during supernova would seed the universe with heavier elements that are formed by nuclear fusion in the hearts of these first giant stars.

It is likely that during these cataclysmic events, some of the first microscopic crystalline minerals, around a dozen or so, would form, including diamond, graphite, and various silicates. These few primordial minerals have been named ur-minerals by Hazen, a reference to the ancient Sumerian city of Ur that marks some of the earliest evidence of complex civilization.

These early ur-minerals would combine, mix, and react over time to form much of the complex world we know today. To understand the mineral story of Earth, though, we have to move quite a way forward in time, to about 9 billion years after the big bang—that's 4.6 billion years ago. This is a time before our familiar planets had formed. In their place was a vast cloud of hydrogen, helium, and dust, the dust probably comprising some of those early ur-minerals. This cloud was contracting and spinning under its own gravity, forming a concentration of material at its center.

This is known as the T Tauri phase of a star's development, the star not yet able to fuse hydrogen and initiate nuclear fusion but still bright and radiant as it collapses under its gravity. Even at this stage, though, it is still energetic and hot, with the young protostar heating up a disk of material that surrounds it. This is the protoplanetary disk and it's from this that the planets will eventually form.

It is thought that around 60 mineral species get cooked in this particular stage of the solar system's development. Wouldn't it be wonderful to glimpse into those early times around 4.6 billion years ago? Well, we can, because some of those early mineral phases have been preserved and occasionally fall to Earth as a class of meteorites called primitive chondrites. These developed as the dust and Sun-bathed minerals started to accrete together, initially due to electrostatic attraction—a bit like dust bunnies under your bed—and later, as they got larger, under gravity. An interesting feature of these meteorites are the small—about 1 millimeter—spherical chondrules. These probably represent molten droplets that were formed by flash heating as the early Sun cooked the materials in its surrounding protoplanetary disk, fascinating echoes of early conditions in that early cloud before the Earth was born.

Over time, these small chondrites would accrete together to form larger bodies. If larger than about 200 kilometers in diameter, heat from decay of radioactive isotopes trapped inside these rock piles, and heat generated by collisions, would cause the interiors of these larger bodies to melt—at least partially melt, anyway—and produce new suites of minerals. These so-called planetesimals would also differentiate under gravity, with heavier components like nickel and iron sinking to the center of the mass. This creates a protoplanet with a basaltic, relatively light lava-like crust surrounding a dense metallic core.

Meteorites called achondrites are thought to represent the shattered crustal fragments of some of these protoplanets. Iron-nickel meteorites are likely their shattered metallic cores. This also tells us that the early solar system was a busy shooting gallery with multiple mergers, titanic collisions, and destruction of some of those early planetary bodies. By the end of this stage and all this activity, we'd see the cumulative count of minerals rise to about 60. We still obviously have quite a long way to go yet, though.

After the Earth had formed and differentiated, the light scum of less dense minerals that remained close to the surface would have cooled to form a skin, a bit like the skin on your cocoa. Its composition would have been similar to basalt like we find today in places like Hawaii. This period is sometimes referred to as the Black Earth, reflecting the blackened basaltic surface that would have characterized our planet at this time. When I visited the big island of Hawaii in 2015, I really got a feeling for what a strange an alien place that early Earth must have looked like.

This black crust would be repeatedly recycled, though, melting and generating magma that would undergo an important process called fractional crystallization. As magma cools, its composition changes. This is because different minerals crystallize out of the melt at different times depending on their melting point. For example, olivine, having a high melting point, will be the first mineral to crystallize and settle to the bottom of the magma chamber. As olivine has now been removed from the remaining magma, the composition of the remaining molten component has been changed. As the magma continues to cool, minerals crystallize out in the order of their melting point, continually changing the composition of the remaining magma and the composition of the minerals it generates. This stage of magma differentiation is probably the level of mineral evolution that the Moon and Mercury reached, but went no further.

It is probably the presence of liquid water on Earth that would let mineral evolution progress even further, though. It is possible our planet may have been cool enough for liquid water we think at around about 4.4 billion years ago. The interaction of minerals with water would raise the number to about 500 in the Earth system. This will also be possibly the stage that the once wet Mars might have also reached.

But another significant development would be the formation of granitic rocks, rocks that contain a lot of quartz and feldspar, which started to form in Earth around about 4 billion years ago. As magmas were continually injected into the early crust, they would partially melt the surrounding crustal rocks, but with only the relatively less dense minerals melting, as it is these that have the lowest melting points. As a result, these magmas had a very different composition of less dense minerals than the parent rocks that were melted to form them. It is this process that would produce the granitic magma that would rise into higher levels of the crust, cool, crystallize, and form granites like you might find in your kitchen countertop.

Because granites are significantly less dense that basaltic rocks—they're about 2.7 versus 3.3 grams per cubic centimeter—they're very buoyant and tend to float on the surface of the dense rocks in the mantle. These accumulations of buoyant granitic rocks would be the seeds of the first continents. Granitic melts would continue to differentiate, helping concentrate rare, mostly lighter elements such as lithium, cesium, beryllium, and boron into granitic rocks like pegmatites, which many rare gems and minerals form from. Through these processes, our mineral count is now around the 1000 mark. Earth, and possibly Venus, reached this stage of granite production and mineral evolution, but our own planet, probably uniquely in the solar system, would have many more stages to pass through before its current final inventory was reached.

The next stage is a significant one. It involves the initiation of a process which, at the moment, we think is only found on our planet, at least in our solar system: that's plate tectonics. Plate tectonics describes the large-scale motions of the fractured plates that make up the Earth's outer surface, what we call the lithosphere. At plate boundaries, lithosphere can slide past each other, as is the case along the San Andreas Fault in California, but can also spread apart, generating new oceanic lithosphere, as occurs along the Mid-Atlantic Ridge.

Some boundaries are marked by the collision of plates either forming large mountains like the Himalaya of Northern India and Nepal or by plates being destroyed as one is physically forced under another in a process that we call subduction, as is occurring along parts of the western edge of North America today, generating the volcanic activity of the Cascade volcanoes like Mount Saint Helens and Mount Rainier.

As I already mentioned, as far as we know, Earth is the only planet to have initiated extensive and prolonged plate tectonics. Plate tectonics is a significant reason for the complexity and diversity of Earth's geology and its biosphere. The temperature and pressure regimes caused by different types of plate movements generated new minerals, as well. Plate tectonics would also elevate mountains, exposing these newly formed minerals to weathering process and generating even more minerals. This is a process still going on today, like you can see here in Scotland, with rocks that were once formed deep in the Earth's crust now exposed and eroding at the surface.

In addition, oceanic water seeping into the crust at those ocean crust-generating, mid-ocean ridge systems, and also, though, taken down into the mantle on subducting slabs of oceanic lithosphere, would alter preexisting rocks, creating even more minerals. This is an ongoing process that is still occurring at mid-ocean ridge hydrothermal vent systems today. It is at vent systems where metals, particularly copper and zinc in combination with sulfur, generate massive sulfide ore deposits.

These processes have concentrated large quantities of metal ores. For example, copper and zinc in the Kidd Mine near Timmins, Ontario was originally a hydrothermal vent system on an ocean floor over 2.7 billion years ago. All this plate tectonics-related activity brings our mineral count now to an impressive 1500—impressive, yes, but still a long way off our current 4400 mineral species. How do we almost triple the diversity of minerals on Earth?

So, in summary, it would appear that 1000 mineral species would be a common tally of minerals for many rocky planets and moons in our solar system, and probably other solar systems, too. Plate tectonics boosted that to 1500 in the early Earth. But it is probably the presence of abundant and very evident life that we need to examine to explain the overwhelming bulk of the 4400 minerals on our planet today. This is the where the stories of mineral and biological-paleontological evolution march forward hand in hand.

We have paleontological evidence of life at around 3.4 billion years ago, bacteria metabolizing sulfur-based compounds living among grains of sand on a beach in what is today Australia. It would appear, however, that life initially, even when we have this fossil evidence, had very little effect on increasing the mineralogical diversity of our planet. That would change, though, and change dramatically about 2.5 billion years ago when we start to see significant numbers of certain microbes spreading across the planet, microbes, like these modern examples, that had developed a neat little photochemical trick: photosynthesis.

Their presence at this time can be seen in specific features that they left behind in the sedimentary rock record called stromatolites that we covered in our lecture on trace fossils. These interesting structures form as mats of photosynthetic bacteria trap and bind fine-grained sediments. The earliest form of photosynthesis used hydrogen sulfide as a hydrogen donor to power this reaction, but later forms of photosynthesis would use water.

The consequence of this would be the release of a dangerous byproduct—well, dangerous to any life that evolved in conditions in the early Earth. And that byproduct was oxygen. It wasn't just photosynthetic bacteria in stromatolites, though; there would be plenty probably of photosynthetic organisms living in the water column, as well. This period of history is known as the Great

Oxygenation Event and is probably the single most important event in the diversification of Earth's mineralogy.

Of the approximately 4400 known mineral species we have today, over half of them are oxidized and hydrated products of other minerals, a situation that can only develop on a planet rich in free oxygen. For example, 256 of the known copper oxide minerals, such as the beautiful malachite and azurite, formed from weathering in an oxygen-rich environment. It is at this point where we see the diversity and complexity of minerals outstripping anything else in the solar system. This is a wonderful example of Earth's system science, the biosphere—life—producing and coevolving with minerals in the geosphere.

Another consequence of this availability of oxygen would be a dramatic change in the chemistry of the oceans. Prior to the Great Oxygenation Event, the Earth's oceans had been largely anoxic—that is, they contained little to no dissolved oxygen. As a consequence, Fe^{2+} iron—that is, unoxidized iron—was the common form found dissolved in seawater. With the introduction of oxygen into this system, that Fe^{2+} iron was oxidized into insoluble minerals like magnetite and hematite. These would effectively form rust in the oceans that would settle out in layers on the ocean floor of the continent shelves, alternating with layers of slightly less iron-rich chert or jasper.

These so-called banded iron formations are some of the most iron rich ores on Earth today, and are the result of this significant change in the Earth's system around 2.3 billion years ago. At around 1.85 billion years ago the deposition of banded iron formation ceases abruptly. The reason for this is greatly debated, but for an interesting possibility we need to travel to the town of Sudbury in Ontario.

Sudbury, home of the Science North interactive museum, the Sudbury Wolves hockey team, *Jeopardy!* host Alex Trebek, and the world's largest nickel. The reason behind the Big Nickel is the association this town has with nickel ore. The nickel ore and other valuable metals such as copper and palladium are concentrated in an oval structure about 62 kilometers by 30 kilometers wide and over 15 kilometers deep that's called the Sudbury Basin. The basin was likely more circular when it formed about 1.85 billion years ago, but has subsequently been deformed by plate tectonic processes. Interpretation of this structure was much debated. Some favored a volcanic origin for this crater, but geologists in 1970 collectively had decided that what they were looking at here was a massive impact crater, a crater now known to be the second largest still present on Earth's surface.

Sudbury, 1.85 billion years ago, lay on the southern margins of a large continental mass called Nena. If you could have travelled back there you would have found a barren landscape crisscrossed by large, braided river systems. But what of that fateful day? The impacting body, probably about 10 kilometers in diameter, would have entered the atmosphere travelling around 8 times the speed of sound. It impacted off the southern shoreline of Nena, digging a vast hole many miles into the Earth's crust.

It's been estimated that around 2 kilometers of crustal material was ejected, forming a crater that was originally 250 kilometers in diameter. Debris hurled out from that crater covered an area of about 1 million 600 square kilometers, with debris being found over 800 kilometers away in Minnesota. It's been estimated that the ocean waters pushed out of the way by this impact would have created an enormous tsunami at the impact site maybe up to 1 kilometer high, and still 100 meters high over 3000 kilometers away.

Some of that material blasted out of the crater, that we call ejecta, returned to Earth and covered the impact area in a layer of fractured rock over 2 kilometers thick. In the impact crater, deep under all that impact debris, magma would slowly rise and cool slowly. As it cooled, heavy minerals started to settle out of the magma, forming a layer at the bottom of the structure. It's here where the richest ore is found today.

All right, but what does this have to do with the end of the unique biological-mineralogical association that resulted in the formation of banded iron formations? An interesting yet still much debated hypothesis was suggested by John Slack and William Cannon of the United States Geological Survey in 2009. They point out that many of the banded iron formations are found directly below the Sudbury ejecta layer, but very different rocks appear to be deposited after the impact. Something, according to the authors, had changed in the ocean system after that impact.

They suggest that the tsunami, falling debris, and large underwater landslides that would have been triggered by the impact could thoroughly mix up the ocean and deliver a pulse of oxygenated surface water into the deep ocean setting. This, they suggest, would have ended the supply of that Fe^{2+}, that dissolved iron, from the deep ocean into the shallow continental shelves where most of the banded iron formations were being deposited, this ending—with a few exceptions—a unique biochemical-mineralogical system.

As I said, this is still a controversial hypothesis. Some note that, although banded iron formations disappear, other iron-rich sediments are deposited after this event. In addition, due to difficulties in correlation, it's still difficult to be certain that banded iron formations further away from Sudbury are not in fact younger than the impact event; as a result, not affected by the impact. Whatever happened, and whatever the exact timing of the end of banded iron formation deposition, on a geological timescale, banded iron formation deposition does end rather abruptly. This change also marks the transformation of the land, as oxygen, now no longer captured to form rust in the oceans, is released to the atmosphere and would start to oxidize minerals on the continents, turning many parts of the surface red, just as it still does in parts of the world today.

What follows, from 1.8 to 1.0 billion years ago, is what is known unofficially as the Boring Billion. This boring billion—well, actually more like 800 million years—sees no new major innovations in life or minerals. But the history of life is a bit like war: long periods of boredom with sudden and short snaps of terror. The terror this time would eventually come in the form of ice.

Between 1 billion and 542 million years ago, the Earth would suffer a series of super glaciations or snowball Earth events. For most of this time, the continents were grouped together around the equator in a supercontinent called Rodinia, but by about 830 million years ago that continent had started to fragment. During fragmentation, shallow seaways opened up into the continental interior. It's been proposed that this situation would have caused intense rainfall in the tropics as these shallow warm seas started to evaporate in the tropical Sun.

As this progressed, the rocks on Rodinia would start to undergo intense tropical weathering—that is, weathering that uses carbon dioxide dissolved in rainwater to convert all those silicate minerals in rocks into clay minerals. These minerals were transferred to the oceans, effectively washing carbon dioxide—the Earth's greenhouse gas—out of the atmosphere.

As the Earth's greenhouse blanket thinned, temperatures dropped and ice advanced from the poles. According to Paul Hoffman of Harvard University, once ice reached a latitude of what is today about Texas, a positive feedback loop would be initiated, with more Sun being reflected back into space than could be absorbed by the planet's surface, causing temperatures to drop even faster. This would generate even more ice that would eventually overrun the equator. At least 2 snowball Earths would develop, each persisting for millions of years. Earth would have resembled a vast, white cue ball from space.

Things would change, though. Volcanoes would occasionally break through the ice and deliver a belch of carbon dioxide into the atmosphere. With no ocean to dissolve into or rocks to weather, carbon dioxide built up to extremely high concentrations. This created a super greenhouse climate, which would cause the snowball to collapse catastrophically and end rapidly.

These snowballs would also have implications for mineral evolution, too. It is thought that the end of each snowball would be associated with extreme weather conditions. These so-called hyper hurricanes would thoroughly mix the oceans, flooding them with nutrients and causing a bloom of oxygen-producing cyanobacteria. The resulting increase in the availability of oxygen provided opportunities for creatures to evolve bigger bodies. This would set the stage for our next leap in the Earth system, the explosion of multicellular life forms. By the time we get to the base of the Cambrian period 542 million years ago, biology would be the main driving force in the formation of new minerals.

The Cambrian would see a proliferation of creatures with biomineralized skeletons composed of calcium carbonate—calcite; phosphate—apatite; silica. The colonization of the planet by organisms, and in particular the movement of plants onto land, would see a vast increase in the amount of clay minerals being produced by biological weathering.

Particularly important would be the effect land plants would have on the development of new types of organic-rich soils and the opportunities for more mineral formation. This expansion of the biosphere and organic carbon production would see the formation of more carbon-rich deposits, too, deposits like black shales, coal, oil, and natural gas.

The explosion of biologically driven mineralogy would increase the total number of mineral species to the current level of 4400, a product of our planet's long and complicated evolution and the prolonged development of its biosphere, a situation that, as far as we know at the moment, is unique in the universe. It is possible, though, that a rich mineral heritage is a fingerprint of life. Could this be used to search for biospheres on other planets?

A complex mineralogy can be seen as evidence of a complex system evolution. As we have seen, the evolution of life on Earth significantly increased its mineralogical diversity. Perhaps in the future, in addition to the search for liquid water, the search for a larger mineral vocabulary will also help us zero in on extraterrestrial life and other complex biospheres.

Lecture 7

Fossil Timekeepers

How do fossils speak to time and cycles of time? They are obviously representatives of times past, but is there more to them than simply being old? This lecture will address several questions: Do we need fossils as clocks? How do fossils act as the time keepers of geology? Do days fossilize? Can fossils record changes in the cycles of the solar system over hundreds of thousands of years, or even longer?

Do We Need Fossils as Clocks?

- Our ability to date our planet and its history is becoming more and more sophisticated. Scientists such as Marie and Pierre Curie and Ernest Rutherford advanced our understanding of radioactivity and radioactive decay and, with it, our ability to date our planet.

- Radiometric dating is based on an understanding of the principles of radioactive decay. It considers the ratio of an unstable radioactive isotope, the parent material, such as uranium 238, to its decay product, the daughter material, which for uranium 238 is lead 206. The uranium doesn't decay entirely into lead all at once but, rather, follows a decay chain with various forms of radiation being emitted as a chain of unstable isotopes is produced along the path to lead 206.

- Because we know the rate at which the parent material decays into the daughter material, we can calculate how long decay has been progressing. The technique assumes that no parent or daughter material has been added to the sample—what is called a closed system.

- Fortunately, crystals in igneous rocks, rocks that cool from a magma, form great closed systems that trap small quantities of radioactive isotopes and, as such, act as clocks, ticking away as time passes by.

- The time it takes for half of the parent to decay into the daughter material is called the half-life. For uranium 238 to lead 206, that is about 4.47 billion years. So, even in Earth's oldest rocks, if there is material to analyze, there should be enough parent material left to work out the ratio and calculate an age.

- Although the vast majority (around 90%) of rocks in Earth's crust are igneous rocks, the vast majority of the rocks that cover the surface of the crust—those that contain the majority of the history of life—are sedimentary rocks. Clastic sedimentary rocks that form from the erosion of older rocks may contain datable crystals from igneous rocks.

- But if you find such a crystal that has not been compromised by the erosion that created the sedimentary rock, it will not provide a date for the sediment or the fossils it contains. It will only provide a date for the igneous rock from which it was derived.

- How do we place fossils in a sequence that makes chronological sense? This was an issue that William Smith solved in the late 1700s. He recognized that various types of fossils followed one another in a predictable order. Once you knew the order, you could place any geological stratum that contained fossils into a time frame relative to another exposure, perhaps at some considerable distance, based purely on the fossils it contained.

- For the first time, scientists had the ability to order the geological strata they found based on the order of the fossils they were finding in them. This also permitted geologists and paleontologists to correlate between areas in time.

- This would allow William Smith to create the first large time-based geological map. This development is the start of the science of biostratigraphy, in which we consider the distribution of a particular fossil species from the time it first

originated to the time it becomes extinct. The time this represents is called a fossil's range.

- Such fossil ranges are collected from many different sections and cross-correlated with other fossils and dating techniques. In this way, we can get a pretty good estimate of the slice of time a particular fossil species represents.

- As such, a species that has traveled far and died young makes the best fossils for dating. This is because they define a focused slice of time over a wide area. Not all fossils are great time keepers, though; some species just existed for much too long and, as a result, don't provide us with sufficient time resolution.

- Given that, some of the best fossils for correlation are fossils that would range far and wide across the oceans, such as free swimmers or planktonic floaters, who are found in many locations and across many environments. Microfossils—a broad group of tiny fossils, generally less than 1 millimeter long—are also great for biostratigraphy, as many were planktonic and distributed widely through the oceans.

Can Days Fossilize?

- As the Earth rotates, a circadian rhythmicity is generated that can impact the behavior and even the anatomy of organisms. For example, the orbital position of the Earth will impact the amount of incoming solar radiation, generating the seasons with various effects on organisms. By careful analysis of certain fossils, it is potentially possible to read these time-related changes recorded in their tissues.

- For example, consider creatures with shells or skeletons that live in shallow marine environments that respond to daily tidal variations. Although care has to be taken to account for other environmental factors, creatures such as bivalves (clams) show growth lines that correspond to daily, monthly, seasonal, or yearly environmental changes. These correspond to packets of different thicknesses of growth bands; collectively, this accounting of time is known as sclerochronology.

- One of the first studies to apply this technique was in 1963, when John Wells of Cornell University interpreted fine ridges on the surface of fossil corals from the Devonian period as being circadian in nature. The ridges were further grouped into regular bands thought to be lunar-monthly breeding cycles. He also identified major annulations that he suggested corresponded to seasonal-yearly environmental changes. From his calculations, Wells estimated that the Devonian year consisted of about 400 days.

- This means that the Earth's rotation about its axis has been slowing down. The Earth's initial spin at the time it formed was due in part to the angular momentum of the initial spinning nebula from which the solar system formed.

- Other factors probably also affected the Earth's rotation, including an impact with Theia, a hypothetical Mars-sized body that collided early in Earth's history and is probably responsible for the formation of the Moon.

Following this event, the Earth may have zipped around on its axis in just 6 hours.

- Since then, the Earth's rotation has been slowing down, mostly due to the Moon's effect on ocean tides. The Moon's gravity is dragging on a tidal bulge in the oceans, slowing the Earth down like a brake on the wheel of a car.

- There are other factors that can affect day length, too. For example, it has been estimated that the devastating 2004 Sumatra-Andaman earthquake in the Indian Ocean effectively shortened the length of the day by about 2.68 milliseconds. This megathrust earthquake saw a large portion of the Indo-Australian plate suddenly shoved under Indonesia and into the planet. In the same way that ice skaters pull their arms into their body, their center of mass, to make them spin faster, the earthquake sped up the planet and shortened, very slightly, the length of our day.

- Fossils provide snapshots through time of the rate of Earth's rotation. For example, by the time of the extinction of the dinosaurs at the end of the Cretaceous, there were 371 days in a year. The Middle Permian year was 390 days long, with around 397 days in the Late Devonian.

- Abundant fossils of animals with mineralized skeletons only really occur after the Cambrian explosion, about 542 million years ago. Can we go any further back with our day-length estimates? We probably can, with a little help from bacterial mats and structures they produce call stromatolites, some of which date back to 3.5 billion years ago.

- Stromatolites are layered structures that form in shallow water. They grow as microbial mats—commonly composed of cyanobacteria—trap, bind, and cement sediments. The bacteria move upward daily, forming a new layer, creating the laminations seen in the fossils.

- These daily laminations have been used by a number of authors to estimate year length in the Precambrian. For example, in 1984, James Vanyo and Stanley Awramik from the University of California, Santa Barbara, estimated that stromatolites studied form the Bitter Springs Formation in central Australia indicate that there were 435 days in a year at 850 million years ago.

What about Longer Cycles in Earth History?

- A particular cycle that has a great influence on global climate over hundreds of thousands of years are Milankovitch cycles, which are caused by 3 properties of Earth's orientation and movement around the solar system: obliquity, precession, and eccentricity.

- Obliquity is the change in the tilt of the Earth's axis, which is never vertical but ranges from 21.1° to 24.5° and back again over a period of about 41,000 years. The tilt of the Earth's axis doesn't always stay pointing at the same place in the, sky though; like a top, it moves in a circular manner that is called precession over a period of around 23,000 years. This "wobble" is largely controlled by the gravitational influences of the Sun and Moon. Eccentricity describes the change in the shape of Earth's orbit over time, from more circular to more elliptical over a period of about 100,000 years. This change is caused by the gravitational influence of Jupiter and Saturn.

- Each of these cycles will affect the amount of solar radiation striking the Earth, but their greatest effects will be felt when these cycles all add together. It is thought that in the current ice age, it is these cycles that are a major influence in the retreat and expansion of ice over time. We are currently in an interglacial time interval.

- During a warmer period of Earth's history, we can still detect these cycles when the Earth doesn't plunge into a glacial period under their influence by using fossils. A good example comes from research of Dr. Brian Huber, a micropaleontologist in the Smithsonian's National Museum of Natural History's Department of Paleobiology.

- Changes in the amount of solar radiation can have impacts on a whole range of Earth systems beyond ice formation, including changes in oxygen distribution in the oceans, sea-level fluctuations, nutrient availability, and temperature. These changes will produce different signals from different fossil communities, but one of the most sensitive are marine microorganisms, such as the foraminifera that Dr. Huber studies.

- Dr. Huber and his collaborators were studying sediments extracted by the Ocean Drilling Program that were deposited during the last stage of the Cretaceous. The Cretaceous was an extremely warm period, with likely little to no ice at the poles. The sediments they recovered showed distinctive variations in color between red and green. Using paleomagnetic data contained within the sediments, they could calibrate these changes with other variables, and they determined that these changes may have been controlled by a 21,000-year precessional cycle.

- Fossils are useful in highlighting cycles over tens, perhaps hundreds, of thousands of years, but what about even longer—perhaps hundreds of millions of years long? Things become a little more difficult when dealing with extended timescales. This is in part due to the incompleteness of the sedimentological record. The older you get, the more incomplete the record becomes.

- One of these long-term cyclical proposals comes from David Raup and Jack Sepkoski of The University of Chicago, who described, based on changes in biodiversity over time, a periodic pattern of mass extinctions with a 26-million-year periodicity. A popular explanation for this was an increase in impacts of comets from a remote zone of the solar system.

- The increased frequency of impacts was explained by Michael Rampino of New York University as being due to the vertical oscillation of the solar system as it periodically passed through the plane of the galaxy. This would disturb these comets and cause them to start to tumble into the inner solar system, some of which would impact the Earth, causing extinction events.

- Raup and Sepkoski's suggested periodicity of mass extinctions met with a lot of criticism, though. Some have claimed that the apparent periodicity was just a statistical artifact. Some, such as Robert Rohde and Richard Muller of the University of California, Berkeley, have proposed an alternate periodicity of 62 million years and another at around 140 million years, with possible causes in comet showers and mantle plume–generated volcanism, among others.

Questions to consider:

1. How much of a record will we leave in Earth's history?
2. Why is radiometric dating not the answer to all of our geological dating needs?

Suggested Reading:

Benton and Harper, *Introduction to Paleobiology and the Fossil Record*, chap. 7.

Winchester, *The Map That Changed the World*.

Lecture 7 Transcript

Fossil Timekeepers

Time—we live by it, spend it, and waste it. One of the earliest clocks produced by humans was a sundial dating to 1500 before the Common Era from ancient Egypt. We've been obsessed with time ever since, but what about natural timepieces?

An awareness of the passage of time is probably imprinted into the fabric of the biosphere. Being able tell day from night, for example, is a useful adaptation if you're a photosynthesizing bacterium adjusting your activity based on a daily rhythm. We humans, and many other organisms, have our own body clocks following the circadian rhythm. Circadian, incidentally, comes from the Latin *circa*, meaning approximately, and *diem*, meaning day. Our circadian clocks tell us when we should be alert, asleep, or hungry. It has a powerful effect on our health too, as anyone who has their body clocks all messed around by a long air flight can tell you.

Of course, life responds to seasonal cycles as well, yearly cycles that govern times of migration, reproduction, hibernation, and many other important events in the biosphere. But there are also lunar cycles that govern the activities of some creatures. For example, in December, on the Great Barrier Reef, corals synchronize the release of egg and sperm in one large coral-based orgy. Other factors probably help trigger the event but it's probably moonlight intensity near a full moon that governs the release.

But how do fossils speak to time and cycles of time? They are obviously representatives of time past, but is there something to them more than just simply being old? So in this lecture let's ask: Do we need fossils as clocks? Can't we just date rocks with radioactivity? How do fossils act as the timekeepers of geology? Do days fossilize? Can fossils record changes in the cycles of the solar system over hundreds of thousands of years? And do fossils hint at an even larger Earth cycle over time?

Our ability to date our planet and its history is getting more and more sophisticated. Scientists like Marie and Pierre Curie and Ernest Rutherford advanced our understanding of radioactivity and radioactive decay and with it our ability to date our planet. Bertram Boltwood of Yale University published the first paper regarding the dating of rocks and minerals using radioactive decay way back in 1907.

Basically, radiometric dating is based on an understanding of the principles of radioactive decay. It considers the ratio of an unstable radioactive isotope—the parent material, like uranium-238—to its decay product, the daughter material, which for uranium-238 is lead-206. The uranium doesn't decay entirely into lead all at once, though, but rather follows a decay chain with various forms of radiation being emitted as a chain of unstable isotopes is produced along the path to lead-206. As we know the rate at which the parent material decays into daughter material, we can calculate how long decay has been progressing. The technique assumes that no parent or daughter material has been added to the sample, what we call a closed system.

Fortunately, crystals in igneous rocks—that's rocks that cool from a magma—form great closed systems that trap small quantities of radioactive isotopes and, as such, act as little hourglasses ticking away as time passes by. The time it takes for half of the parent to decay into the daughter material is called the half-life. For uranium-238 to lead-206, that is about 4.47 billion years. So even in Earth's oldest rocks, if there is material to be analyzed, there should be enough parent material left to work out the ratio and calculate the age. This is not the only radioactive isotope scheme, though. There are many, each of which can be used to cross-check the results of the other.

Perhaps the most familiar form of radiometric dating is carbon-14 dating, though. This is a technique commonly used by archaeologists. Carbon-14 has a very short half-life, though, of about 5730 years, meaning that by 50,000 years there is very little material left to date. Even so, it's a useful technique for dating recent geological events such as the last glacial episode of the Pleistocene ice age.

So, do we really need fossils for dating? Surely all you need is an expensive lab and, as long as you choose your samples carefully, off you go—date away. Of course, there is a fundamental problem. Although the vast majority—around 90%—of the rocks in Earth's crust are igneous rocks, the vast majority of rocks that cover the surface of the crust, and those that contain the majority of the history of life, are sedimentary rocks.

Clastic sedimentary rocks—for example, sandstones—that form from the erosion of older rocks may contain datable crystals from igneous rocks. But if you find such a crystal that has not been compromised by the erosion that created the sedimentary rock, it will of course not provide a date of the sediment or the fossils it might contain; it will only provide a date of the igneous rock from which it was derived.

We can get a bit of a break, though, with volcanic eruptions, particularly ones that produce a lot of ash. These ashes can travel considerable distance and also contain datable igneous crystals that formed close to the time of the eruption of the volcano. For example, *Ardipithecis ramidus*, a famous human ancestor from the Afar Desert in Ethiopia, was conveniently found sandwiched between two ash layers, giving Ardi a very precise age of around about 4.4 million years old.

But ash horizons are not present everywhere, and there can be a vast sequence of fossil-rich strata that contains no ash beds at all. Therefore, how do we place fossils in a sequence that makes chronological sense so that we can order the pages of life's history book? This is an issue William Smith, son of a blacksmith, solved for us back in the late 1700s, and whose trials and tribulations—and believe me, he had many—were covered in my other series, "A New History of Life." William was an engineer—if alive today he may have been considered to be a geological engineer—and had been charged to build canals to carry coal from the Somerset coalfields.

What he recognized, while he was cutting his way through the English countryside, was that various types of fossils followed one another in a predictable order. Once you knew the order, you could place any geological stratum that contained fossils into a time frame relative to another exposure, perhaps at some considerable distance, based purely on the fossils that it contained. At a very gross level, it means in a sequence of rocks that have not been faulted, folded, or otherwise overturned by tectonic processes, this fossil will always be found below this fossil. At a finer scale, it means that this trilobite is always found below this trilobite.

This might not sound like a revolutionary concept to us today, but back then this was powerful stuff. For the first time scientists had the ability to order the geological strata that they found based on the order of the fossils they were finding in them. But more than that, it also permitted geologists and paleontologists to correlate between areas in time. This would allow William Smith to create the first large time-based geological map, and also meant that the history of the book of life on Earth could really start to be written, the pages ordered and controlled by fossils.

This development is the start of the science of biostratigraphy and, of course, only works because Darwin's evolution by natural selection has been continually updating the characters in the long-running story of life. In biostratigraphy, we consider the distribution of a particular fossil species from the time it first originated to the time it becomes extinct, what we call its first and last occurrence datum. The time this represents is called the fossil's range.

Such fossil ranges are collected from many different sections and cross-correlated with other fossils and dating techniques. In this way we can get a pretty good estimate of the slice of time a particular fossil represents. As such, a species that has travelled far but died young makes the best fossils for dating. This is because they define a focused slice of time over a wide area. Now, not all fossils are great timekeepers, though. Some species just existed for too long, and as a result don't provide us with sufficient time resolution.

Given that, some of the best fossils for correlation are fossils that would range far and wide across the oceans, free swimmers like some groups of ammonites or planktonic floaters like the graptolites who are found in many locations and across many environments. Microfossils—that's a broad group of fossils generally less than 1 millimeter long—are also great for biostratigraphy, as many of these forms were planktonic and distributed widely through the oceans.

My own research focused on one such group from the Paleozoic called the chitinozoa. These are enigmatic fossils that, as of yet, have not been tied down to any group of creatures either living or extinct. The best bet so far is that they might represent the egg cases of some sort of marine animal. Their unknown biological affinities, though, doesn't impact their usefulness to biostratigraphy. Whatever they were, they evolved fast and produced many species that were widely distributed.

As microfossils, they have great advantages over larger macrofossils, in that 25 grams of rock may contain hundreds, sometimes thousands of chitinozoa. This makes them a rich source of readily available biostratigraphic information, something very important to an oil company, for example, who may need to date a core of rock, or chips of rock, from a borehole in which the chance of finding a large macrofossil would be pretty slim.

Chitinozoa have now been used to correlate areas at a regional, national, and, for those forms that have been found to have a particularly global distribution, even international levels. This table is taken from a paper that I had some small part in writing back in the 1990s when we were using commonly occurring species of chitinozoa to chop up the Silurian time period based on the fossil ranges that we talked about earlier. From this biostratigraphy, for example, if you ever find *Eisenakitina phillipi* here, you can know you are specifically in the Ludfordian stage of the Ludlow series of the Silurian period.

Dr. Brian Huber in the Department of Paleobiology is also a micropaleontologist, but the group he works with are the foraminifera, a protist group that secretes a shell, usually out of calcium carbonate, that is an important part of the microplankton today but also of the geological past, too. Dr. Huber's research interests are focused on a more recent time in Earth's history than my own, specifically the Cretaceous period and the younger Cenozoic era. The advantage that Dr. Huber has is a far more complete sedimentological record than is available to people like me who work with older rocks. Basically, the further you go back in time, the less material you have to study due to the destruction of older rocks by erosion and plate tectonic processes. Fortunately for Brian, he has available sediments at the bottom of today's ocean basins that date back to the Jurassic period with even a few slivers of older Permian sediments dating back to around 280 million years.

Forams have been an important part of the ocean ecosystem for many hundreds of millions years, and are abundant in diverse environments, including the ocean floor, that's where we find benthic forams, and at various layers in the water column, that's where we find planktonic forams. The period that Brian is studying has some superb, almost continuous sections of sediment extracted by organizations such as the Ocean Drilling Program by ships like the *JOIDES Resolution*.

But there is another advantage to working in ocean sediments, and it has a lot to do with the Earth's core. As we covered in another lecture, we live on a highly differentiated planet. There is a thin atmosphere surrounding an equally thin crust that sits upon the hot, dense rock of the mantle. At the center of our planet is a solid metallic core composed mostly of iron and some nickel. This is surrounded by a liquid outer core composed of iron, nickel, and some lighter trace elements.

The differentiated structure probably developed very early in our planet's history when heating of the early Earth melted the planet and caused heavier elements like iron and nickel to sink to the center under gravity. This was a vital event in our planet's history, as it is from Earth's core that Earth's magnetism is generated. This magnetic field, or the magnetosphere, extends far into space and protects us from harmful solar radiation by deflecting most of the charged particles that would otherwise strike the surface of the planet. This may explain, in part, why we have managed to hold onto our thick atmosphere and why a planet like Mars didn't due to it losing its magnetic shield, probably because of the cooling of that planet's core. With its magnetosphere gone, the solar winds were able to strip away the atmosphere of Mars into space.

But something strange happens to Earth's magnetic field every now and again. At the moment, if you take a compass, the needle will point toward the North Pole. But this has not always been the case. At times in Earth's past, that needle would have pointed to the south. This has occurred throughout Earth's history with the last one about 780,000 years ago. The switch probably took less than the length of a human lifetime to complete, too.

What causes these geomagnetic reversals? Well, at the moment we're not really sure. Some have suggested that the shock of an impact could cause the magnetic poles to reverse, or perhaps slabs of oceanic lithosphere descending into the mantle and colliding with the core. Whatever the reason, these reversals have an interesting effect on sediments deposited in the oceans.

Iron minerals falling through the deep ocean eventually settle on the ocean floor. The fine sediments that form in these environments are particularly useful as iron minerals contained in them take up the orientation of the Earth's magnetic field, both its inclination to the surface and its magnetic orientation, either normal or reversed. These magnetic reversals act like the rings across a tree trunk that can be read in sediments, providing a magneto-stratigraphy.

A similar process occurs with magnetic minerals occurring in lava. As the lava cools, it also takes up a magnetic signal. The advantage here, though, is that these rocks are igneous rocks and can be radiometrically dated. This allows us to match specific magnetic reversals to actual dates, which can then be applied to the patterns of reversals we find in the ocean sediments. This gives scientists like Dr. Huber, who work with such oceanic sediment cores, a very powerful tool. Not only do these cores have an excellent record of changing species of microfossils through time, but they also allow us to cross-correlate their ranges with the geomagnetic reversals also contained in the sediments.

So fossils are great for ordering the pages of Earth's history, especially when we can correlate them with other dating schemes, but what about insights into timescales at a more human scale? Let's ask: Can days and years fossilize?

You might not have noticed it, but on June 30, 2015, at 23:59 and 60 seconds, an extra leap second was added to what is called Coordinated Universal Time. This addition was to account for a discrepancy in the way we calculate time based on two systems. Firstly, International Atomic Time, which is a time based on a number of very precise atomic clocks; and secondly, Universal Time, sometimes called Astronomical Time, which is calculation based on the Earth's spin around its axis—essentially, the length of a day. The reason we had to add a second to Universal Time is because Earth is running slow. Since 1972, a total of 26 leap seconds have been added to Universal Time, meaning the Earth has slowed down 26 seconds since then. This presents an interesting question: If the Earth is slowing down, how long was a year in the geological past? And is there any hard geological evidence to support it?

As we noted earlier, as the Earth rotates, a circadian rhythmicity is generated that can impact the behavior and even the anatomy of organisms. For example, the orbital positions of the Earth will impact the amount of incoming solar radiation generating the seasons with various effects on organisms. By careful analysis of certain fossils, it is potentially possible to read these time-related changes recorded in their tissues.

Here's an example. Consider creatures with shells or skeletons that live in shallow marine environments that respond to daily tidal variations. Although care has to be taken to account for other environmental factors, creatures like bivalves, or clams, show growth lines that correspond to daily, monthly, seasonal, or yearly environmental changes. This corresponds to packets of different thicknesses of growth bands. Collectively, this accounting of time is known as sclerochronology.

One of the first studies to apply this technique was back in 1963 and used corals. Professor John Wells of Cornell University interpreted fine ridges on the surface of corals from the Devonian period as being circadian in nature. The ridges were further grouped into regular bands thought to be lunar monthly breeding cycles. He also identified major annulations, and he suggested that these corresponded to seasonal yearly environmental changes. From his calculations, Wells estimated that the Devonian year consisted of about 400 days.

This means that the Earth's rotation about its axis has been slowing down. The Earth's initial spin, at the time it formed, was due in part to the angular momentum of the initial spinning nebula from which the solar system formed. Other factors probably affected the Earth's rotation as well, including an impact of Theia, a hypothetical Mars-sized body that collided early in Earth's history and is probably responsible for the formation of the Moon. Following this event, the Earth may have zipped around on its axis in just 6 hours. Since then, the Earth's rotation has been slowing down, mostly due to the Moon's effect on ocean tides The Moon's gravity is dragging on a tidal bulge in the oceans, slowing the Earth down like a brake on the wheel of a car.

There are other factors that can affect day length, too. For example, it's been estimated that the devastating 2004 Sumatra-Andaman earthquake in the Indian Ocean effectively shortened the length of the day by about 2.68 milliseconds. The megathrust earthquake, as it's called, saw a large portion of the Indo-Australian plate suddenly shoved under Indonesia and into the planet. This is a bit like an ice skater pulling their arms into their body, which is their center of mass, which makes them spin faster. In the same way, the earthquake sped up the planet and shortened—very slightly—the length of our day.

Fossils provide snapshots through time of the rate of Earth's rotation. For example, by the time of the extinction of the dinosaurs at the end of the Cretaceous, there were 371 days in a year. The Middle Permian year was 390 days long, and around 397 days in a late Devonian year. But how about earlier than that? Abundant fossils of animals with mineralized skeletons only really occur after the Cambrian explosion about 542 million years ago. Can we go any further back with our day length estimates?

We probably can, with a little help from bacterial mats and structures that they produce call stromatolites that we met in an earlier lecture, some of which date back to 3.5 billion years ago. If you remember, stromatolites are layered structures that form in shallow water. They grow as microbial mats commonly composed as cyanobacteria trap, and bind, and cement the sediments. The bacteria move upward daily, forming a new layer, creating the laminations seen in the fossils. These daily laminations have been used by a number of authors to estimate year length in the Precambrian. For example, J. P. Vanyo and S. M. Awramik from UC Santa Barbara who in 1984 estimated that stromatolites studied form the Bitter Springs Formation in central Australia indicated that there were about 435 days in a year at around 850 million years ago.

So how about longer cycles in Earth's history? A particular cycle that has a great influence on global climate over hundreds of thousands of years are Milankovitch cycles. These cycles are caused by 3 properties of Earth's orientation and movement around the solar system. First is obliquity. This is the change in the tilt of the Earth's axis, which is never vertical but ranges from about 22.1° to 24.5° and back again over a period of about 41,000 years. We're currently at about 23.44°, somewhat more than halfway between the two extremes.

The tilt of the Earth's axis doesn't always stay pointing in the same place in the sky, though. Like a top, it moves in a circular manner that is called precession over a period of about 23,000 years. Currently, our axis in the Northern Hemisphere points to the pole star. In the past, though, it has pointed at stars like Vega in the constellation of Lyra. This wobble is largely controlled by the gravitational influences of the Moon and Sun. And there is eccentricity, which describes the change in shape of Earth's orbit over time from more or less circular to more elliptical over a period of about 100,000 years. This change is caused by the gravitational influence of Jupiter and Saturn.

Each of these cycles will affect the amount of solar radiation striking the Earth, but their greatest effects will be felt when those cycles all add together. It is thought that in the current ice age it is these cycles that are a major influence in the retreat and expansion of ice over time. We are currently just in an interglacial time interval. But what about these cycles during the warmer period of Earth's history? Can we still detect them when the Earth doesn't plunge into a glacial period under their influence? Well, yes we can, by using fossils. A good example comes from some more of the research that's conducted by Dr. Huber at the Smithsonian.

Changes in the amount of solar radiation can have impacts in a whole range of Earth systems, beyond ice formation. For example, it could affect changes in oxygen distribution in the oceans, sea-level fluctuations or nutrient availability, temperatures, and probably other conditions, as well. These changes will produce different signals from different fossil communities, but one of the most sensitive are marine microorganisms like the foraminifera that Dr. Huber studies.

Dr. Huber and his collaborators were studying sediments extracted by the Ocean Drilling Program that were deposited during the last stage of the Cretaceous, in the Maastrichtian stage. The Cretaceous was an extremely warm period with likely little or no ice at the poles. The area of study, the Blake Nose off the coast of Florida, was in a tropical to subtropical location at around about 30°N latitude at this time. The sediments they recovered showed distinctive variations in color between red and green.

Using paleomagnetic data contained within the sediments, they could calibrate these changes with other variables, variables such as the paleotemperature of the ocean recorded isotopically in the foraminifera shells, the mineral character of the sediments, and the relative abundance of the forams. From this, they determined that these changes may have been controlled by a 21,000-year precessional cycle.

How does this system work? Well, that's the crunch, isn't it? But one possible explanation is that the climatic variation caused by the precessional cycle was affecting the upwelling of cold water in this particular part of the Atlantic. The redder, more oxidized sediments may have been associated with low rates of upwelling, lower rates of oceanic productivity, and an association of particular foram species.

The greener parts of the sediment core that they found were associated with colder surface water temperatures and higher rates of oceanic productivity, probably a response to increased nutrient availability in that cold, upwelling water. The green color could probably be explained by the increased delivery of organic material to the ocean floor. This would rapidly use up any oxygen, favoring reducing conditions which typically produce these green-coloured sediments.

So, fossils are useful in highlighting cycles over tens, perhaps hundreds of thousands of years. How about even longer, perhaps hundreds of millions of years long? Things get a little more difficult when dealing with these extended timescales. This is, in part, due to the incompleteness of the sedimentological record. As I said earlier, the older you get, the more incomplete the record becomes. One of these long-term cyclical proposals of David Raup and Jack Sepkoski of The University of Chicago described, based on changes in biodiversity over time, a periodic pattern of mass extinctions with a 26-million-year periodicity.

A popular explanation of this was an increase in impacts of comets from a remote zone of the solar system called the Oort Cloud, a spherical shell of comet debris that orbits our sun at the far distant edge of near interstellar space. This increased frequency of impacts was explained by Mike Rampino of New York University as being due to the vertical oscillation of the solar system as it periodically passed through the plane of the galaxy. This would disturb the comets and cause them to start to tumble into the inner solar system, some of which would impact the Earth causing extinction events.

Raup and Sepkoski's suggestion of periodicity of mass extinction met with a lot of criticism, though. Some have claimed that the apparent periodicity was just a statistical artifact. Some, such as Robert Rohde and Richard Muller of the University of California, Berkeley have proposed an alternate periodicity of 62 million years and another of around 140 million years with possible causes in comet showers and mantle plume-generated volcanism, amongst others.

Another possible mechanism to support a 62-million-year extinction periodicity is increased cosmic ray exposure when the solar system moves to the northern side of the galactic plane. This increased exposure occurs as our galaxy is moving toward the Virgo Cluster, a collection of perhaps 2000 galaxies and the source of these cosmic rays. In this hypothesis, increased cosmic ray exposure generates energetic particles which are dangerous to organisms. It was also postulated that cosmic rays could have helped destroy the ozone layer and, through interactions with the atmosphere, caused smog and global dimming and cooling effects. It's important to note, though, that all these long-term extinction cycles are what you might call a slippery fish scientifically, and should be regarded with caution.

We've come a long way, though, from William Smith and his fossil observations in canal excavations in the gentle hills of England. The science of biostratigraphy is now a critical tool both to scientists and the resources industry, but fossils as clocks now are also revealing the possibility of deep-seated cycles in the history of life, the solar system, and possibly beyond. The further back we look, the muddier and more out of focus the view is, but the benefit of science is that with every passing year of research our sight gets deeper and a little clearer.

Lecture 8 Fossils and the Shifting Crust

Exotic fossil assemblages can be set adrift on continents and continental fragments to beach thousands of miles away in a completely different part of the world. In doing so, they leave a story of their origin and journey through time. In this lecture, you will learn what paleobiogeography is, what the fossils in Alfred Wegener's jigsaw puzzle were, how fossils can time the closing of an ocean, and how fossils trace the dance of continental fragments through time.

Paleobiogeography

- Why are creatures where they are? We obviously don't live on a planet where life-forms are spread in a homogeneous manner; different types of animals and plants have distributions and concentrations. Basically, all creatures have a geographical range, some broad and some narrow. Endemic species are only found in a specific area, while cosmopolitan species are found in a range of environments.

- On a very broad scale, life can be divided up into several biogeographical provinces, or ecozones, which are geographical areas of the world that have characteristic communities of species. The Nearctic ecozone includes North America and Greenland. Europe, Asia, and North Africa are in the Palearctic. Others include the Neotropic, Afrotropic, Indo-Malaya, Australasia, and Arctic.

- Ecozones can also be recognized from Earth's geological past, but in this case, we have to consider the additional complication that wandering continents add to the story. The first thing we need to consider is the manner in which diversity changes, very broadly, across our planet. To do that, we also need to think about how our planet's magnetic field intersects with the ground.

- It has been suggested that the movement of liquid metal in the outer core around the solid metal inner core, due to convection and the Coriolis effect of the spinning Earth, produces electric currents that, in turn, generate Earth's magnetic field. The Earth is like a giant bar magnet, with lines of force running from the North Pole to the South Pole. Just like a bar magnet, on our planet, the magnetic field is inclined toward the vertical at the poles, and at the equator it will be parallel to the surface of the ground. This means that magnetic inclination and latitude are linked.

- This signal can be locked into certain fine-grained sediments and basalt lava when iron-rich minerals take up the magnetic inclination at the time of their formation. So, if you can record the inclination of the magnetic field in the rocks, you can also estimate the paleolatitude of that rock at its time of formation. There is a relationship between latitude and the diversity of organisms, too: The diversity of organisms is highest at the equator and drops off toward the poles.

- Both the magnetic inclination and the diversity data provide potentially useful information about where on the surface of the planet, in a latitudinal sense, a particular rock was when it formed. These are important clues that we can use when trying to recreate the history and movement of areas of our restless planet.

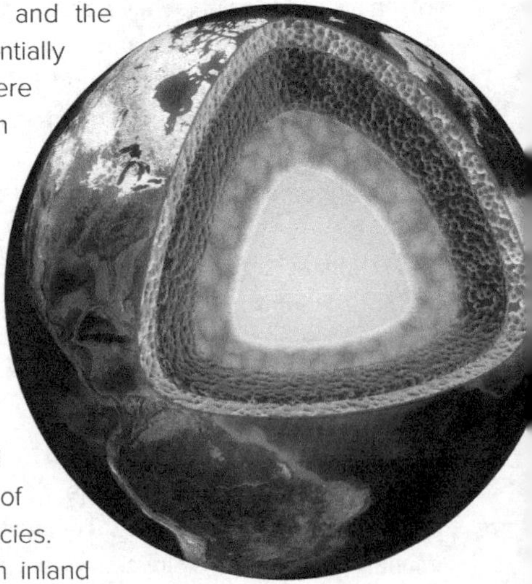

- The second point we need to consider is the concept of barriers to the migration of species. On land, barriers could be an inland sea, mountains, or even dense forest. In the ocean, barriers can include swift currents or deeper parts of the ocean, where food resources may be limited. Barriers could also be due to different temperature and climatic regimes.

- Paleontologists are also concerned with a dimension beyond the currently geographical one: They want to know what happens to the distribution of plants and animals over time, what is called dispersal biogeography.

- One of the first people to consider such migrations was American paleontologist George Gaylord Simpson, who imagined species originating at a central location and then dispersing over time. Dispersal would vary depending on the ease of movement of creatures.

- Simpson referred to a corridor as a place where creatures can mix fairly easily. Other migration routes are more selective, allowing the passage of only a restricted selection of creatures. Simpson termed this type of feature a filter bridge. The third dispersal mechanism Simpson termed sweepstakes, which describes migration due to the relatively rare but still important effect that luck has in the movement of organisms from one place to another. Simpson also recognized that dispersal, and associated isolation, is a powerful force in evolution.

- When considering the distribution of species, we have to consider that it is not only creatures that migrate—continents do, too. Simpson was no fan of the idea of drifting continents, so his view of the dispersal of organisms through time was essentially a static one, with the continents and the ocean basins occupying their current locations for more than hundreds of millions of years.

- But with the dawn of plate tectonics, and the dynamic movements of continents over time, a whole new way of looking at the dispersal of fossil species came to light.

Wegener and Continental Drift

- German climate scientist Alfred Wegener challenged the static view of continents. He amassed a wealth of data to suggest that the continents were once joined, including similarities of the stratigraphic record on distant continents, evidence of glaciations that once covered a united supercontinent, and the fit of coastlines on either side of the Atlantic.

- In 1915, he proposed the existence of a supercontinent called Pangaea that existed more than 250 million years ago. Accordingly, the current continents represent the fragments of that united landmass that have subsequently drifted to their current locations.

- But perhaps some of his most compelling evidence for continental drift came from fossils, many of whose current distribution is puzzling in the context of a static planet. For example, fossils of *Cynognathus*, a meter-long predator from the early Middle Triassic period, have been found in South Africa and China. It could have walked to those locations based on earlier views of static, immobile continents.

- But specimens were also found in Argentina and Antarctica. Physiologically, these creatures were not adapted to swimming, so how they could be found on such distant continents separated by enormous oceans? This distribution only makes sense once the continents are drawn back together.

- But there was a problem with the mechansim Wegener proposed to explain how the continents drifted. He suggested that the gravitational pull of the Sun and Moon and the spin of the Earth were "dragging" the continents around the planet. This was a very difficult pill for many scientists to swallow, because these forces are nowhere near what would be needed to move a continent. The hypothesis of drifting continents was largely ridiculed.

- The distribution of Wegener's fossils was explained away by the rafting of creatures, the presence of land bridges, or island hopping. It is difficult, though, to see how these mechanisms could operate over large oceans and how they could account for the distribution of so many fossil species.

The History of the Iapetus Ocean

- Eventually, the idea of drifting continents would be revived, but this time with the more plausible mechanism of seafloor spreading. Scientists such as Harry Hess, Marie Tharp, Bruce Heezen, and John Tuzo Wilson would pull together information from ocean-floor topography, seismic records, and ocean-floor magnetism to give us the theory of plate tectonics that we are familiar with today.

- Wilson proposed—on the basis of fossils and his understanding of tectonic plate motions—the existence of a large ocean in the Northern Hemisphere. He called this ocean the proto-Atlantic and claimed that this ocean, later called Iapetus, closed during the Silurian and Devonian periods.

- Although of the same geological age, Cambrian trilobites of western Newfoundland are different from those of eastern Newfoundland. The association of fossils in the west are called the Laurentian fauna, and those in the east are called the Avalonian fauna.

- The western Newfoundland faunas have more in common with those of Scotland, northwestern Ireland, and most of the rest of North America. By contrast, the eastern Newfoundland faunas, which also occur in New Brunswick, Nova Scotia, and Massachusetts, share more fossil biogeographical connections with most of Europe, including England, Wales, and southeastern Ireland.

- Before the advent of plate tectonics, this mismatch of trilobite faunas across the Atlantic was explained away by a bunch of geological and oceanographic gymnastics. But in a plate tectonics context, this tells us that Newfoundland is a sutured landmass. In other words, the 2 halves were once associated with different continents on either side of an ancient ocean that subsequently have been brought together and fused by plate tectonics.

- The fossils are different because the western and eastern parts of Newfoundland were in different climatic zones during the Cambrian on opposite sides of the Iapetus Ocean, with the ocean being sufficiently wide at that point to even prevent the mixing of marine species. The 2 faunas were brought together when the ocean closed. Where the 2 faunas now meet represents the line along which the eastern (Laurentian) and western (Avalonian) terranes were sutured together in what today is Newfoundland.

- When the Atlantic Ocean opened up, splitting the continents apart again, fragments of the Avalonian or Laurentian faunas were stranded on either side of the ocean. Fossils would not only uncover the presence of this ancient ocean; they would also help document its closure over time.

- It is thought that Iapetus opened in the Late Precambrian as an older supercontinent, called Rodinia, fragmented. British paleontologists Stuart McKerrow and Leonard Robert Morrison Cocks would record changing faunas found on either side of the Iapetus Ocean as this body of water narrowed. They found that faunas in Europe and North America were most different during the Cambrian and Ordovician, with only planktonic species—which floated in the ocean and would have been able to mix relatively freely across a large body of water—being found on both sides of the ocean. It is estimated at its widest point that Iapetus would be about 4000 kilometers wide.

- As the ocean started to close and the distance between North America and Europe was reduced, creatures that had a planktonic larval stage started to mix on either side. In addition to planktonic forms, nektonic, free-swimming organisms were also able to make the crossing. As the ocean narrowed even more, less mobile forms were able to make it across. By the end of the Devonian, the Iapetus Ocean was sufficiently narrow that even freshwater fish were similar in western Europe and eastern North America.

Exotic Terranes

- Since the early days of plate tectonics, our understanding of the complexity of the wandering continents has increased considerably. One of the ways our understanding has become more complex is an appreciation of what are called exotic terranes. In addition to large continental masses lumbering around the planet, it was realized that small fragments of continents have also been rifting off larger parent bodies, zipping around the Earth like marbles, colliding with other areas, potentially thousands of miles from where they originated.

- Exotic fragments can generally be recognized by geologists in the field when mapping highlights major fault zones that represent lines of disjunction between the rock units to either side. The exotic blocks themselves differ dramatically from the surrounding geology, with paleontology and even sometimes paleomagnetic inclination very different from the surrounding geology, suggesting a more exotic, perhaps significantly distant, origin.

- Fragments can be composed of ancient volcanic islands, oceanic ridges, various ocean-floor volcanic features, and fragments of other continents.

- This so-called accretion tectonics is probably a common process through geological time but is most easily recognized in relatively recent rocks with relatively less deformation, where traces of these fragments can still be uncovered.

- In considering how fossils help us in our understanding of the movement of terranes, we need to appreciate how populations, or species, of organisms differ in relation to the geographic distance between them. In other words, the farther away they are from their original source, the more dissimilar they become.

- The similarity between populations can be calculated using the Simpson coefficient, in which the higher the coefficient, the greater the similarity between 2 populations. This is a powerful tool when attempting to reconstruct the movement of terranes.

- Diversity gradients and paleomagnetism are useful tools in determining the ancient latitude that a particular terrane occupied, but there is an obvious drawback: We have no idea about the longitude of the terrane, and this is where fossils and similarity coefficients come into play.

- Studies of exotic terranes have helped us unravel a picture of what is today a fairly complicated geology, the product of the collision between various bits and pieces zooming across the Pacific and colliding with the main North American continental landmass.

Questions to consider:

1. Should we expect greater biodiversity during times of continental amalgamation or fragmentation?
2. How much of the ancient history of the dancing continents and continental fragments is lost to us?

Suggested Reading:

Keary, Klepeis, and Vine, *Global Tectonics*.

Plummer, Carlson, and Hammersley, *Physical Geology*.

Paleogeographic and Tectonic History of Western North America, http://cpgeosystems.com/wnampalgeog.html.

Lecture 8 Transcript: Fossils and the Shifting Crust

Why are creatures where they are? We obviously don't live on a planet where life-forms are spread in a homogeneous manner—different types of animals and plants have distributions and concentrations. Basically, all creatures have a geographical range; some broad, some narrow.

For example, this is a pebble toad, *Oreophrynella nigra*. It lives specifically on a type of mountain called a tepui in Venezuela: steep-sided, flat-topped mountains that rise majestically out of the jungle like islands in the sky. This toad is what we call an endemic species; it is only found in this specific highland area of Venezuela. Incidentally, this little toad has developed an interesting defense strategy against anything that might fancy toad for dinner. If it feels threatened, it can tuck itself into a ball and throw itself down a hillside like a rolling stone. I must admit, I've been in some parties when I wish I could've done that.

The opposite of an endemic species is a cosmopolitan species. For an example, well, just go look in a mirror. From the steaming tropics to the ice and snow of the arctic, from the deepest jungles to the tops of mountains and baking deserts, you will find *Homo sapiens*.

Since we evolved in Africa around 200,000 years ago, we have spread to every corner of the globe and are certainly the most cosmopolitan primate on Earth today, although I'm quite sure that our primate cousins would claim that that was nothing to be really proud about. On a very broad scale, though, life can be divided up into several biogeographical provinces or ecozones. These are geographical areas of the world that have characteristic communities of species.

For example, I am currently in the Nearctic ecozone that includes North American and Greenland. Europe, Asia, and North Africa are in the Palearctic.

Others include the Neotropic, Afrotropic, Indo-Malaya, Australasia, and Antarctic at the bottom of the world. Ecozones can also be recognized from Earth's geological past, but in this case we have to consider the additional complication that wandering continents add to the story.

Some things we need to consider, then. First is the manner in which diversity changes very broadly across the plant. To do that, although it might sound a little off-topic, I also need you to think about how the planet's magnetic field intersects with the ground. It has been suggested that the movement of liquid metal in the outer core around the solid metal inner core, due to convection and the Coriolis effect of the spinning Earth, produces electric currents, which in turn generate Earth's magnetic field.

At a very simple level, though, you can imagine the Earth to be like a giant bar magnet, with lines of force running from the North Pole to the South. Just like a bar magnet, our planet, the magnetic field is inclined toward the vertical at the poles, and at the equator it would be more parallel to the surface of the ground. Practically, this means magnetic inclination and latitude are linked.

This signal, as we've seen, can be locked into certain fine-grained sediments in basalt lava when iron-rich minerals take up the magnetic inclination at the time of their formation. So, if you can record the inclination of the magnetic field in the rocks, you can also estimate the paleolatitude of the rock at the time of its formation.

There is a relationship between latitude and the diversity of organisms, too. The diversity of organisms is highest at the equator and drops off toward the poles. Have a look at this diagram adapted from a 2006 paper written by my colleague Paul Smith at the University of British Columbia. It shows a clear relationship between increasing magnetic inclination of magnetic minerals in rocks and decreasing diversity of organisms, in this case bivalves. Both the magnetic inclination and the diversity data provide potentially useful information about where on the surface of the planet, in a latitudinal sense, a particular rock was formed. These are important clues that we can use when trying to recreate the history and movement of areas of our restless planet. The second point we need to consider, though, is the concept of barriers to the migration of species.

On land, barriers could be an inland sea, like existed in North America during the late Cretaceous period. They could be mountains, or even a dense forest. In the oceans, barriers can include swift currents or deeper parts of the ocean where food resources may be limited and can be as much of a barrier to shallow marine creatures as a mountain is on land. Barriers could also be due to different temperature and climatic regimes. Of course, as paleontologists, we are also concerned with a dimension beyond the current geographic one. We want to know what happens to the distribution of plants and animals over time, what we call dispersal biogeography.

One of the first people to consider such migrations was American paleontologist George Gaylord Simpson. Simpson imagined species originating at a central location and then dispersing over time. Dispersal would vary depending on the ease of movement of the creatures. For example, if you consider mammals at the taxonomic level of orders, the largest groupings of the class Mammalia across North America, 100% of orders found in New York State are also found in Oregon. The degree of similarity will decrease as you consider the finer taxonomic resolutions, like the smaller groupings of family, genus, and species, but in general you can regard North America as being fairly homogeneous. This is what Simpson referred to as a corridor, a place where creatures can mix fairly easily.

Other migration routes are more selective, allowing for the passage of only a restricted selection of creatures. For example, during the last ice age the Bering Strait became dry land as sea levels fell. This allowed migration of animals between Asia and North America. But possibly, due to climatic factors, not all animals exploited this new opportunity. Creatures that would migrate include the mammoths and bison that would spread from Europe to North America, and horses and camels that would migrate from North America to Europe and beyond. Simpson termed this type of feature that imparts a selectivity of what can migrate a filter bridge.

Possibly the most famous example of a filter bridge comes from the development of the Isthmus of Panama, the thin ribbon of east-west trending land that today connects North and Central America with South America. At about 3 million year ago, volcanic islands formed by the subduction of the Cocos Plate below the

Caribbean Plate were starting to be connected as sediments filled in the gaps in between them. As a result of this, a migration of faunas occurred between North America—which eventually was linked up with tectonic plates forming Central America—and South America, which up till that point had been pretty well geographically isolated. To use Simpson's evocative term, South America, until linked with the rest of the world, has spent considerable time in splendid isolation.

Because of this isolation, a whole range of exotic marsupials had evolved in South America, like Thylacosmilus, a marsupial predator resembling a saber-toothed cat. And there was fast-footed, camel-like Macrauchenia, and the Glyptodon, not a marsupial but a relative of the armadillo, with adults about the size of a VW Beetle.

Overall, there would be a greater net migration of creatures from the north into the south, an event we call the Great American Biotic Interchange. But this was no simple mixing of species, as many of those exotic creatures in South America would be driven into extinction. But why mostly the South American forms? The answer may in part be due to climate. In general, during this time, much of South America was very warm and moist all year round. When large terrestrial animals of the South American fauna moved north across the land bridge and into North America, they rapidly encountered more extreme variations in climate. This effectively placed a lid on most of the southern critters migration, but not on the dispersal of the hardier North American animals who were able to cross this filter bridge and start to replace those wonderful, and now, regrettably, extinct beasts of South America.

The third dispersal mechanism Simpson termed sweepstakes. This describes migration due to the relatively rare but still important effect that plain dumb luck has in the movement of some organisms from one place to another. For example, creatures that are rafted to oceanic islands, perhaps after major floods when large mats of vegetation are swept from large continental river systems out into sea. This is possibly the manner in which the ancestor of the Galapagos marine iguana landed on the island about 8 million years ago.

Simpson also recognized that dispersal and associated isolation is a powerful force in evolution. The isolation of South America, for example, had permitted the development of its unique fauna. The same is true for the weird and beautifully wonderful fauna of present day Australia.

Something to consider, though, about the development of land bridges is that, although they permit the mixing of some groups of creatures, they may also act as barriers to others. For example, the formation of the Panama isthmus separated 2 oceans that were once linked through the Caribbean, an event called the Great Schism. Populations that were once in contact were now isolated on either side of this newly emplaced landmass. This may account for the diversification of mollusks on both sides of the isthmus due to the isolation of populations and their independent evolution in these new marine environments.

This evolution of new species may have been caused by the mollusks on the Caribbean side having to adapt to more nutrient-poor conditions once the nutrient-rich water of the deep Pacific was now blocked. As paleontologists and geologists, though, we have to account for another factor when considering the distribution of species, for it is not only creatures that migrate, continents do, too.

Simpson, however, was no fan of the idea of drifting continents, so his view of the dispersal of organisms through time was essentially a static one, with continents and the ocean basins occupying their current locations for over hundreds of millions of years. But with the dawn of plate tectonics and the dynamic movements of continents over time, a whole new way of looking at the dispersal of fossil species came to light.

One of those concepts was developed by Malcolm McKenna, a curator of virtual paleontology at the American Museum of National History in New York City. He called his idea Noah's arks, and it describes how isolated landmasses could drift across an ocean and run aground on another continent, releasing any unique endemic species that had evolved on the ark during its isolation. India may have played just such an ark role. About 170 million years ago, India and

Madagascar were locked in the middle of Gondwanaland, a southern landmass comprising South America, Africa, India, Madagascar, Antarctica, and Australia.

By 150 million years ago, Madagascar, India, Antarctica, and Australia split from Africa, almost separated by a narrow ocean. And by 105 million years ago, India and Madagascar had split from Antarctica and were drifting northward. Here, at 66 million years ago, at the time of the death of the dinosaurs, India had left Madagascar behind and was zipping in glorious isolation northward on a collision course with the underbelly of Asia.

By the early Eocene, groups of mammals like the even-toed ungulates, which include modern animals such as pigs, giraffes, and hippos, and up-toed ungulates, such as modern horses, tapirs, and rhinos, suddenly appear in Asia. It's been proposed that it is from animals like this, *Cambaytherium thewessi*, found recently by researchers from Johns Hopkins University, that all the even- and odd-toed ungulates would evolve, carried by Noah's ark India across the ocean, evolving in isolation, till they—and possibly other mammals—were released when India finally docked with Asia. This is an intriguing possibility, although it has been suggested from research in amber and arthropods that these deposits may have formed after docking had occurred. As such, the Gondwana-India ark fauna, if they exist, may have to be sought from older deposits.

But for the rest of this lecture, I'd like to consider another type of geological ship: a Viking funeral ship. For just as Viking chieftains were sent to Valhalla on a ship that was set afire and pushed out to sea, so too can exotic fossil assemblages be set adrift on continents and continental fragments to beach thousands of miles away in a completely different part of the world. In doing so, they leave a story of their origin and their journey through time.

So for the rest of the lecture, let's ask: Who were the fossils in Alfred Wegener's jigsaw puzzle? How can fossils time the closing of an ocean? And what are exotic terranes? And how do fossils trace the dance of continental fragments through time?

The story of the German climate scientist Alfred Wegener who challenged the static view of continents has been told many times. To summarize, though, Wegner amassed a wealth of data to suggest that the continents were once joined. These included similarities of the stratigraphic record on distant continents, evidence of glaciations that once covered a united supercontinent, and the apparent fit of coastlines on either side of the Atlantic. In 1915, he would propose the existence of a supercontinent called Pangaea that existed over 250 million years ago. Accordingly, the current continents represent the fragments of that united landmass that have subsequently drifted to their current locations.

But perhaps some of his most compelling evidence for continental drift came from fossils, many of whose current distribution is puzzling in the context of a static planet. Consider, for example, Cynognathus, a meter-long predator from the early Middle Triassic period. Cynognathus is a member of a group of reptiles, the synapsids, that were probably very diverse during the Permian period, but took a severe hammering at the Permo-Triassic extinction event 252 million year ago. The survival of this group of reptiles is important to our own story too, as synapsid reptiles would give rise to the true mammals later in the Triassic period.

It is the paleogeographical distribution of Cynognathus that is interesting for our story, though. Fossils have been found in the Karoo Basin of South Africa and Lesotho, and also in China. Nothing unusual there, it could have walked to those locations based on earlier views of static, immobile continents. But what about specimens found in Argentina? And now also from the Fremouw Formation in Antarctica? That would be an impossible journey even for such a powerful land creature as Cynognathus.

It is the same story for another synapsid, but this time a herbivore, Lystrosaurus, that evolved in the Permian but would become one of the most common vertebrate elements in Early Triassic faunas, and possibly prey for carnivores like Cynognathus. Once again, it is the distribution of these animals that is of interest. Like Cynognathus, Lystrosaurus has now been found in South Africa, Antarctica, China, and also India. Physiologically, neither of these creatures was adapted to swimming, so how could it be that they could be found on such distant continents separated by enormous oceans?

The same is true for this creature, Mesosaurus, a Permian marine reptile about 3.3 meters long. It had a streamlined body and webbed feet, long, narrow jaws and nostrils on the top of its head, probably swimming a bit like a crocodile by undulations of its long, flexible tail. Like crocodiles, though, it is unlikely that this creature was able to cross a major ocean. This makes its distribution on either side of the Atlantic, in Africa and South America, once again difficult to explain with a static Earth.

And, famously, there is Glossopteris, an extinct order of woody, tree-like seed ferns—some reached 30 meters tall—with tongue-shaped leaves and associations with other floral elements such as horsetails and lycopods. Glossopteris evolved during the early Permian and became the overwhelmingly dominant species in the Southern Hemisphere during that time, where it was an important coal producer. Fossils of Glossopteris are found in Australia, South Africa, South America, Madagascar, India, and Antarctica, certainly a bizarrely wide distribution for any plant with the continents fixed in their current locations.

These distributions of course only make sense once the continents are drawn back together, as you see here. The distribution of plants and animals now make sense in terms of a geographical range across a large landmass. No one here is having to leap across oceans.

This might sound like a bit of a slam dunk for continental drift. There's a problem, though, a problem with the mechanism Wegener proposed to explain how the continents drifted. He suggested that the gravitational pull of the Sun and Moon, and the spin of the Earth, were dragging the continents around the planet. This was a very difficult pill for many scientists to swallow, as these forces are nowhere near what would be needed to move a continent. For this reason, and probably because he was challenging the received view of the old school, the hypothesis of drifting continents was largely ridiculed.

The distribution of Wegener's fossils was explained away by the rafting of creatures, the presence of land bridges, or island hopping. It is difficult, though, to see how these mechanisms could operate over large oceans and how they could account for the distribution of so many fossil species.

Eventually, the idea of drifting continents would be revived, but this time with the more plausible mechanism of seafloor spreading. Scientists like Harry Hess, Marie Tharp, Bruce Heezen, and J. Tuzo Wilson would pull together information from ocean floor topography, seismic records, and ocean floor magnetism to give us the theory of plate tectonics that we are familiar with today.

It would be J. Tuzo Wilson who, on the basis of fossils like trilobites, brachiopods, and graptolites, and his understanding of plate tectonic motions, would propose the existence of a large ocean in the Northern Hemisphere. He called this ocean the proto-Atlantic and claimed that this ocean, later called Iapetus, closed during the Silurian and Devonian periods.

To illustrate what Professor Wilson noticed about these fossils that would allow him to make such a brave prediction, let's travel to Canada and in particular to the beautiful province of Newfoundland. Newfoundland lies off the coast of eastern Canada along the Atlantic Ocean and has a long and fascinating geological history. Of particular interest to our story are the Cambrian trilobites and the fact that, although of the same geological age, trilobites of western Newfoundland are different from those of eastern Newfoundland The association of fossils in the west are called the Laurentian fauna and those in the east are termed the Avalonian fauna.

What is particularly odd is that the western Newfoundland Laurentian faunas have more in common with those of Scotland, northwestern Ireland, and most of rest of North America. By contrast, the eastern Newfoundland Avalonian faunas that also occur in New Brunswick, Nova Scotia, and Massachusetts, share more fossil biogeographical connections with most of Europe, including England, Wales, and southeast Ireland.

Before the advent of plate tectonics, this mismatch of trilobite faunas across the Atlantic was explained away by a bunch of geological and oceanographic gymnastics, including the existence of land barriers or really odd oceanic currents. But, in a plate tectonic context, what it tells us is that Newfoundland is a sutured landmass. In other words, the 2 halves were once associated with different continents on either side of an ancient ocean that subsequently have been brought together and fused by plate tectonics. You can see the 2 halves here, around 455 million years ago during the Ordovician, when eastern and western Newfoundland was separated by the Iapetus Ocean.

The fossils are different because the western and eastern parts of Newfoundland were in different climatic zones during the Cambrian, on opposite sides of the Iapetus Ocean, with the ocean being sufficiently wide at that point to even prevent the mixing of marine species. The 2 trilobite faunas were brought together when the ocean closed. Where the 2 faunas meet represents the line along which the eastern Laurentian and the western Avalonian terranes were sutured together in what is today Newfoundland.

When the Atlantic Ocean opened up, splitting the continents apart again, fragments of the Avalonian or Laurentian faunas were stranded on either side of the ocean. The collision of these continents, though, would create an enormous mountain chain not unlike the Alps today, the much-reduced fragments of which now exist in the highly eroded mountains of Scotland and the modern Appalachians.

Fossils would not only uncover the presence of this ancient ocean, though, but would also help document its closure over time. It is thought that the Iapetus Ocean opened in the late Precambrian as an older supercontinent called Rodinia fragmented. British paleontologists like Stuart McKerrow and L. R. M. Cocks would record changing faunas on either side of the Iapetus Ocean as this body of water narrowed. McKerrow and Cocks found that faunas in Europe and North America were most different during the Cambrian and Ordovician, with only planktonic species like the Cambrian graptolite Dictyonema and the Ordovician form *Didymograptus bifidus* being found on both sides of the ocean.

Graptolites are colonies of little, filter-feeding zooids that floated in the ocean and would've probably been able to mix freely across a large body of water like this. It is estimated, at its widest point, Iapetus would be about 4000 kilometers wide. As the oceans started to close and the distance between North American and Europe was reduced, creatures that had a planktonic larval stage, like trilobites and brachiopods, started to mix on either side, first at the level of genera, but later with particular species found on either side of the ocean. In addition to planktonic forms, nektonic free-swimming organisms like the conodonts were also starting to make the crossing.

As the ocean narrowed even more, less mobile forms were able to make it across, like tiny, benthic ostracods, which were now able to just crawl from one side to the other. And by the end of the Devonian, the Iapetus Ocean was sufficiently narrow that even freshwater fish were similar in western Europe and eastern North America.

This classic study shows how fossil funeral ships like the Avalonian and the Laurentian trilobites on Newfoundland alerted geologists to the possibility of a lost, ancient ocean, and then how, charting the dispersal of fossil communities across that ocean, it closed over time. And if you travel to the Isle of Man off the northwest coast of England, you can go and visit where continents actually touched, now represented by the Niarbyl Fault, all that remains of the once mighty Iapetus Ocean and the mountains that formed by this closure. To me, this is pure geological poetry.

Since the early days of plate tectonics, when the theory was just starting to make sense of so many puzzling features of paleontology and geology, our understanding of the complexity of the wandering continents has increased considerably. One of the ways our understanding has become more complex is in an appreciation of what are called exotic terranes. In addition to large continental masses lumbering around the planet, it was realized that small fragments of continents have also been rifting off larger parent bodies and zipping around the Earth like marbles, colliding with other areas potentially thousands of miles away from where they originated.

A classic example of this comes from western North America: the North American Cordillera. Exotic fragments can generally be recognized by geologists in the field when mapping highlights major fault zones that represent lines of disjunction between the rock units to either side. The exotic blocks themselves differ dramatically from the surrounding geology, like a jigsaw piece in the wrong place, with paleontology and sometimes paleomagnetic inclination very different from the surrounding geology, suggesting a more exotic, perhaps significantly distant origin.

More than 50 exotic terranes have been identified in the Western North American Cordillera, and they may constitute about 70% of the total region, producing what could only be called a tectonic collage. Fragments can be composed of ancient volcanic islands, oceanic ridges, various ocean floor volcanic features, and fragments of other continents. This so-called accretion tectonics is probably a common process through geological time, but is most easily recognized in relatively recent rocks with relatively less deformation, where traces of those fragments can still be uncovered.

In considering how fossils help us in understanding of the movement of terranes, we need to appreciate how populations of species of organisms differ in relation to the geographic distance between them; or, in other words, the further away they are from their original source, the more dissimilar they become. The similarity between populations can be calculated using the Simpson coefficient, in which the higher the coefficient, the greater the similarity between 2 populations.

For example, consider the Simpson coefficient recorded for either side of the Pacific and Atlantic from the Jurassic period to the present day. Basically, as the Pacific shrinks due to the widening of the Atlantic Ocean, there is an increasing similarity between Pacific species and an increasing dissimilarity between Atlantic species—just what you'd expect, right? This is a powerful tool when attempting to reconstruct the movement of terranes. As we mentioned earlier, diversity gradients and paleomagnetism are useful tools in determining the ancient latitude that a particular terrane occupied, but there is an obvious drawback. We have no idea about the longitude of the terrane, and this is where fossils and similarity coefficients come into play.

A good example of the use of fossils in this way comes from Paul Smith's work on a number of exotic terranes in western Canada, and in particular, 2 specific terranes called Wrangellia and Stikinia. Paul was able to compare the similarity of certain endemic Jurassic ammonites and bivalves found in these terranes versus species of the same groups found on the North American mainland.

The comparisons were made by examining fossil assemblages, in which he was able to determine that the Wrangellia terrane and Stikinia terrane, during their drift across the ancient Pacific Ocean, maintained a fairly close biogeographical association, not separated by much distance, as suggested by the similarity of fossil invertebrate assemblages from both terranes. It was also possible to determine that the terranes were in the eastern Pacific during the Early Jurassic, due to the presence of North and South American taxa present in the assemblages. Paul could also establish the relative positions of the terranes, with Wrangellia being outboard of Stikinia, as Stikinia shows a higher similarity with the North American craton than does Wrangellia, and due to the minor presence of western Pacific fossil species on Wrangellia that are absent on Stikinia.

These and other observations have permitted a fairly detailed accounting of these 2 blocks through this time interval and beyond. Other studies like Paul's have helped us unravel a picture of what today is a really complicated geology, the product of collisions between various bits and pieces zooming across the Pacific and colliding with the main North American continental landmass.

I think Wegener would have been astounded at the level of detail we have now uncovered about the movement of the fractured surface of our planet. The history of the changing geography of planet Earth is still being uncovered, though, aided by our friends the wandering fossils.

Lecture 9 Our Vast Troves of Microfossils

Micropaleontology is a world of paleontology that often gets overlooked—quite literally—because it is the world of the very small. In this lecture, you will learn about microfossils. You will also learn about foraminifera, including how these fossils chart global climate over 120 million years and what they tell us about the death of the dinosaurs. You will also discover what microfossils tell us about how evolution works.

Microfossils

- There is no fixed definition of what a microfossil is. Basically, if it's very small and needs a microscope to be seen, you can call it a microfossil. Most microfossils are less than 1 millimeter in size, but some are much larger. Given such a broad definition, microfossils can come from a wide variety of sources. They are also the first fossils we find in the geological record, given that first life was probably microbial.

- In addition to microbes, microfossils also include many important components of microplankton, which form the base of almost all aquatic food chains. Microplanktons' sensitivity and reaction to events in the wider Earth system are critical in any narrative we are trying to develop about the evolution of life on our planet.

- Planktonic microfossils include those with organic walls, such as the dinoflagellates, a group of marine protists that move around using a whiplike flagellum. Many dinoflagellates are photosynthetic, but some are tiny predators that feed on other protozoa. As fossils, dinoflagellates are mostly known from those species that have an encystment stage as part of their life cycles.

- Not all microplankton have a test, or shell, made of organic material, though; many secrete mineralized skeletons. For example, protozoa called radiolaria secrete beautiful silica skeletons. Together with the diatoms, one of the most important components of microplankton today, they are important sediment producers, covering the deep ocean floor with a fine sedimentary rock called a siliceous ooze.

- Larger organisms can also contribute to the microfossil assemblage. Spores and pollen from plants are an important component of the paleontologist's toolkit, both for environmental analysis and for the correlation and dating of rocks. We find the first plant spores at about 470 million years ago, possibly produced by the first plant colonizers of the land.

- By the time we get to some of the first plant macrofossils, we start to see an increasing diversity of spores, reflecting the spread and diversification of plants across the coastal landscape.

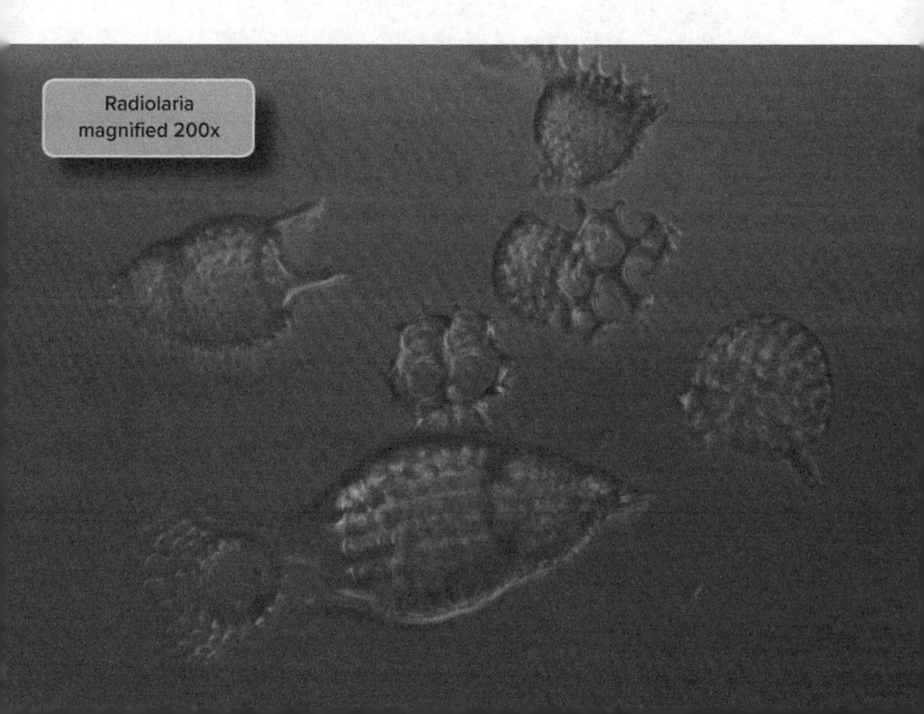

Radiolaria magnified 200x

- But larger animals are in on the microfossil game, too. Perhaps some of the most famous are the conodonts. Initially, the conodont animal was known only from its conodont elements: tiny teeth-like objects composed of calcium phosphate. They are found from the Cambrian period all the way through the end of the Triassic, which is associated with a probable extinction event at about 200 million years ago.

- Conodonts are diverse and evolved many species, some of which had quite short geological ranges, making them very useful in biostratigraphy. However, even though the conodont elements were discovered in the mid-1800s, we still didn't know what conodonts actually were.

- Because conodonts were commonly found in assemblages of paired elements, it was assumed that they formed some sort of articulated apparatus. But it would not be until the early 1980s that the soft parts of an eel-like chordate were found with conodonts arranged as teeth at the feeding end of the animal. Conodonts turned out to be chordates—not a direct ancestor of us, but certainly one of our early cousins.

Foraminifera

- A particularly long-lived group of microfossils that have become invaluable in our understanding of the Earth system over time is foraminifera, or forams for short. Forams are a group of protists that secrete a test. Forams are found in many diverse environments, from the shallow to the deep ocean, with some even in moist terrestrial environments. They are found in climates that range from the tropics to the poles.

- Forams are an extremely powerful tool in paleontology and Earth history. Because they are widely distributed by ocean currents, they can be found in many different sediment types, making them great for correlation between different areas. And, importantly, they have a relatively continuous fossil record in ocean basins since the Jurassic.

- They also occur in high numbers—sometimes in the tens of thousands for a relatively small volume of sediment. And, very importantly, they evolved rapidly, producing short ranging forms that define short packages of geological time. This makes them excellent biostratigraphic tools.

- Perhaps one of their most significant contributions is in our understanding of long- and short-term climate change in the past. Forams are excellent paleothermometers. Their usefulness in this regard comes from the isotopic composition of the calcium carbonate shells that forams secrete, which gives us a proxy for the temperature of seawater at the time of formation of a particular shell.

- Oxygen isotopes data gained from forams have provided us with insights into the climate change in the most recent era of Earth's history, the Cenozoic, which runs from the end of the Cretaceous 66 million years ago to the present day.

- At around 55 million years ago, the record of forams shows a period of warming known as the Paleocene-Eocene thermal maximum. This warming matches the time of migration of many tropical mammals toward the poles. This warming was possibly triggered by carbon dioxide emissions related to volcanism associated with the breakup of the supercontinent of Pangaea and perhaps the release of methane, another greenhouse gas, from the oceans.

- Forams also record a major cooling event at the Eocene-Oligocene boundary, about 34 million years ago. This signals the development of the first major ice sheet in Antarctica as the continent drifted to the South Pole and started to become isolated by the Antarctic circumpolar current.

- There is another drop in temperatures at about 14 million years ago, called the Middle Miocene climate transition, which by 8 million years ago would see temperatures drop to levels that would establish ice cover present at the current levels on Antarctica.

- That is the picture from the Cenozoic, but can we take this back any further in time? This is just what Dr. Brian Huber of the Smithsonian's National Museum of Natural History's Department of Paleobiology has been doing—pushing the record of temperature changes back 120 million years, well into the last period of the Mesozoic era, the Cretaceous, when dinosaurs still ruled the land.

- One of the reasons Dr. Huber can make this trip back in time is due to the nature of the geological materials that he samples. One of the largest depositories of Earth's sediments, forams, and therefore climatic records, are the oceans. But oceans are continually being opened up at mid-oceanic ridges and destroyed at subduction zones, destroying or altering the record such that sensitive isotopic information is lost.

- Fortunately, Dr. Huber has found a number of locations where Cretaceous sediments are present and the level of preservation of forams within them is excellent. From the forams he has collected, Dr. Huber has extended the climate record for the Cenozoic, starting at 66 million years ago, back another 55 million years to 120 million years ago. Dr. Huber and his colleagues have uncovered a remarkable record of climatic changes that can be described as varying from warm to very warm over 55 million years.

- The end-Cretaceous extinction event is probably one of the most well-known crises is Earth's history, probably because of its link to the death of the dinosaurs and the massive impact centered on the present-day village of Chicxulub on the Yucatán Peninsula 66 million years ago. Because forams were, and still are, an extremely important part of the ocean system, they are useful during events like this, both for timing and for providing insight into the causes of the extinction.

- One of the questions regarding the Cretaceous-Paleogene extinction concerns the state of the biosphere prior to the Cretaceous-Paleogene boundary. The vast majority of scientists accept that an impact occurred at the end of the Cretaceous and that it had a severe and detrimental effect on the biosphere. But was this the only cause?

- An additional finger of blame is often pointed toward India and a sequence of rocks found on the Deccan plateau called the Deccan Traps. These are a vast outpouring of basalt lava that occurred at the end of the Cretaceous thought to last around 30,000 years, producing lava flows that might have originally covered around 1.5 million square kilometers. It is not lava that was the problem, though; creatures can always migrate away from centers of volcanism. It is the carbon dioxide and associated global warming that could potentially cause the most serious impact to the biosphere.

- Dr. Huber's research shows that warming started at 65.9 million years ago. This trend began just before the impact occurred at Chicxulub. This temperature increase reversed a long, slow cooling that had been progressing throughout the Late Cretaceous. Forams don't record any major extinction throughout this interval of time.

- With regard to the meteor impact at Chicxulub, forams tell an interesting story. Cores taken from the Pacific and Atlantic Oceans will often tell a familiar story across the Cretaceous-Paleogene boundary: a dramatic change in strata from white, chalky sediments into a much darker horizon that contains molten material sprayed out of the impact site, some of which fell into the oceans. These strata are capped by a thin, rusty fireball layer, representing fine debris and soot that rained down out of the atmosphere after the main event.

- Forams appear to be doing fine before the impact occurred, but after it, they register a 90% extinction of the group. For forams, it was likely darkness that would be the killer. Fine ash and soot thrown high into the atmosphere would cut off the Sun and shut down photosynthesis both in the oceans and on the land. And once the food-web support was removed, the Mesozoic biosphere collapsed.

Microfossils and Evolution

- Microfossils hold great potential in detailing changes in climate and recording the progression of major events in Earth history, such as mass extinction events. This is in part due to their shear abundance when compared to large (macro) fossils. This abundance also allows us to investigate some of the fundamental processes of evolution, an opportunity that was taken up by Dr. Gene Hunt of the Department of Paleobiology at Smithsonian's National Museum of Natural History.

- Dr. Hunt studies ostracods, tiny crustaceans that are typically around 1 millimeter in size. They secrete a bivalved organic or calcareous shell and are found from the Ordovician period to the present day.

- When Darwin first proposed his theory of evolution, it was criticized by some as not being supported by evidence from the fossil record. Since Darwin's time, the numbers of fossils in the collections of museums, universities, and research institutes has expanded considerably, leaving no doubt that evolution has occurred, but with the question of how evolution occurs still up for debate.

- It is in part this question that Dr. Hunt has been trying to answer using ostracods. Does evolution occur in a gradual linear manner, often called phyletic gradualism, or does evolution progress in a series of rapid pulses separated by periods of apparent stasis with little change? The latter is a hypothesis proposed by Niles Eldredge and Stephen Jay Gould in 1972 that they called punctuated equilibrium.

- There is another process, though, called random walks. This describes how trends in various features of a fossil group can develop that are not necessarily driven by natural selection. In random walks, a particular characteristic of a feature—for example, its size or complexity—may increase or decrease randomly if there is no selective pressure.

- Random walks are very rarely completely random. There will often be a bounding wall, which will prevent a certain feature from varying above or below a particular value. For example, consider the size of ostracods. Although they are very small, there will be a size below which it would be impossible for these little crustaceans to exist, bounded by such things as the functional size of organs or the ability to efficiently respire.

- Dr. Hunt tested 251 data sets of the morphological characteristics from 53 different evolutionary lineages of fossils, including ostracods, and found that directional trends only best fit about 5% of examples. About 50% could be described as random walks and 45% as stasis, with no appreciable trend. This fits nicely with what you would expect from punctuated equilibrium, with most fossils demonstrating either stasis or random walks between punctuated bursts of rapid evolution, with just a minor component of what could be described as directed phyletic gradualism.

Questions to consider:

1. Why are microfossils so valuable in correlating rocks (biostratigraphy) and in elucidating environmental change in the oceans?
2. Why are microfossils the oldest fossils we will likely ever find?

Suggested Reading:

Armstrong and Brasier, *Microfossils*.

Knell, *The Great Fossil Enigma*.

Lecture 9 Transcript

Our Vast Troves of Microfossils

When you think of a natural history museum, you think of this, right? Giant skeletons of dinosaurs, mammoths, and perhaps a whale or two? Now don't get me wrong, these are wonderful exhibits, but they do tend to dominate, don't they, perhaps taking our gaze away from other important fossil groups such as the invertebrates, like the trilobites, and, of course, the fossil record of plants, although the National Museum of Natural History has wonderful displays of both.

But there is another world of paleontology that often gets overlooked, quite literally, as this is the world of the very small: micropaleontology. In this lecture I would like ask: What are microfossils? I want to introduce you to the foraminifera. I'd like to see how these fossils chart global climate change over 120 million years, and what they tell us about the death of the dinosaurs, and find out what microfossils tell us about the way that evolution actually works.

Now, there's no fixed definition of what a microfossil actually is. Basically, if it's very small and needs a microscope to be seen, then you can call it a microfossil. Most microfossils are less than 1 millimeter in size, but some are a lot larger. For example, the limestone blocks that make up the pyramids in Egypt are packed full of the shells of Eocene microfossils called nummulites that belong to a large group of microfossils called the foraminifera. Some of these microfossils can reach sizes of 2 inches—that's 5 centimeters. Not bad for a single-celled protist.

Given such a broad definition—basically, small fossils—microfossils can come from a wide variety of sources. They're also not surprisingly the first fossils we find in the geological record, given that first life was probably microbial. At present, the oldest fossils and microfossils come from the Strelley Pool Formation in Australia. The fossils there were originally sulfur-metabolizing microbes living in between the sand grains of a tidal beach over 3.43 billion years ago in what is today the Pilbara region of western Australia.

In addition to microbes, microfossils also include many important components of the microplankton, which forms the base of almost all aquatic food chains. Microplankton's sensitivity and reaction to Earth events in the wider Earth system are critical in any narrative we're trying to develop about the evolution of our planet.

Planktonic microfossils include those with organic walls like the dinoflagellates, a group of marine protists that move around using a whip-like flagellum. Many dinoflagellates are photosynthetic but some are tiny predators that feed on other protozoa. As fossils, dinoflagellates are mostly known from those species that have an encystment stage as part of their life cycles. They vary in size from about 15 to 80 microns—a micron is 1 millionth of a meter. It's kind of difficult to imagine the length of a micron, so consider this: a human hair is about 50 to 70 microns in diameter. Many pollen, such as the ragweed pollen, the cause of much summer allergy misery, are often 20 to 50 microns in diameter.

Not all microplankton has a test made of organic material, though; many secrete mineralized skeletons. For example, the coccolithophores are photosynthetic protistan algae that secrete a series of plates, or coccoliths, composed of calcium carbonate. Coccoliths have been so productive in the past that they have become significant producers of sediments in carbonate rocks, and today can still form enormous blooms that can be seen from space. In this image, you can see a bloom occurring off the coast of Cornwall in southwest England, close to where I went to university at the University of Plymouth.

A good example of very high coccolithophore productivity comes from the famous White Cliffs of Dover that are almost 100% coccolith. These vast accumulations were produced in the warm shallow seas of tropical Europe during the Cretaceous period. These, though, from an aesthetic point of view, are really some of my favorites: the radiolaria. These protozoa secrete beautiful silica skeletons, often looking more like Christmas tree ornaments than products of microscopic biology. Together with the diatoms, one of the most important components of microplankton today, they are another important sediment producer, covering the deep ocean floor with a fine sedimentary rock called a siliceous ooze.

Larger organisms can also contribute to the microfossil assemblage. Spores and pollen from plants are an important component of the paleontologist's tool kit, both for environmental analysis and for the correlation and dating of rocks. We find the first plant spores at about 470 million years ago, possibly produced by the first plant colonizers of the land, perhaps something like modern liverworts. These also provide some of the earliest evidence of land plants that were being consumed by herbivores. As we can see here, this specimen, from the Middle Devonian of New York State, has a hole and surface feeding trace made by a small arthropod. By the time we get to some of the first plant macrofossils, like Cooksonia at 415 million years ago during the Silurian, we start to see an increasing diversity of spores reflecting the spread and diversification of plants across the coastal landscape.

But animals are in on the microfossil game too, and perhaps some of the most famous are the conodonts. Initially, the conodont animal was only known from its conodont elements, these teeth-like objects commonly about 250 to 2000 microns in size. They're composed of calcium phosphate. They're found in the Cambrian period all the way through to the end-Triassic, which is associated with a probable extinction event at about 200 million years ago.

Conodonts are diverse and include many species, some of which had quite short geological ranges, making them, again, useful in biostratigraphy. However, even though the elements were discovered in the mid-1800s, we still don't know what conodonts actually were for a long time.

As conodonts were commonly found in assemblages of paired elements, it was assumed that they formed some sort of articulated apparatus, but it would not be until the early 1980s that the soft parts of an eel-like chordate were found with conodonts arranged as teeth at the feeding end of the animal. Conodonts turned out to be chordates, not a direct ancestor of us but certainly one of our early cousins.

But let's focus in a little here. I'd like to introduce you to a particularly long-lived group of microfossils that have become invaluable in our understanding of the Earth's system over time. Let's consider the foraminifera. The foraminifera, or forams for short, are a group of protists that secrete a shell, or test. A mobile part of the foram's cytoplasm called the ectoplasm—no, that's not the stuff that was reportedly excreted by Victorian spiritualists—can develop tentacle-like structures, pseudopodia, that reach outside the shell and capture prey like diatoms and other microscopic creatures.

Many forams also contain photosymbiotic algae. In the planktonic form *Globigerinoides sacculifer*, the pseudopodia contain symbiotic photosynthetic dinoflagellates. These dinoflagellates are transported to the distal parts of the pseudopodia in the morning so they can bask in the sunlight, and then retract back into the shell in the evening. Some non-planktonic benthic forms with streamlined shapes actively burrow through sediment, processing organic detritus, while others crawl around on the ocean floor using their pseudopodia for locomotion.

Forams are found in many diverse environments, from the shallow to the deep ocean, with some even in moist terrestrial environments. They are found in climates that range from the tropics to the poles. This is a diverse group of very successful creatures.

The shells of forams are composed of various materials. Some are made of organic material called tectin while others are agglutinated, or glued, composed of grains either randomly, or very selectively, taken from the environment and stuck on the creature to form its shell. For example, the foram technitella has a particular fancy for making a test out of sponge spicules that it collects from its local neighborhood. Other forams secrete minerals to form a shell, most commonly out of calcium carbonate, but some species also produce silica.

But forams are not just an interesting, tiny curiosity; they are a powerful tool in paleontology and Earth history. As they are widely distributed by ocean currents, especially the planktonic forms, they can be found in many different sediment types, making them great for correlation between different areas. And, importantly, they have a relatively continuous fossil record in ocean basins since the Jurassic. They also occur in high numbers, sometimes in the tens of thousands, for a relatively very small volume of sediment. And, very importantly, they evolved rapidly, producing short-ranging forms that define short packages of geological time. This makes them excellent biostratigraphic tools, and are used extensively by the petroleum industry.

Perhaps one of their most significant contributions is in our understanding of long- and short-term climate change in the past. Forams are excellent paleothermometers. Their usefulness in this regard comes from the isotopic composition of the calcium carbonate shells that the forams secrete, and in particular the stable isotopes of oxygen that get incorporated into the calcium carbonate. We are especially considering here the ratio between 2 isotopes: oxygen-16 and the heavier oxygen-18 that has 2 extra neutrons in its nucleus.

And how does this system work? When the foram is secreting its shell, oxygen atoms from the surrounding seawater go into the newly formed solid mineral. Oxygen-18, being heavier and not as energetic, gets incorporated more readily into the crystal structure. The colder the seawater is the more oxygen-18 gets included, therefore giving us a proxy for the temperature of the seawater at the time of formation of that particular shell. A change in 1 per mill in oxygen-18 levels represents a change in temperature of about 4° to 5° centigrade.

Oxygen isotopes data gained from forams have provided us with insights into the climate change in the most recent era of Earth's history, the Cenozoic, that runs from the end of the Cretaceous 66 million years ago to the present day. Let's examine a few points on this graph. At around 55 million years ago an event that causes a negative spike in the oxygen-18 records of forams records a warming known as the Paleocene-Eocene thermal maximum where global temperatures soar 5° centigrade—that's 9° Fahrenheit. This warming matches the time of migration of many tropical mammals toward the poles. It also explains the presence of palm trees in Greenland and Patagonia, and the mangrove swamps in southern Australia, which at that time was well south, at about 65° south.

This warming was possibly triggered by carbon dioxide emissions related to volcanism associated with the breakup of the supercontinent of Pangaea, and perhaps the release of methane, another greenhouse gas, from the oceans. As forams are distributed vertically through the water column, they also provide a valuable picture of what parts of the ocean were warming. In this particular warming event, planktonic and benthonic forams show that the entire water column and not just the surface waters were experiencing warm conditions.

The forams also record a major cooling event at the Eocene-Oligocene boundary about 34 million years ago, with a sudden positive excursion in the oxygen-18 isotopic record. This signals the development of the first major ice sheet in Antarctica as the continent drifted to the South Pole and started to become isolated by the Antarctic circumpolar current.

There is another drop in temperatures here at about 14 million years ago called the Middle Miocene climate transition, which, by 8 million years ago, would see temperatures drop to levels that would establish ice cover present at the current levels on Antarctica. As you can see, the overall trend is one of cooling and increasingly positive oxygen-18 oxygen values, all faithfully recorded by our tiny paleothermometers, the foraminifera.

So that is the picture from the Cenozoic, but can we take this back any further? This is just what Dr. Brian Huber of the Department of Paleobiology has been doing, pushing the record of temperature changes back 120 million years, well into the last period of the Mesozoic era, the Cretaceous, when dinosaurs still ruled the land and ammonites swam through the oceans keeping a wary eye out for the scary, giant aquatic reptiles of the time.

One of the reasons Dr. Huber can make this trip back in time is due to the nature of the geological materials that he samples. One of the largest depositories of Earth's sediments—forams, and therefore climatic records—are the oceans, faithfully recording information layer by layer like the pages of a book. As we know, we live on an active, dynamic planet, and oceans are continually being opened up at mid-ocean ridges but also destroyed at subduction zones. When the latter happens, the record is destroyed or so altered by heat and pressure that sensitive isotopic information is lost.

The oldest piece of active ocean, one that is still generating new crust, comes from the western Pacific and northwestern Atlantic dating to about 180 million years during the Jurassic period. You can see it here on this image, with the young, newly formed ocean crust, colored in red, concentrated around the mid-ocean ridge systems; and the older crust, coloured in blues and greens, further away, demonstrating how they have been moved from the ridge systems to their current locations. For the Atlantic Ocean, this operates at about the speed that your fingernails grow.

The older you go back in time, though, the fewer sedimentary sequences in the oceans you're going to find. And when you do find them, there will be a greater chance that, over time, these sediments, and the fossils they contain, may have been altered, destroying the precious isotopic signals you're looking for. Fortunately, Dr. Huber has found a number of locations where Cretaceous sediments are preserved and the level of preservation of the forams is excellent. Much of the data he has retrieved for his studies of the Cretaceous has come from the Integrated Ocean Drilling Program, the IODP, an international marine research program that specializes in taking samples of sediments from below the seafloor by drilling, using ships like this, the *Resolution*, sponsored by JOIDES, the Joint Oceanographic Institutions for Deep Earth Sampling.

From the forams he has collected, Dr. Huber has extended the climate record we were looking at for the Cenozoic, starting at 66 million years ago, back another 55 million years to about 120 million years ago. Just look at the oxygen isotope data and those low to negative readings. Remember, the more negative the oxygen-18 signal, the warmer the climate. So if you thought that we hit warm times during the Cenozoic, you ain't seen nothing yet. Get ready for the Cretaceous super greenhouse.

Here is the world around 90 million years ago. Sea levels were far higher and there was probably no ice at the poles. As a consequence of those high sea levels, many of the interiors of the continents were flooded. See how much of Europe is reduced to a series of tropical islands, and that much of what is today the Sahara Desert was occupied by a warm shallow sea. North America at this time was split in two by a body of water running north to south called the Western Interior Seaway. That explains why you can find rocks, originally marine sediments, dating between 87 to 82 million years ago called the Niobrara Chalk that is found as far apart as Kansas and British Columbia. Marine creatures like this mosasaur could have swam from what it today the Arctic Ocean all the way to the Gulf of Mexico.

Dr. Huber and his colleagues have uncovered a remarkable record of climatic changes that you can describe as varying from warm to very warm over 55 million years. Let's focus in on some specific points that illustrate just how useful these microfossils are in paleoceanography and climatology.

One of the events that Brian was interested in occurs between two stages in the Cretaceous called the Albian and the Aptian at around 112.5 million years ago. Sediments taken from the ocean off the coast of present-day Florida, called the Blake Plateau, provided a record of sedimentation across this interval. The boundary between the Albian and the Aptian is recorded by a sudden color change that you can see in this sediment core.

At the same level, there is an extinction in forams and a general change in their size across the boundary, from large in the Aptian to small in the Albian, with the diversity of planktonic forms reduced, as well. At the same time, there is a negative shift in the oxygen-18 isotope data in the foram tests, indicating that the oceans were getting warmer. As well, there is a corresponding shift in the carbon-13 isotopes, which is generally associated with environmental problems in the biosphere, something that we covered in another lecture in this series.

That sudden color change is actually a little worrying. It could mean that a period of erosion had occurred, removing sediment, and accounting for the sudden changes in the fossil assemblage and isotopic data across the horizon. Such a gap in the geological record is called an unconformity. In fact, when analyzed, it was found that there was at least a 1-million-year gap in deposition, which might account for the sudden changes we were recording there.

Fortunately, there were other sites to study—for example, on the Falkland Plateau down in the Antarctic, where a complete section without an unconformity confirmed the extinction event in the foram fossils and their isotopic shifts. Given that the same signal was also found in the Southern Hemisphere, we can determine that we have something of a global event here, something impacting the entire biosphere, forcing microplankton into extinction and warming the oceans. As of yet, we don't know what caused this event, but it does show how the study of forams is steadily adding more detail to our story of our planet.

Another extinction was recorded around 94 million years ago when a number of well-established forams disappear. The deeper ocean benthics from off the coast of Florida record an increase in temperature of ocean bottom waters of up to 19° centigrade. That is very warm, indeed. Closely associated with these extinctions is evidence of substantial volcanic activity related to the formation of the Caribbean Plate, a large igneous platform and a major outpouring of basalt thought to have erupted around 4×10^6 km^3 of material, and likely is the source of the warming through carbon dioxide emissions and the cause of the extinctions.

The forams provide more detail about this extinction, though. In addition to recording warming temperatures and extinction of species, the pattern of this extinction has a tale to tell, too. It would appear that the forams that suffered the most were deep-dwelling planktonic forms. Foram species are often distributed based on bathymetry—that's water depths—with certain species occupying certain depths in the water column.

The loss of an entire group of species associated with a particular level in the ocean suggests that there is no longer a sufficient ocean temperature gradient to support as many diverse, depth-related species associations. That is, the ocean is becoming more homogeneous in temperature, leaving no room for these particular deep-dwelling, warm-sensitive forams. This demonstrates the power of these amazing little fossils, for not only are they recording shifts in temperature and the timing of the extinction, but also the detailed ecological factors involved in their extinction, too.

But thinking about extinction events, how do forams speak to one of the most famous, the one that killed off the dinosaurs? The end-Cretaceous extinction event is probably one of the most well-known crises in Earth's history, most likely because of its link with the death of the dinosaurs and the massive impact centered on the present-day village of Chicxulub on the Yucatan Peninsula 66 million years ago.

The story of this extinction is far from being completely told, though. As forams were, and still are, an extremely important part of the ocean system, they are useful during events like this, both for timing and providing insight into the causes of the extinction. One of the questions regarding the Cretaceous-Paleogene extinction concerns the state of the biosphere prior to the Cretaceous-Paleogene boundary. The vast majority of scientists accept that an impact occurred at the end of the Cretaceous and that it had a severe and detrimental effect on the biosphere, but what was the cause? Was that the only cause?

An additional finger of blame is often pointed toward India and a sequence of rocks found on the Deccan plateau called the Deccan Traps—*trappa* means "step" in Swedish. There are vast outpourings of basalt lava that occurred at the end of the Cretaceous, thought to last around 30,000 years, producing lava flows that might have originally covered around 1.5 million square kilometers. It's not lava that's the problem though—creatures can always migrate away from centers of volcanism. It is the carbon dioxide and associated global warming that could potentially cause the most serious impact to the biosphere.

So what do the forams tell us? Dr. Huber's research shows that warming started at around about 65.9 million years ago. This trend began just before the impact occurred at Chicxulub, with a definite shift in oxygen-18 values toward the negative. This temperature increase reversed a long, slow cooling trend that had been progressing throughout the late Cretaceous. The forams don't record any major extinction throughout this interval of time, although sites close to the Antarctic do record temperatures rising perhaps as much as 15° centigrade. Species that previously existed at low latitudes had started to migrate to higher latitudes as ocean temperatures warmed, but there does not appear to be any major global stress in the foram populations at the time.

This warming is also is registered on land, too. For example, in North Dakota, which today has an average temperature of around 43° to 45° Fahrenheit, had an average temperature of around 65° to 68°. Even more impressive is when you realize that North Dakota at this time was at about the same latitude as Quebec City is today. Imagine Canada with alligators and palm trees. Overall, although there were definitely shifts in the distribution of species, it does not appear that the biosphere was particularly severely impacted during this warming.

But what about the meteor impact at Chicxulub? Again, the forams tell an interesting story. Cores taken from the Pacific and Atlantic Ocean will often tell a familiar story across the Cretaceous-Paleogene boundary, a dramatic change in strata from white, chalky sediments into a much darker horizon that contains molten material that was sprayed out of the impact site, some of which fell into the oceans. These strata are capped by a thin, rusty fireball layer representing fine debris and soot that rained down out of the atmosphere after the main event.

As we noted, the forams appear to be doing fine before the impact occurred, but after it they register a 90% extinction of the group, particularly in the planktonic forms. Before the impact, there is a diverse fauna of various types and sizes of foram, and after the event, in the earliest Paleogene, only a few small, less ornate forms survive.

For the forams, it was likely darkness that would be the killer. You can imagine fine ash and soot thrown high into the atmosphere would cut off the sun and shut down photosynthesis, both in the oceans and on the land. Like a house of cards, once you take away the food web support at the bottom of the pile, the whole structure—in this case, the Mesozoic biosphere—came tumbling down.

So, microfossils hold great potential in detailing changes in climate and recording the progression of major events in Earth history like mass extinction events. This, in part, is due to their sheer abundance when compared to large macrofossils. This abundance also allows us to investigate some of the fundamental processes of evolution, an opportunity that was taken up by Dr. Gene Hunt of the Department of Paleobiology. Gene studies ostracods, tiny crustaceans around about 1 millimeter in size that secrete a bivalved organic or calcareous shell—it makes them look a bit like tiny pistachios—and are often found from the Ordovician period right through to the present day.

When Darwin first proposed his theory of evolution, it was criticized by some as not being supported by evidence from the fossil record. Since Darwin's time, the numbers of fossils in the collections of museums, universities, and research institutes has expanded considerably, leaving no doubt that evolution has occurred, but with the question of how evolution occurs still up for debate.

It is this question that Gene Hunt has been trying to answer using ostracods. Does evolution occur in a gradual linear manner, often called phyletic gradualism? Or does evolution progress in a series of rapid pulses separated by periods of apparent stasis with little change; a hypothesis proposed by Niles Eldredge and Stephen Jay Gould in 1972 that they called punctuated equilibrium.

There is another process, though, called random walks. This describes how trends in various features of a fossil group can develop that are not necessarily driven by natural selection. In random walks, a particular characteristic of a feature—for example, its size or complexity—may increase or decrease randomly if there is no selective pressure. As an analogy, consider flipping a coin. You can get runs of heads or runs of tails, even though, over a long time period, you would tend to get an average of 50/50 heads and tails.

To add just one more layer of complexity to this, though, we must realize that random walks are very rarely completely random. There will often be a bounding wall, which will prevent a certain feature varying above or below a particular value. For example, consider the size of ostracods. Although very small, there will be a size below which it would be impossible for these little crustaceans to exist, bounded by such things as functional size of organs or the ability to efficiently respire.

This concept has often been illustrated with an analogy called the drunkard's walk. Imagine a paleontologist emerging from a bar, a little worse for wear from too much celebrating after their paper just got published in a prestigious scientific journal. They emerge from the bar and start to walk along the sidewalk, but with one side of the sidewalk bound by a wall. They will randomly wobble along the pathway, but any increase will be away from the bounding surface, until the paleontologist in question either gets home or winds up in a ditch somewhere.

Gene Hunt tested 251 data sets of the morphological characteristics from 53 different evolutionary lineages of fossils, including ostracods, and found that directional trends only fit about 5% of examples. About 50% could be described as random walks and 45% stasis with no appreciable trend. This fits nicely with what you would expect from punctuated equilibrium, with most fossils demonstrating either stasis or random walks between punctuated bursts of rapid evolution, with just a minor component of what could be described as phyletic gradualism.

Hopefully, these examples demonstrate how micropaleontology provides high-resolution data for our continuing story of the evolution of Earth and the biosphere. Sure, the big stuff is cool, but very often the pages of the story they appear on are written on the pages of the mighty microfossils.

Lecture 10

Ocean Fire and the Origin of Life

This lecture will examine an intriguing hypothesis regarding the origin of life, and ultimately the origin of the science of paleontology, in the ocean depths. The lecture will address these questions: How do we explore the Earth's mid-oceanic mountain chain? Do we have geological and paleontological evidence for ancient undersea volcanic ecosystems? Why are oceanic volcanoes a good candidate for life's origins? Could life have arisen in a similar manner on other worlds?

HMS *Challenger* and *Trieste*

- Interest in the ocean floor is not a recent development. People have speculated about "what is down there" for a long time. But the first systematic survey of the ocean floor would have to wait for the HMS *Challenger*, which would sail out of Portsmouth, England, on December 21, 1872. The HMS *Challenger* would travel about 70,000 miles (130,000 kilometers), taking ocean-floor dredges, recording the temperature of the ocean at various depths, performing open-water trawls, and, in the process, discovering 4700 new species of marine life.

- As recently as the late 19th century, knowledge of the ocean was basically restricted to the topmost few fathoms, about 18 feet. The *Challenger* would perform one of the first systematic surveys of the ocean floor, using 181 miles (291 kilometers) of Italian hemp and a lead weight. On March 23, 1875, between Guam and Palau in the southwestern Pacific, the line they tossed overboard just kept on going down, eventually recording around 4475 fathoms—about 5 miles, or more than 26,000 feet, deep.

- In the 1930s, Swiss physicist, inventor, and explorer Auguste Piccard, whose first interest was the upper atmosphere, constructed pressure spheres that he attached to high-altitude balloons to measure cosmic rays. Later, he realized that he could modify his sphere to withstand pressure at depth, too.

- He invented the bathyscaphe *Trieste*, which was launched on August 26, 1953, operated by the French Navy but later purchased by the U.S. Navy. The dive began on January 23, 1960. Shadowed by the USS *Lewis*, the *Trieste* descended toward the Challenger Deep—the location that the HMS *Challenger* had sampled about 85 years earlier. At 4 hours and 47 minutes, they reach the ocean floor at 35,814 feet (10,916 meters). Just before touchdown, they spotted a flat fish swimming by—quite a surprise, as it was not known that fish could survive at these great depths and pressures.

- This mission ran at a time when the paradigm about how the world looks and operates was changing. We now know that magma oozes up at ridges in the ocean crust, forming new material and pushing the older oceanic lithosphere away to either side. Continents are carried as the plates spread away from the ridges. Ultimately, oceanic lithosphere descends into the mantle at the ocean trenches.

- This project provided vital information about one of those trenches. The Challenger Deep is just part of the Mariana Trench, a feature 2550 kilometers (1580 miles) long with an average width of 69 kilometers (43 miles). This is just one of many trenches surrounding the Pacific Ocean, marking the point where the Pacific Plate is being subducted into the Earth's mantle.

- The groundbreaking work of the *Trieste* would pave the way for exploration of another feature of the newly resolved ocean floor—the plate-generating ridges that traverse the Earth's oceans—and with the exploration of these features, a new possibility regarding the origin of life would emerge, too.

Earth's Mid-Oceanic Mountain Chain

- Ocean ridges produce new ocean crust, and as such, they are a hot, active, dynamic feature of our planet. They form a chain of volcanic mountains about 31,000 miles (50,000 kilometers) long, rising an average of about 2.7 miles (4500 meters) above the seafloor. Although the global mid-oceanic ridge system is mostly hidden beneath the ocean's surface, it is the most prominent topographic feature on the surface of our planet.

- By the 1970s, sonar and magnetic mapping of the ocean floor had pinpointed the location of Earth's ocean ridge systems, but no one had ever seen them up close. Even so, some had speculated that they might be the site of hydrothermal activity, areas where ocean water would sink into cracks in the newly formed crust, become heated by magmatic fluids and the still-warm rocks, and get expelled again as hot water. These underwater hot springs may hold the key to the origin of life on Earth.

- We had hints of the existence of these hot springs going back as far as the early 1880s. A Russian ship, the *Vital*, was sailing in the Red Sea when it sampled water at 200 feet that appeared to be warmer than water at the surface. The presence of hot, mineral-rich water in this area was confirmed by later exploration. In 1965, the research vessel *Atlantis II* recorded water temperatures at 133° Fahrenheit.

- The U.S. National Science Foundation sent the research vessel *Chain* to take more readings. They took sediment cores of the ocean floor. The sediment they retrieved was bizarre—rich in metals such as copper, zinc, and manganese. By now, the idea of spreading ocean floors was being widely accepted, with the Red Sea identified as a young and newly formed oceanic rift.

- The hunt was on for other oceanic hydrothermal sites. In 1972, a promising site in the Galapagos rift zone was selected by the presence of hot water found on earlier expeditions and was explored by the research vessel *Thomas Washington*. Robotic and submarine-mounted cameras recorded curious mounds encrusted with minerals around 15 to 75 feet high sticking above the ocean-floor sediments about 10 to 20 miles south of the Galapagos rift. They also detected hot fluids rising from the ridge and recorded bursts of earthquakes where the water temperature was particularly high.

- In 1977, the DSV *ALVIN* visited the Galapagos rift. In addition to finding hydrothermal vents, they also found a rich and bizarre biological community. The inhabitants included mussels and white clams, some more than a foot long, and bacteria-laden beard worms, many times larger than their shallow-water relatives, that covered the lava rock.

- Something other than the Sun must be powering life down here. The water collected by *ALVIN* contained hydrogen sulfide, which is produced by primitive microscopic microbes called archaea living in and on the hot rocks and sediments. They take sulfate that occurs in seawater and reduce it by chemically removing oxygen, producing energy and releasing hydrogen sulfide as a waste product. Higher organisms feed on archaea, making this a chemosynthetic-based ecosystem rather than a photosynthetic-based one.

Ancient Undersea Volcanic Ecosystems

- There is evidence of these hydrothermal vent systems in the geological past. In fact, these systems are an extremely important source of metal ores. The vents form as cold ocean water descends into cracks in the ocean floor, where it is heated by hot rocks still close to the magmatic source at the ridge.

- The seawater starts to alter, and get altered by, minerals present in the surrounding rocks. The altered seawater is then expelled as a superheated metal-rich brine through hydrothermal vents, such as so-called black smokers. These volcanogenic massive sulfide deposits are important sources of ores containing copper, zinc, lead, gold, and silver. But these metal sulfides are not the only indicators of these ancient ecosystems; we also find whole fossilized vent communities.

- One of the oldest fossil vent systems we currently have comes from northeastern China. In 2007, an ancient Precambrian community was described by Jiang-Hai Li of Peking University and Timothy Kusky of Saint Louis University, who discovered evidence of a volcanogenic massive sulfide deposit dating to 1.43 billion years ago—well before the diversification of multicellular creatures. These sections preserved some of the black smoker hydrothermal vent chimneys, just like the ones we find under the ocean today.

- In addition, within these ore deposits were the fossilized remains of microbes that were living in and on this ancient hydrothermal system. The microbial community was probably sulfate-reducing, just like microbes in modern vent settings.

A Candidate for Life's Origins

- Given that the oldest vent fossils we find are much younger than those at Strelley Pool in western Australia, why are scientists still so keen on hydrothermal vents as the location for the origin of life?

- There are a lot of raw materials—all those metals to act as catalysts for the generation of useful organics—in these hydrothermal pressure cookers. And it appears that the last universal common ancestor of living things today was an extreme thermophile, a microbe that liked the heat, just like we find in modern oceanic hydrothermal settings.

- Our job now is to come up with a hypothesis that bridges the gaps from an inorganic environment to organic molecules to the first living cells on Earth. Some ideas center around the production of self-replicating molecules as a precursor for life, although probably not DNA, which requires enzymes to reproduce themselves, which are encoded on DNA—a chicken-and-egg scenario. Perhaps a simpler self-replicating molecule, such as RNA, was the earliest form of life.

- Recently, scientists have speculated that a metabolism-first rather than genetic-first model makes more sense. Work at University College London by chemists Nora de Leeuw and Nathan Hollingsworth has shown how the mineral greigite, found inside hydrothermal vents, might be acting like enzymes in living organisms, providing a catalytic site for carbon dioxide dissolved in seawater. In addition, vent systems also provide a lot of heat to power chemical reactions that can generate complex organic molecules.

- But it has also been suggested that, although hydrothermal systems are our best bet for the location of the origin of life, perhaps we have been looking at the wrong type of hydrothermal system. In 2000, a National Science Foundation–funded project found an area called the Lost City in the Atlantis Massif, 62 miles (100 kilometers) west of the Mid-Atlantic Ridge. They found a field of hydrothermal vents that are very different from those sitting on the spreading ridges.

- In 2003, the submersible vessel *ALVIN* found white-colored chimneys composed of calcium carbonate rising from 30 to 60 feet off the ocean floor. Unlike the dark-colored black smokers of a ridge axis, these structures are not releasing significant amounts of carbon dioxide or hydrogen sulfide. Instead, they are producing high quantities of hydrogen and methane, with some hydrogen sulfide, and in alkaline rather than acidic waters.

- Areas containing white smokers could be a better location for the generation of life. They have a proven record of producing important quantities of organic molecules and have the energy and catalysts present to power an interesting biochemistry. In addition, the chimneys at the Lost City have been forming for about 30,000 years—much longer than most black smoker systems, which will only be active while they are over the hot magmatic rocks of the spreading center. This system, therefore, may provide a longer-term site for the evolution of a complex biochemistry and perhaps life.

- The rocks of both hydrothermal systems also possess an interesting microstructure: tiny pockets where organic chemistry could be concentrated and perhaps develop other features, such as a cell membrane. At a certain point in time, these primitive cells may have become sufficiently resilient to leave the vent system and start to populate the ocean.

Other Worlds

- Vent discoveries have also opened up possibilities for the search for life elsewhere in our solar system. In particular, astrobiologists are interested in moons like Europa, which orbits Jupiter, and Enceladus, which orbits Saturn.

- It is thought that Europa may have a liquid water ocean below its icy crust that would massage the interior of the planet, generating heat and perhaps allowing for the existence of hydrothermal vents, around which life might

Jupiter with Europa and Io

develop. These ideas might be tested if a NASA mission is launched in the 2020s.

- We also have spectacular evidence of a similar ocean on Enceladus from 2005, when NASA's *Cassini* satellite detected jets of water being released from the moon. *Cassini* was able to fly through these plumes and detected not only water vapor but also nitrogen, methane, and carbon dioxide—all useful building blocks for the formation of interesting organic chemistry, perhaps leading to the biochemistry of the simplest organisms.

- It is possible that there may be hydrothermal vents deep in Enceladus, too—and, if so, perhaps life as well. This is an exciting possibility—not only in the search for extraterrestrial life, but also for the history of life on our planet. If life is found around such vent systems on other worlds, then perhaps anywhere we have liquid water and hydrothermal vents we should expect life processes to initiate.

Questions to consider:

1. Is plate tectonics a vital component of planets that might develop life?
2. On how many other worlds might life have started in our solar system?

Suggested Reading:

Corfield, *The Silent Landscape*.

Knoll, *Life on a Young Planet*.

Lecture 10 Transcript

Ocean Fire and the Origin of Life

Life has literally changed the way the world works. We are just a small part of the Earth's biosphere but, to our best knowledge, we're the first to spread Earth's biosphere beyond its cradle. For me, this is what comes to mind when I consider one of humankind's most significant endeavors, initiated in 1961 and sponsored by the National Aeronautics and Space Administration—NASA—this was Project Apollo. And by 1969, well, this kind of speaks for itself, doesn't it?

But in this lecture I'm not going to be looking at the future of our species, or even of our biosphere; I want to look at the origins of life. And for that, I would like to start with another monumental project, but this time from 1960, a year when Kennedy would debate with Nixon, British prime minister Harold Macmillan warns the apartheid regime in South Africa about the wind of change they would be facing, and Alfred Hitchcock would terrify moviegoers with *Psycho*.

And then there was Project Nekton, an event of great importance that would receive no coverage at all. This was a project so risky that the navy thought it best to keep it top secret; a project, just like Apollo, to explore another frontier, but this time to a frontier still on planet Earth; a mission to the greatest depths of the ocean: the Challenger Deep.

Interest in the ocean floor is not a recent development; people have speculated about what's down there for a long time. But the first systematic survey of the ocean floor would have to wait for HMS *Challenger* that would sail out of Portsmouth on December 21, 1872. The HMS *Challenger* would travel some 70,000 miles, taking ocean-floor dredges, recording the temperature of the oceans at various depths, performing open-water trawls, and in the process discovering 4700 new species of marine life, life that included microplankton like these radiolarian—collected, categorized, and beautifully illustrated.

As recently as the late 19th century, knowledge of the ocean was basically restricted to the topmost few fathoms, about 8 feet. The *Challenger* would perform one of the first systematic surveys of the ocean floor using 181 miles of Italian hemp and a lead weight; simple but effective. But on March 23, 1875, at sample section 225 between Guam and Palau in the southwestern Pacific, the line they tossed overboard just kept on going, eventually recording around 4475 fathoms—about 5 miles deep.

To really come to grips with this famous deep, we need to wait for this man, Swiss physicist, inventor, and explorer Auguste Piccard. Actually, Auguste's first interest was the upper atmosphere. He constructed pressure spheres that he attached to high altitude balloons to measure cosmic rays for his old buddy Albert Einstein. Later, he realized that he could modify his sphere to withstand pressures at depth, too. And, yes, if you're wondering, he was the inspiration for Jean-Luc Picard in *Star Trek* and probably also for the look-alike Professor Calculus in *Tintin*.

What he invented was the bathyscaphe *Trieste*. It was launched on August 26, 1953, operated by the French Navy, but later purchased by the U.S. Navy. The bulk of the vessel consists of the float, filled with 22,000 U.S. gallons of gasoline—not for power, but as a useful, noncompressible liquid for buoyancy. Lead weights were located in ballast hoppers to help the vessel sink, and would be released by electromagnets. This is actually a very clever design as, during a power failure, all the lead weights would be released and the *Trieste* would simply float to the surface. This buoyancy control allowed for a free-dive capability, not relying on a cable to lower and pull the vessel back up. The pilots would get into the circular pressure vessel via this small tunnel. The sphere itself had steel walls that were 5 inches thick, and was provided with a very small Plexiglas window through which all of the observations were going to be made.

The dive occurred in complete secrecy, given the possibility of failure, and began on January 23, 1960. The *Trieste* would carry Jacques Piccard—that's the son of Auguste—and Lt. Don Walsh of the U.S. Navy. Shadowed by the USS *Lewis*, the *Trieste* descended toward the Challenger Deep, the location that HMS *Challenger* had sampled about 85 years earlier. They entered complete darkness by 3000 feet, with only an occasional bioluminescent organism to break the gloom.

And after 2 hours, at 30,000 feet—bang! The whole ship shudders and an ominous crack appeared in the Plexiglas window. Fortunately, the window was made in 2 halves, and only the outer pane was damaged. Bravely, Jacques and Don decide to carry on. Personally, I would have been heading back to the surface for cocktails.

At 4 hours and 47 minutes they reach the ocean floor at 35,814 feet. Just before touchdown, they spot a flatfish swimming by—quite a surprise, as it was not known that fish could survive at these great pressures and depths, pressures close to 16,000 pounds per square inch. That's 1000 times the standard atmospheric pressure at sea level. And if you want an idea as to how deep the Challenger Deep is, consider this. If you could drop Mount Everest into the sea at this location, it would sink to the seafloor with 1.28 miles of ocean still above the top of the peak.

This mission ran at a time when the paradigm about how the world looks and operates was changing. People like Harry Hess, Bruce Heezen, and Marie Tharp, and a whole host of others, were piecing together what would become our modern understanding of plate tectonics. Significantly, the ocean floor turned out to be full of topography and not largely flat as many had previously assumed. Rather than a static planet, the Earth was found to be an active, dynamic world with oceans opening and closing, and continents zipping around the surface of the planet. We now know that magma oozes up at the ridges in the ocean crust, forming new material and pushing the older oceanic lithosphere away to either side. Continents are carried as the plates spread away from the ridges. Ultimately, oceanic lithosphere descends into the mantle at the ocean trenches.

Project Nekton provided vital information about one of those trenches. The Challenger Deep is just part of the Mariana Trench, a feature that's around about

2550 kilometers long—that's about 1580 miles—with an average width of 69 kilometers—that's 43 miles. This is just one of the many trenches surrounding the Pacific Ocean, marking the point where the Pacific Plate is being subducted into Earth's mantle. The groundbreaking work of the *Trieste* would pave the way for exploration of another feature of the newly resolved ocean floor, those plate-generating ridges running like a series of baseball seams traversing the Earth's oceans. And with the exploration of these new features, a new possibility regarding the origin of life itself would emerge, too.

Ocean ridges produce new ocean crust and as such they are hot, active, dynamic features of our planet. They form a chain of volcanic mountains about 31,000 miles long, rising an average of about 2.7 miles above the seafloor. For comparison, the volcano Mount Rainier in Washington State is about the same height above sea level. Although the global mid-ocean ridge system is mostly hidden beneath the ocean's surface, it is the most prominent feature on the surface of our planet.

So, in this lecture, let's build on the pioneering work of Project Nekton as we look into the intriguing hypotheses regarding the origin of life, and, ultimately, the origin of the science of paleontology in the ocean depths. Let's ask: How do we explore the Earth's mid-ocean mountain chain? Do we have geological and paleontological evidence for undersea volcanic ecosystems? Why are deep oceanic volcanoes a good candidate for life's origins? And could life have arisen in a similar manner on other worlds?

By the 1970s, sonar and magnetic mapping of the ocean floor had pinpointed the location of Earth's ocean ridge systems, but no one had ever seen them up close. Even so, some had speculated that they might be the site of hydrothermal activity, areas where ocean water would sink into cracks in the newly formed crust, become heated by magmatic fluids and the still warm rocks, and get expelled again as hot water, just like Old Faithful in Yellowstone National Park. It is these submarine geysers that may hold the key to the origin of life on Earth.

In truth, we had hints of the existence of these underwater hot springs going as far back as the early 1880s. A Russian ship, the *Vital*, was sailing in the Red Sea close to Mecca when it sampled water at 200 feet which appeared to be warmer than the water at the surface. The presence of hot mineral-rich or salty brines in this area was confirmed by later expeditions. In 1965, the research vessel *Atlantis II* recorded water of temperatures around 133° Fahrenheit and recovered a sample of oozy sediment that was reported as being too hot to touch.

Intrigued by this development, the U.S. National Science Foundation sent the research vessel *Chain* to take more readings, but this time sediment cores of the ocean floor would be taken, too. The sediment they retrieved was bizarre, rich in metals like copper, zinc, and manganese. The sedimentary strata were also extremely colorful, with Woods Hole Oceanographic Institute researcher Egon Degens describing their colors as being as bright as an Indian sand painting or a Mexican serape. By now, the idea of spreading ocean floors was being widely accepted, with the Red Sea identified as a young and newly formed oceanic rift.

The hunt was on for other oceanic hydrothermal sites during the 1970s, using elevated seafloor water temperatures as a proxy for hydrothermal activity. In 1972, a promising site was selected by the presence of hot water found on earlier expeditions. This site was explored by the research vessel *Thomas Washington* on the Southtow expedition to the Galapagos Rift zone.

The expedition included a rig called Deep-Tow that was outfitted with sonar, magnetometers, and cameras to take pictures of the ocean floor. The project was funded in part by the U.S. Navy who were motivated by the awful loss of the submarine USS *Thresher* in 1963 in the North Atlantic. Such experimental technology, it was hoped, could perhaps be of use in future emergencies at sea. Robotic and submarine-mounted cameras recorded curious mounds encrusted with minerals around 15 to 75 feet high sticking above the ocean floor sediments about 10 to 12 miles south of the Galapagos Rift. They also detected hot fluids rising from the ridge and recorded bursts of earthquakes where the water temperature was particularly high. Another very strange observation was the sudden appearance of dead bottom-dwelling fish at the surface, like this cusk eel, above the area where the small earthquakes were detected.

What was happening here would remain a mystery until 1977 when the DSV *Alvin*, the descendant of Piccard's *Trieste*, would visit the Galapagos Rift. The team on this mission included experts in many fields of geoscience, but no one thought biologists would be needed. That proved to be an oversight. Although they found wonderful hydrothermal vents spewing hot mineral-rich water, they also found a rich and bizarre biological community associated with these hot springs. The inhabitants included mussels and white clams, some over a foot long, and bacteria-laden beard worms many times larger than their shallow water relatives that covered the lava rock. White crabs and purple octopuses scuttled around this shelly community, too.

One of the odd creatures they found they called a sea dandelion. They're actually a member of the phylum Cnidaria that includes jellyfish and corals. Called the siphonophores, they're closely related to the surface-dwelling Portuguese man-of-war. Like all siphonophores, the sea dandelion is not one animal but a colony composed of zooids performing different functions like defense, feeding, and reproduction.

At one spot that was named the Garden of Eden, they found 1.5-foot-long tube worms swaying in the warm waters close to the scalding—about 750° Fahrenheit—water of the vents. The Garden of Eden was a relatively new volcanic site and was at the exact location where the earthquakes and dead fish were recorded in 1972. What the scientists had unknowingly recorded back then was a volcanic eruption on the ridge causing the earthquakes and killing off the fish.

These wonderful and bizarre creatures were completely unexpected, and as a result of this abundance of riches, there was hardly any formaldehyde to preserve the specimens they brought back. Indeed, they had to resort to some fairly powerful Russian vodka when the formaldehyde ran out. But how could creatures live down there? The common view at the time was that all biological systems were based on food chains powered at their base by the Sun—a photosynthetic pyramid.

It is possible that the marine detritus or snow—a continual shower of organic material mostly composed of decomposed organisms from the surface waters—

could be supporting this distinctive ecosystem. However, it was unlikely that this could support the vast biomass observed on the rift. Something else other than the Sun must be powering life down here. Hints to the source came from the smell that came from the water collected by *Alvin*, the distinctive smell of rotten eggs: the smell of hydrogen sulfide.

Hydrogen sulfide is produced by primitive microscopic microbes called archaea, such as *Archaeoglobus fulgidus*, living in and on hot rocks and sediments. They take sulfate that occurs in seawater and reduce it, chemically removing oxygen, producing energy and releasing hydrogen sulfide as a waste product. Higher organisms feed on the bacteria, making this a chemosynthetic- rather than a photosynthetic-based ecosystem.

OK, so this is a unique ecosystem. What does that have to do with the origin of life? Well, before we get there, let's first look at vent systems through a paleontological lens. Do we even have evidence of these hydrothermal vent systems in the geological past? We do. In fact, a lot of attention has been focused on these systems recently, as they are an important source of metal ores. The vents form as cold water descends into cracks in the ocean floor, where it is heated by hot rocks still close to the magmatic source of the ridge.

The seawater starts to alter, and get altered by, minerals present in the surrounding rocks. The altered seawater is then expelled as a superheated, metal-rich brine through hydrothermal vents like these so-called black smokers. These volcanogenic massive sulfides, or VMS deposits, are important sources of ores containing copper, zinc, lead, gold, and silver.

But these metal sulfides are not the only indicators of these ancient ecosystems. We find whole fossilized vent communities, too. For example, researchers, including Crispin Little of the University of Leeds, were investigating the sulfide deposit in the San Rafael Mountains of California near Santa Barbara. In addition to typical metal sulfide minerals, they also found a number of fossil creatures that were part of this vent community. Using fossil plankton called radiolarians, this ecosystem was dated to the early Jurassic period between 190 and 182 million years ago.

Using the inclination of magnetic minerals within the rocks, the researchers were able to determine the paleolatitude of this material at time of formation. It originally formed somewhere along the equator on the Farallon Plate, a slab of ancient oceanic crust which, over many millions of years, has been in collision with western North America. As collision occurred, parts of the plate that didn't get subducted have been accreted to the western edge of North America and elevated by crustal movements, so that we now have what was once an ancient ocean vent system high and dry in the San Rafael Mountains.

The fossils they found included tube worms like those found in modern vent systems, a gastropod, and a species of brachiopod, a creature resembling a clam but from a different group of animals. These creatures were living around a vent system and, not unlike modern vent systems today, likely had microbes living in their tissues—what we call endosymbionts—like some of their modern counterparts.

Geologically speaking, though, these ancient vent communities are pretty recent, and if we're interested in the origin of life, we need to go way back in time. One of the oldest fossil vent systems we currently have comes from northeastern China, just 150 kilometers east of Beijing. This ancient Precambrian community was described in 2007 by Jianghai Li of Peking University and Timothy Kusky of St. Louis University. What they discovered was evidence of a volcanogenic massive sulfide deposit dating to 1.43 billion years ago. That is well before the diversification of multicellular creatures. These sections actually preserved some of the black smoker hydrothermal vent chimneys as well, just like the ones we find under the ocean today.

But there was more, for within these ore deposits were the fossilized remains of microbes that were living in, and on, this ancient hydrothermal system. Some of these microbes were forming biofilms and producing layered domed structures on the nearby ocean floor and around the vents. The microbial community was probably sulfate reducing, just like microbes in modern vent settings, and these ecosystems are almost a billion and a half years old.

As ancient as this is, though, it is not as old as the oldest fossils; we have already met them in another lecture in this series. They come from the Pilbara area of western Australia in a section called Strelley Pool dated to 3.43 billion years old, way older than the vent fossils in China. Today, the area is dry and desolate, but beautifully desolate in a way. However, at 3.43 billion years ago, it was very different.

Back then, Strelley pool would have been on the edge of an ocean, and the fossils in question were microbes living between grains of sand on a tidally influenced beach. They were also living in an oxygen-free environment. It would be many hundreds of millions of years before significant quantities of oxygen would accumulate on the Earth. This shallow, sunlit, and relatively low temperature and pressure environment is very different, though, to any of the vent systems—modern or fossils—we've described earlier.

So, given that the oldest vent fossils we find are much younger than those at Strelley Pool, why are scientists so keen on hydrothermal vents as being the location for the origin of life? Well, for a start, there are a lot of raw materials down there. All those metals can act as catalysts for the generation of useful organics in these hydrothermal pressure cookers.

But something else, too. When you consider the tree of life—moving down from the tips of the twigs, down the branches, and eventually to the main trunk—it would appear that the last universal common ancestor of living things today was what we call an extreme thermophile. Or, in other words, it was a microbe that liked the heat, just like we find in modern oceanic hydrothermal settings.

Our job now is to come up with a hypothesis that bridges the gaps from an inorganic environment to organic molecules to the first living cells on Earth. Some ideas center around the production of self-replicating molecules as the precursors for life, although probably not DNA, which itself requires enzymes to reproduce themselves which are encoded on DNA—a classic chicken and egg scenario. However, perhaps a simpler, self-replicating molecule like single stranded RNA, a molecule present even in very simple life-forms, was the earliest form of life. This formed the basis of RNA world, a phrase coined by Nobel laureate Walter Gilbert in 1986 to describe how the first biosphere on Earth may have been RNA- and not DNA-based.

But, more recently, scientists have speculated that a metabolism-first model would make more sense. Work at the University College of London by chemists like Nora de Leeuw and Nathan Hollingsworth have shown that the mineral greigite, found inside hydrothermal vents, might be acting like enzymes in living organisms, providing a catalytic site for carbon dioxide dissolved in seawater. In addition, vent systems also provide a lot of heat to power chemical reactions that can generate complex organic molecules like formic acid, acetic acid, methanol, and pyruvic acid, an important component in the metabolism of modern organisms.

But it has also been suggested that, although hydrothermal systems are our best bet for the location of the origin of life, perhaps we've been looking at the wrong type of hydrothermal system. In the year 2000, a National Science Foundation funded project was exploring an area about 2300 miles east of Florida and 62 miles west of the Mid-Atlantic Ridge. This topographic high is called the Atlantis Massif. Located on this 1.2–1.5-million-year-old ocean floor was found an area called the Lost City, and you can see why.

What they found was astounding: a field of hydrothermal vents very different from those that are found sitting on the spreading ridges themselves. In 2003, the submersible vessel *Alvin* once again descended to the depths, this time finding white, cream-colored chimneys composed of calcium carbonate rising 30–60 feet off the ocean floor.

Unlike the dark-colored black smokers of a ridge axis, these structures are not releasing significant amounts of carbon dioxide or hydrogen sulfide. Instead, they are producing high quantities of hydrogen and methane, with some intermixed hydrogen sulfide—and in alkaline rather than the acidic waters of the black smokers. The local environment is cooler as well—still hot, but not the scorching 760° Fahrenheit of a black smoker. The white city pillars of the Lost City emit water closer to 200° Fahrenheit.

The heat driving these hydrothermal systems, though, is not from young hot rocks freshly formed at mid-ocean ridge systems. The heat in this system comes from a geological process called serpentinization. At the Atlantis Massif, mantle rocks called peridotite have been brought close to the surface by faults.

These rocks, rich in the mineral olivine, are only stable at great depths and high temperatures.

Bringing peridotite rocks close to the surface and causing them to interact with seawater moving through cracks in the seafloor results in a major transformation. In this transformation, olivine changes into other minerals like magnetite and serpentine, forming a rock called serpentinite. This transformation is an exothermal reaction that generates heat, enough to raise the temperature to about 550° Fahrenheit—that's 260° centigrade. It is this heat that helps drive the Lost City system, very different to the black smoker systems that are driven by magmatic heat. The serpentinite produced is also lower in density than peridotite, which causes the immediate area to writhe and fracture. This introduces more seawater into the crust and continues the process, a kind of a positive feedback loop.

So could areas containing white smokers be a better location for the generation of life? They certainly have a proven record of producing important quantities of organic molecules, and have the energy and catalysts present to power an interesting biochemistry. They have another advantage, too. The chimneys at the Lost City have been forming for about 30,000 years. That's around 2 orders of magnitude longer than most black smoker systems, which will only be active while they are over the hot magmatic rocks of the spreading center. This serpentinite-powered system, therefore, may provide a longer-term site for the evolution of a complex biochemistry and perhaps life itself.

The rocks of both hydrothermal systems also possess an interesting microstructure: tiny pockets where organic chemistry could be concentrated and perhaps develop other features such as a cell membrane. At a certain point in time, these primitive cells may have become sufficiently resilient to leave the vent system and start to populate the ocean.

But let's pause just for a second. Just what is life? Perhaps we can define it by the 3 common components to all life on Earth. Firstly, metabolism: that's chemical reactions that power life. Encapsulation: something to isolate the creature from the external environment, like a cell membrane. And reproduction, which in the earliest organisms would have consisted of fission or budding and other asexual replicating processes. Once this was achieved and life could pass copies of itself into the future, the biosphere was born, perhaps deep in the dark ocean depths in a hot, rocky nursery.

The vent discoveries we have been discussing have also opened up the possibilities for the search for life elsewhere in the solar system. In particular, astrobiologists are interested in moons like Europa that orbits Jupiter, and Enceladus that orbits Saturn. It is thought that Europa may have a liquid water ocean below its icy crust, perhaps maintained by the gravitational tidal action of Jupiter that would massage the interior of the planet, generating heat and perhaps allowing for the existence of hydrothermal vents around which life might develop. These ideas might be tested if a NASA mission, such as a Europa multiple-flyby mission satellite, is launched sometime in the future.

We also have spectacular evidence of a similar ocean on Enceladus, where NASA's *Cassini* satellite detected jets of water being released from the moon in 2005. *Cassini* was actually able to fly through these plumes and detected not only water vapor but also nitrogen, methane, and carbon dioxide, all useful building blocks for the formation of interesting organic chemistry, and perhaps leading to the biochemistry of the simplest organisms.

It is possible that there may be hydrothermal vents deep in Enceladus too, and if so, perhaps life as well. This is an exciting possibility, not only in the search for extraterrestrial life, but also for the history of life on our planet. If life is found around such vent systems on other worlds, then perhaps, anywhere where we have liquid water and hydrothermal vents, we should expect life processes to initiate. Confirming this will probably come well after I have been recycled into the Earth's biosphere, but what a wonderful confirmation that would be.

Lecture 11

The Ancient Roots of Biodiversity

Although we have likely had life on Earth for around 4 billion years, the spectacular biosphere we see all around us today may be relatively new. This lecture will examine how we get the first indications of a diverse biosphere. What was Darwin's dilemma? What are the first stirrings of an enlarged biosphere? How would these new organisms develop? Why did this first explosion of life occur, and what happened to it?

Darwin's Dilemma

- Charles Darwin's theory of evolution by natural selection elegantly accounts for all the wonderful diversity we see all around us today. His theory predicts that, through time, there should be a lineage of creatures eventually ending with what is today called the last universal common ancestor.
- Darwin was well aware that fossils were useful indicators of past life and ecosystems but also understood that the record was incomplete. Even so, the fossil record should still demonstrate increasing complexity from simple forms following the Early Precambrian dawn of the last universal common ancestor to the more complex biosphere we have today.
- There was a problem, though: Close to the base of the Cambrian period, today dated around 542 million years ago, the fossil record appears to indicate that a diverse array of large, complex creatures apparently materialized out of thin air. This appearance occurred even though life appeared to follow Darwin's predictions after the Cambrian period. This emergence of a complex biosphere geologically in an instant was contrary to what Darwin had predicted.

- Darwin suggested that as paleontological exploration continued, simpler fossil forms would likely turn up in older strata, but for a while this was a problem. Today, we are aware that Darwin's dilemma is related, in part, to the fact that the pace of evolution did not follow the traditional "slow, steady rate" views of the theory that many held at that time.

- Indeed, complex life and all the major plant and animal phyla that we know today did appear in the record rapidly. Essentially, the large biomineralized arthropods, such as the trilobites, that we find at around 521 million years ago, just 20 million years into the Cambrian period, arose geologically very quickly.

- Today, this rapid evolution of large animals is called the Cambrian explosion. Our insights into the world's biosphere just following this explosion of life were greatly expanded by the discovery of an extremely important fossil treasure—the Burgess Shale deposit in British Columbia, Canada—by one of the Smithsonian's most famous secretaries and director of the National Museum of Natural History, Charles Walcott.

- The discoveries of Walcott and later discoveries of Burgess-type deposits span the Early and Middle Cambrian. Together, the Burgess-type faunas paint a picture of a wonderfully diverse biosphere with the majority of all the major phyla represented.

- But was this the only big boom for complex life? Is it possible that there was an earlier explosion of life, a precursor to the explosion represented in the Burgess Shale?

A Bigger Biosphere

- To answer that, we stay in Canada but travel to the other side of the country, Newfoundland, which lies off the coast of eastern Canada in the Atlantic Ocean. In 1968, at a location known as Mistaken Point, Shiva Balak Misra, a graduate student at Memorial University of Newfoundland, discovered an entire ancient world—an extensive ecosystem preserved on the surface of a series of gently dipping rocks. Many believe the exosystem to be complete, incorporating all of the life-forms that were present at that time, what is called a biocoenosis, or life assemblage.

- What Misra revealed was an ancient deep ocean floor, complete with the creatures that were living on it. The rocks are now mudstone but were originally muddy sediments. The creatures he found living in these deepwater, low-oxygen conditions are from the latest Precambrian on what is now the Avalon Peninsula. At that time, this area was located between 40° and 65° south latitude, very different from its current location at 46.6° north.

- The quiet, low-oxygen conditions these creatures lived in probably aided in their preservation, but another important feature is their location: close to an ancient volcano. The volcano has long since been eroded away, but the ash that it spewed settled down through the water column, burying the creatures where they lived. The ash helped cast the fossils by forming an external mold, but it also provided an absolute date for the entombing sediments. From radiometric dating, we know that the fossils are about 565 million years old, from a period called the Ediacaran.

- The creatures found here are not like any we see today. Many of these fossils are collectively called rangeomorphs, frond-like creatures that are composed of simple budding elements that divide and repeat over 4 levels of organization in a simple fractal manner. This type of reproduction is a simple yet effective solution to build large bodies from small self-repeating elements.

- These, and related Ediacaran organisms, don't have a mouth or a gut, and some have suggested that they were osmotrophs, absorbing nutrients and organic material directly from the seawater through their bodies. They were certainly not photosynthetic, as they lived well below the photic zone, the depth to which light can penetrate into water.

- This ecosystem from Newfoundland is known as the Avalon Assemblage, and this initial burst of large creatures has been termed the Avalon explosion. But this is not the only assemblage of creatures known from the Ediacaran period. There are 2 others that are found in different environments: the White Sea and Nama Assemblages.

The Development of New Organisms

- The White Sea Assemblage was named for a typical occurrence of the assemblage found in northwestern Russia. The Russian Assemblage was not the first group of these particular Ediacaran creatures to be found, though. The first discovery of these creatures came from the Flinders Ranges of Australia in an area called the Ediacara Hills. In fact, it is this area that lends its name to this latest interval of the Precambrian: the Ediacaran period.

- Like the Mistaken Point fossils, these were discovered by a young geologist, Reginald Sprigg, who observed the impressions of the fossils on a rock surface. The fossils are dated at around 550 million years ago, younger than those from Mistaken Point.

- The environment that the White Sea creatures lived in was very different from those discovered in Newfoundland. The White Sea fossils lived on shallow, sandy sediments in sunlit waters in temperate conditions. Storm events occasionally smothered entire communities and preserved them more or less in place.

- The Nama Assemblage consists of forms that are interpreted as being more tropical in their distribution. They have been recovered from sections in Namibia, southern Africa.

- The temporal relationships between each of these assemblages is disputed by some, but in general, the Avalon-type cluster of species appears to be the first pulse of innovation followed by a second wave represented by the White Sea and Nama clusters of species.

- The second wave Ediacaran creatures still contains the frond-like rangeomorphs that we see in Newfoundland. It is possible that the Avalon-type assemblages still existed in the deep, dark waters surrounding the continents, but in the second wave, there are now other creatures, too.

- In the Ediacaran period, there is evidence that the biosphere was no longer static and was beginning to show glimpses of the wonderful animals that were to follow—animals that would differentiate their bodies to perform specialized functions and would move and interact with their environment in a variety of diverse ways. But why did this event occur at this point in time? What was driving the Avalon explosion?

What Caused the Explosion of Life?

- Around 2.5 to 1.85 billion years ago, it is believed that the photosynthetic bacteria had started to deliver significant quantities of oxygen to the surface sediments, oceans, and atmosphere of the Earth system. Following that interval, the Earth enters into a period called the boring billion between 1.85 and about 1 billion years ago, where nothing much appears to change in the Earth's geochemical or biological systems.

- Things would change, though, with a series of global glaciations, called snowballs—between 850 and 635 million years ago, within a period called the Cryogenian—that were probably related to the breakup of a supercontinent called Rodinia about 850 million years ago.

- A possible explanation may be related to increased weathering rates following a snowball event. This could fertilize the oceans and cause a bloom of photosynthetic microplankton and a release of atmospheric oxygen. The increased erosion rate and delivery of sediment to the ocean would also increase the rate at which organic material was buried. Removing the organic material in this manner reduces the amount of oxygen that would usually be used up in its oxidation, further contributing to the buildup of free oxygen.

- Following the snowballs, oxygen levels would have increased to such a level that it would be possible for creatures composed of multiple cells to exist. Prior to this, oxygen concentrations were only sufficient to power a microbial level of biological organization. This would allow larger creatures to evolve, as oxygen could now diffuse through layers of cells. It has also been suggested that a more oxygenated ocean system would also make available trace elements that would be key in the development of more complex metabolisms.

- That is one hypothesis. The story of oxygen in the Earth system is a complex one and is changing very rapidly. It would appear that oxygen levels fluctuated dramatically through the Ediacaran period and into the Cambrian period.

- Some researchers, such as Douglas Erwin of the Smithsonian's National Museum of Natural History's Department of Paleobiology, have suggested that the actual roots of current diversity may even lie earlier than we thought, perhaps within the Cryogenian period, the age of the great snowballs. He suggests that the snowball events may have been the proving ground where many animals developed their genetic toolkits, priming the fuse of the diversification of the Avalon and Cambrian explosions.

- But what happened to the Ediacarans? At the end of the Ediacaran period, there is a large shift in carbon isotopes recorded in the geological record. With each of the 5 mass extinctions, starting with the extinction at the end of Ordovician about 443 million years ago, a similar perturbation is recorded and often relates to a severe disruption in the biosphere.

- So, does the Ediacaran biota disappear because of an extinction event caused by some unknown perturbations in global geochemical cycles? Or is it possible that their removal from the fossil record may simply be a result of changing conditions of preservation?

- Many Ediacaran organisms are partly preserved as the result of microbial mantling, a "mask" of microbes that grew over dead Ediacarans that aided in their preservation. With the advent of more sophisticated grazing and burrowing organisms, the unique conditions that preserved these creatures ended and, with it, the record of the Ediacarans. But there is another possibility: one that evokes a replacement of the Ediacarans by other animals.

- In 2015, Erwin and a number of his colleagues from various universities released a paper exploring the possibility that the disappearance of the Ediacarans represents the first mass extinction event. In fact, the paper proposes that life itself may have caused a crisis for the Ediacarans.

- The Early Ediacaran animals, although a fantastic leap forward in complexity, were essentially immobile, probably passively absorbing nutrients from seawater. By the time we see complex animals, the biosphere is starting to actively interact with the rest of the Earth system. Animals have become ecosystem engineers. Is it possible that these new bioengineers changed conditions so much that it made life untenable for the Ediacarans?

- What is suggested is an Ediacaran Assemblage that was becoming increasingly outcompeted and marginalized by a developing Cambrian fauna. In physically interacting with the oceanic substrate, these new Cambrian forms would have competed for resources, increased the delivery of carbon to the ocean floor, and more effectively mixed and oxygenated ocean sediments. For the Ediacarans, the world was changing beyond their ability to adapt, and they faded away, leaving the stage set for the Cambrian explosion.

- Whatever their fate, by the time we see the first large fauna of the Cambrian, the Ediacarans are gone, either suddenly or gradually replaced by other creatures. The putative ancestors of later organisms are gone, too, their ancestors evolving into the wonderful creatures from the Cambrian explosion.

Questions to consider:

1. Following the appearance of the Ediacaran animals, was the evolution of even more complex life inevitable?
2. How important is mobility to the development of our complex biosphere?

Suggested Reading:

Erwin and Valentine, *The Cambrian Explosion*.

Fedonkin, Gehling, Grey, Narbonne, and Vickers-Rich, *The Rise of Animals*.

Lecture 11 Transcript

The Ancient Roots of Biodiversity

Life is divided into species, but how many species are there? It was over 250 years ago that Swedish biologist Carl Linnaeus, a man of short stature but imbued with a grand mission—and an ego to match—set out to classify all of God's Creation. He decided that, in all likelihood, there were only about 10,000 species of plants on our planet. Our current estimate for flowering plants alone is about 400,000 known species. It is estimated that we have only described around 1.2 to 1.5 million of all the species on our planet. Just how many species are there in total?

Now, this is an important question, a question that goes beyond simple bean counting, as knowing the total number of species and their ecological roles and relationships may have serious implications for how we estimate what the minimum number of species may be needed to maintain a healthy biosphere. Estimates of the total number of species on Earth have varied wildly between 3 million and 100 million, depending what you're counting and how the number of unknown species was calculated.

One approach, taken by Camilo Mora of the University of Hawaii and fellow researchers at Dalhousie University in Nova Scotia, may help bring that number into a little more focus. Their estimates only consider eukaryotic organisms, those single-celled or multicellular organisms that are more complex than a prokaryote, like a bacterium—that is, everything from a paramecium to a penguin.

They considered the taxonomic arrangement of species to make an estimate. They looked at the way that species are ordered into various hierarchical groupings and suggested that there is a consistent numerical trend in the number of species in each level of this hierarchy. They validated their hypothesis by testing it against well-known groups of taxa and came up with an estimate of around 8.7—plus or minus 1.3—million species on Earth. If this is the case, it would mean that around 86% of land species and 91% of marine species may remain undescribed. Other studies are more conservative, however, placing the number around 3 to 5 million species in total. But importantly, both estimates are comparatively close, within a factor of about 2.

Whatever the total number of species, the diversity of life on our planet is staggering. But how long have we had this situation? Well, although we have likely had life on planet Earth for around 4 billion years, the spectacular biosphere we see all around us today may be relatively new, possibly representing as little as 13% of the total history of the biosphere.

For most of Earth's history, life was simple and single-celled. It could pull off some pretty amazing tricks like photosynthesis and by 1.2 billion years ago we do have evidence of some of the first multicellular algae. By 542 million years ago, skeletonized invertebrates have made an appearance, but to be honest, most of the creatures during the first 3.5 billion years of life's story would not make a particularly awe-inspiring zoo exhibit, nothing quite like what Richard Dawkins rightfully calls the greatest show on Earth.

So, in this lecture, let's look at how we get the first indications of a diverse biosphere and ask: What was Darwin's Dilemma? What are the first stirrings of an enlarged biosphere? How would these new organisms develop? And why did this first big bang of life occur and what happened to it?

Charles Darwin's theory of evolution by natural selection elegantly accounts for all the wonderful diversity we see all around us today. His theory predicts that, through time, there should be a lineage of creatures eventually ending with what we call today LUCA, the Last Universal Common Ancestor. As he stated: "Therefore I should infer from analogy that probably all the organic beings which have ever lived on this earth have descended from some one primordial form, into which life was first breathed."

Darwin was well aware that fossils were useful indicators of past life and ecosystems, but also understood that the record was incomplete, stating that it had only here and there a short chapter preserved. Even so, the fossil record should still demonstrate increasing complexity from simple forms following the distant, early Precambrian dawn of LUCA to the more complex biosphere we have today. There was a problem, though, because close to the base of the Cambrian period, today dated around 542 million years ago, the fossil record appears to indicate that a diverse array of large, complex creatures apparently materialized out of thin air. This appearance occurred even though life appeared to follow Darwin's predictions after the Cambrian period. This emergence of a complex biosphere, geologically in an instant, was contrary to what Darwin had predicted.

Darwin himself suggested that, as paleontology exploration continued, simpler fossil forms would likely turn up in older strata, but for a while this was a problem. Today, we are aware that Darwin's Dilemma is related, in part, to the fact that the pace of evolution did not follow the traditional slow, steady state views of the theory that many held at that time. Indeed, complex life and all the major plant and animal phyla that we do know today did appear in the record rapidly. Essentially, the large biomineralized arthropods such as trilobites that we find around 521 million years ago, just 20 million years into the Cambrian period, arose geologically in a few short breaths.

Today, this rapid evolution of large animals is called the Cambrian Explosion and has been covered in detail in books, in magazines, and on television. Our insights into the world's biosphere just following this explosion of life were greatly expanded by the discovery of an extremely important fossil treasure, the Burgess Shale deposit in British Columbia, Canada by one of the Smithsonian's most famous secretaries and director of the National Museum of Natural History, Charles Walcott. The discoveries of Walcott and later discoveries of Burgess type deposits span the Early and Middle Cambrian, with the oldest being found in China, dating to about 515 million years ago. Together, the Burgess type faunas paint a picture of a wonderfully diverse biosphere with the majority of the major phyla represented.

But was this the only big boom for complex life? Is it possible that there was an earlier explosion of life, a precursor to the explosion so wonderfully represented in the Burgess Shale? To answer that, we stay in Canada, but this time travel to the other side of the country, to the beautiful province of Newfoundland.

Newfoundland lies off the coast of eastern Canada in the Atlantic Ocean and has a long and fascinating geological history. It was on Newfoundland that the birth and death of an ancient ocean called Iapetus is recorded that we covered in an earlier lecture. Newfoundland also has a long history of habitation of indigenous peoples, and also Vikings, but it would start to become more well known to Europeans when explorer John Cabot sailed into the area in 1497. Cabot was particularly startled by the richness of the fish around its shores. John Cabot's crew reported: "The sea there is full of fish that can be taken not only with nets but with fishing baskets."

The fish stocks, now unfortunately much depleted in Newfoundland, were in part one of the great draws to the area. Ships had to be careful, though. In the foggy and rainy conditions, very common in those parts, a ship heading toward Cape Race could turn north prematurely and hit these unforgiving rocks. This is why this area is called Mistaken Point, a site of many shipwrecks over the years.

But these rocks tell another story. This story, though, would have to wait till 1968, when a young graduate student from a very different part of the world moved to Memorial University at St. John's, the capital city of Newfoundland, to start his master's degree in geology.

Shiva Balak Misra was born in a small village near Lucknow, in India. As a boy, he had to walk over 15 miles to go to school each day. After winning a scholarship to study in Canada, he arrived in a very different world to his native India. Little did he know that he would also ultimately discover an entirely lost world, as well. He was tasked with completing a geological survey of the area that includes Mistaken Point, and the production of a geological map of the area shown here. What the young geologist would discover was amazing.

Now, if you're lucky enough to receive a permit to visit Mistaken Point, you will find a series of gently dipping rocks like giant steps across the landscape. If you take a look on the surface of those rocks, particularly if the Sun is low on the horizon and casting high-contrast shadows, an entire ancient world is revealed. Preserved in these rocks is an extensive ecosystem many believe to be complete, incorporating all of the life-forms that were present at that time, what we call a biocoenosis, or life assemblage.

What Misra revealed was an ancient deep-ocean floor, complete with the creatures that were living on it. The rocks are now mudstone but were originally muddy sediments. The creatures he found living in these deepwater, low-oxygen conditions are from the latest Precambrian on what is now the Avalon Peninsula. At that time, this area was located between 40° to 65° south latitude, very different to its current location at 46.6° north. The quiet, low-oxygen conditions these creatures lived in probably aided in their preservation, but another important feature is their location, close to an ancient volcano.

Of course, the volcano has long since been eroded away, but the ash that it spewed settled down through the water column, burying the creatures where they lived, like an ancient version of Pompeii. The ash helped cast the fossils by forming an external mold, but it helped in another way too, by providing an absolute date for the entombing sediments by radiometric dating. We covered this in an earlier lecture, but let's review the basic principles again.

Volcanic ash contains crystals that start to form in the magma chamber of a volcano. As that magma fragments in an eruption and is violently ejected as ash, some of those crystals are incorporated in the ejecta cloud. Many crystals, particularly zircons, contain small amounts of radioactive material. Once zircon crystals are formed, the radioactive components start to decay, yielding stable, nonradioactive daughter products.

As the decay rate is known from laboratory experiments, the ratio between parent-to-daughter material acts a bit like a ticking clock used to date the sediments entombing these fossils. From this, we know that the fossils at Mistaken Point are about 565 million years old in a period we call the Ediacaran.

The creatures you find here are not like any we see today. Many of these fossils are collectively called rangeomorphs, frond-like creatures that are composed of single, budding elements that divide and repeat over 4 levels of organization in a simple, fractal-like manner. This simple type of reproduction is a simple yet effective solution to build large bodies from small self-repeating elements, a bit like toy bricks. To build a body like this you would probably only need about 7 or 8 genetic commands. For comparison, we consist of over 25,000. Although these creatures were fixed in one location, they should not be mistaken for plants or seaweed—they were something very different.

These and related Ediacaran organisms don't have a mouth or a gut and some have suggested they were osmotrophs, absorbing nutrients and organic material directly from the seawater through their bodies. Their simple, fractal, self-repeating body plans helped them increase their surface area for absorption of material from the surroundings. They were certainly not photosynthetic as they lived well below the photic zone, the depth to which light can penetrate into water. Reproductive organs appear to be missing, too. Perhaps they just budded or released one of their cactus-like reproductive elements into the ocean that would sink to the ocean floor and start to germinate, thus growing another fractal-like creature elsewhere on the bottom of the ocean.

It is very difficult to find anything currently living that looks anything like these very strange creatures. Many suggestions have been made, ranging from obscure groups of primitive invertebrates to groups of very large protists. Guy Narbonne of Queens University has suggested they might represent the primitive ancestors of a large modern group of creatures, perhaps animals, or even fungi. Or perhaps, as suggested by researchers like Dolf Seilacher, they represent a completely separate and distinct group of life, the Vendozoa that flourished during the Ediacaran but didn't make it into the next eon, the Phanerozoic.

The most common type of rangeomorph found here are spindle creatures like *Fractofusus misrai*, named in honor of Professor Misra. The largest of these creatures were about a foot long and found on the ocean floor. From either side of a central rib was a series of subordinate ribs that were further subdivided into fronds in a typical rangeomorph manner. Some of these organisms were more elevated and upstanding. This is Bradgatia, a bushy kind of cabbage looking-like creature that sat on the ocean floor. Large specimens of this species stood about 8.5 inches—that's 21.5 centimeters tall.

Occupying higher levels of the water column, swaying in the water like modern kelp, were creatures such as Charnia, long and ribbon-like and about as thin your fingernail, some reaching lengths of about 6.5 feet in length, although most were pretty much smaller, possibly reflecting the increased scarcity of nutrients away from the sediment-water interface.

So this is the picture of life, the first attempt that life really had to get big: thin creatures with no mouths or guts, simple building block-like creatures fixed onto the substrate in darkness. Collectively, this ecosystem from Newfoundland is known as the Avalon Assemblage, and this initial burst of large creatures has been termed the Avalon Explosion.

But this is not the only assemblage of creatures known from the Ediacaran period. There are 2 others that are found in different environments: the White Sea and the Nama Assemblage. Both of these biotas should also be considered, for they have an important story to tell, too.

The first of these, the White Sea Assemblage, was named for a typical occurrence of the assemblage found in northwestern Russia. The Russian assemblage, though, was not the first group of these particular Ediacaran creatures to be found. The first discovery of these creatures came from the Flinders Ranges of Australia in the area called the Ediacara Hills. In fact, it is this area that lends its name to this latest interval of the Precambrian—the Ediacaran Period.

Like the Mistaken Point fossils, these too were a surprise when discovered by a young geologist, one Reginald Sprigg, who had been sent to this area by the South Australian government to see if some old mines in the area should be reopened. The story goes that he was eating lunch when he observed the impressions of fossils on a rock surface. The fossils are dated at around 550 million years ago, younger than those from Mistaken Point in Newfoundland.

The environment that the White Sea creatures lived in was very different from those discovered in Newfoundland. The White Sea fossils lived on shallow, sandy sediments in sunlit waters in temperate conditions. Storm events would occasionally smother entire communities and preserve them more or less intact.

The third assemblage type, the Nama Assemblage, consists of forms that are interpreted as being more tropical in their distribution. They have been recovered from sections in Namibia, southern Africa. The temporal relationships between each of the assemblages is disputed by some, but in general the Avalon-type cluster of species appears to be the first pulse of innovation, followed by a second wave represented by the White Sea and Nama cluster of species.

The second wave Ediacaran creatures still contain the frond-like rangeomorphs that we see in Newfoundland, like Charnia, a sample of which, from the Ediacara Hills in Australia, can be seen in the National Museum of Natural History. It is possible that the Avalon-type assemblages still existed in the deep, dark waters surrounding the continents, but in the second wave there were other new other creatures, too. There are triradially-arranged organisms like Tribrachidium, and pentaradial forms like Arkarua, suggested by some to be ancestors of echinoderms from which we may get urchins and starfish, although this link is somewhat tenuous.

Something else had evolved, though—strange looking creatures like Dickinsonia. Dickinsonia is an animal that looks like a large, segmented flatworm and could range in size from a penny to several feet in length. What it actually is, though, is still very much in dispute. Is it possible in Dickinsonia, though, that we find evidence for a low-energy type of mobility in the biosphere. The floors of the oceans in these shallow marine environments were encrusted with microbial mats, some of which have Dickinsonia-shaped impressions, perhaps suggesting that it was adopting a strategy of absorbing the microbes in a particular area of the mat, and after a feeding event, picking up its skirts and then moving on to another location. Of course, this doesn't prove that Dickinsonia was moving under its own steam, but it does show that not all large creatures were fixed to the substrate.

And consider Spriggina, a creature that appears to have a segmented body specialized into definite functional areas, the possibility of a head and a tail, as well. Was this a creature that was going places and probing the sediment or its surface for food? Spriggina has been touted as a possible forerunner of the arthropods, and it does look a little like some of the trilobites that would be a common component of the Cambrian period. We will look more closely at this in another lecture.

But if you want more convincing evidence that the biosphere was starting to move around, you should consider Kimberella, a slug-like creature that could reach lengths of about 6 inches. Kimberella shows definite signs that it was interacting with the sediment, probably feeding on microbes on the sediment-water interface. Scratches on the sediment may represent feeding marks left as Kimberella raked the surface for microbes. This is evidence that the biosphere is no longer passive; it is actively interacting with the rest of the Earth system.

Kimberella is significant for another reason, too. It is bilaterally symmetrical and consequently represents one of the earliest fossils of the most species-rich group of animals: the Bilateria. The Bilateria is a major group of animals that excludes sponges and corals. Some have suggested an even more refined relationship, suggesting that Kimberella might be an ancestral mollusk, an ancestor of calamari and escargot everywhere.

So, if you were snorkeling, preferably in one of the more tropical Ediacaran locations, you may have seen a biosphere that was no longer static, a biosphere beginning to show glimpses of the wonderful animals that were to follow. Animals with a definite front and a definite back end, animals that would differentiate their bodies to perform specialized functions, animals that would move and interact with their environment in a variety of diverse and wonderful ways. But why did this event occur at this point in time? What was driving the Avalon Explosion?

Around 2.5 to 1.85 billion years ago, it is believed that the photosynthetic bacteria had started to deliver significant quantities of oxygen to the surface sediments, oceans, and atmosphere of the Earth system. Following that interval, the Earth enters into a period called the Boring Billion between about 1.85 and about 1 billion years ago, where nothing much appears to change in the Earth's geochemical or biological systems.

Things would change, though, with a series of global glaciations called snowballs between 850 to 635 million years ago within a period called the Cryogenian that were probably related to the breakup of a supercontinent called Rodinia about 850 million years ago. A possible explanation may be related to increased weathering rates following the ends of a snowball event. This would fertilize the oceans and cause a bloom of photosynthetic microplankton and a release of atmospheric oxygen. The increased erosion rate and delivery of sediment to the ocean would also increase the rate at which organic material was being buried. This would remove the organic material and reduce the amount of oxygen that would usually be used up in its oxidation, further contributing to the buildup of free oxygen.

Following the snowballs, oxygen levels would have increased to such a level that it would be possible for creatures composed of multiple cells to exist. Prior to this, oxygen concentrations were only sufficient to power a microbial level of biological organization. This would allow larger creatures to evolve, as oxygen could now diffuse through layers of cells. It has also been suggested that a more oxygenated ocean system would also make available trace elements that would be key in the development of more complex metabolisms.

Well, at least that's one hypothesis. To be honest, the story of oxygen in the Earth system is a complex one and changing very rapidly. It would appear that oxygen levels fluctuated dramatically through the Ediacaran period and into the Cambrian period. Some researchers, like Doug Erwin of the Department of Paleobiology at the Smithsonian, have suggested that the actual roots of current diversity may lie even earlier than we thought, perhaps within the Cryogenian period, the age of the great snowballs. He suggests that the snowball events themselves may have been a kind of proving ground where many animals developed their genetic tool kits, priming the fuse for the diversification of the Avalon and Cambrian Explosions.

But what happened to the Ediacarans? At the end of the Ediacaran period there is a large shift in carbon isotopes recorded in the geological record. With each of the 5 mass extinctions, starting with the extinction at the end of the Ordovician about 443 million years ago, a similar perturbation is recorded and often relates to a severe disruption in the biosphere. So, does the Ediacaran biota disappear because of an extinction event caused by some maybe unknown perturbations in global geochemical cycles?

Or is it possible that their removal from the fossil record may simply be a result of changing conditions of preservation. Many Ediacaran organisms are partly preserved as a result of microbial mantling, a mask of microbes that grew over dead Ediacarans and aided in their preservation. With the advent of more sophisticated grazing and burrowing animals, the unique conditions that preserved these creatures ended, and with it the record of the Ediacarans, too. But there is another possibility, one that evokes a replacement of the Ediacarans by other animals.

In addition to his interest in Ediacaran biota, Doug Erwin has made many contributions to our understandings of mass extinction events. In 2015, he and a number of his colleagues from various universities in the United States released a paper exploring the possibility that the disappearance of the Ediacarans represents the first mass extinction event. I must admit that, in general, I've always regarded geophysical causes like volcanic activity, glaciations, or meteor strikes as generally the triggers of mass extinction events. What Doug Erwin's paper proposes, though, is that life itself may have caused a crisis for the Ediacarans.

As we have already seen, the early Ediacaran animals, although a fantastic leap forward in complexity, were essentially immobile, probably passively absorbing nutrients from seawater. By the time we see complex animals, things have started to change. The biosphere is starting to actively interact with the rest of the Earth system: animals have become ecosystem engineers. Is it possible that these new bio-engineers changed conditions so much that it made life untenable for the Ediacarans?

Doug and his colleagues examined some of the youngest Ediacaran assemblages from locations in Namibia called Farm Swartpunt. They hypothesized that if this biotic replacement model was correct, then the extinction mechanism there should have 2 effects. Firstly, there should be an increase in the size and variety of trace fossils left behind by increasingly complex animals interacting with the ocean sediments; and secondly, that the younger Ediacaran assemblages should be relatively depauperate—that means reduced in species—when compared to older assemblages like those in Canada, Russia, and south Australia.

The presence of relatively high-diversity trace fossils in many sections of uppermost Ediacaran and lowermost Cambrian rocks, including those at Farm Swartpunt, supports the idea of increased ecosystem engineering at this time. They also noted a drop in diversity of the Ediacarans at Farm Swartpunt. Obviously, this is still only one location, and it was located at relatively high latitudes at time of deposition, but even so, diversity was much lower than what was expected, and there is no evidence of environmental stress to account for it.

What is suggested is an Ediacaran assemblage that was becoming increasingly outcompeted and marginalized by developing Cambrian fauna. In physically interacting with the oceanic substrate, these new Cambrian forms would have competed for resources, increased the delivery of carbon to the ocean floor, and more effectively mixed and oxygenated ocean sediments. For the Ediacarans, the world was changing beyond their ability to adapt and they faded away, leaving the stage set for the Cambrian Explosion.

Whatever their fate, by the time we see the first large fauna of the Cambrian, the Ediacarans like the rangeomorphs and similar creatures are gone, either suddenly or gradually, replaced by other creatures. The putative ancestors of later organisms such as Kimberella are gone too, their ancestors evolving into the wonderful creatures from the Cambrian Explosion, thus laying the foundations for the greatest show on Earth.

Lecture 12

Arthropod Rule on Planet Earth

The Department of Paleobiology at the Smithsonian's National Museum of Natural History has had a long and important association with the study of fossil arthropods. This lecture will examine some of the past history and collections, as well as some of the current research that is being undertaken on this extremely important group of animals. In this lecture, you will learn about the origins of the arthropods and how our perception of arthropods would change after the explosion of life.

The Origins of Arthropods

- The last common ancestor of the Arthropods is out there, somewhere, in rocks that are more than half a billion years old. By removing all the recent modifications in arthropod design, we can figure out the basic characters of this time-distant creature—or ur-arthropod, as it is sometimes called, after the cradle of human urbanization, the ancient Sumerian city of Ur in Iraq.

- The ur-arthropod is imagined as a segmented, bilaterally symmetrical, highly appendaged creature with each segment covered by its own armor plate, or sclerite. Each undifferentiated segment would be provided with a pair of biramous, or branched, limbs with a mouth positioned underneath the body at the head end. The head would have eyes, often compound, and probably one or more pairs of antennae. Given the nature of some of the very early arthropods, it would most probably feed by processing sediment for organics with quite complicated mouthparts.

- Some of the earliest arthropod-looking creatures in the fossil record come from the Ediacara Hills in Australia. The fossils are found in a geological formation called the Rawnsley Quartzite that were originally sands deposited in shallow tidal waters around 555 million years ago during the Late Precambrian Ediacaran period.

- One of the members of this diverse, and sometimes strange, fauna of the Ediacaran Hills is a wormlike fossil called *Spriggina*, named for Reginald Sprigg, who discovered the Ediacaran fauna in 1946. *Spriggina* was around 1 to 2 inches (3 to 5 centimeters) long and appears to be segmented, supplied with rows of plates along its back. In addition, unlike many other creatures in the Ediacaran fauna, it has an obvious head, or cephalon, not unlike the head shields we find in trilobites, which are definitive arthropods occurring later during the Cambrian period.

- If it does represent an earthly arthropod, then *Spriggina* is a problematic fossil. First, one of the key features of arthropods, jointed legs, have not been found on any specimen thus far. In addition, although the creature appears to be symmetrical, *Spriggina* actually has a special form of symmetry called glide symmetry, where the segments running down the center line of the creature are imbricated, forming a steplike pattern.

Fossil trilobite imprint in the sediment

- Another contender for the arthropod ancestor from this time is *Parvancorina*, which has also been compared to trilobites. This is a fairly simple creature with a shield-like body and blunt head. It is bilaterally symmetrical, but no segments or limbs have been found.

Changing Perceptions of Arthropods

- The Burgess Shale is located in the Rocky Mountains of British Columbia. It was discovered by Charles Walcott, former Secretary of the Smithsonian, at the end of his 1909 field season and named for nearby Mount Burgess. He would return a number of times and amass a wonderful collection of fossils. Walcott's fossil quarry is now in part of Yoho National Park.

- In total, Walcott would recover more than 65,000 specimens that he would faithfully record, extract, and return to the Smithsonian, forming one of the most important collections of Cambrian fossils in the world. Although its full importance was not really realized until the 1970s, Walcott's discovery sheds light on an incredible ecosystem—a world that had recently gone through the Cambrian explosion and that would see the relatively simple animals from the Ediacaran diversify into all the body plans of animals we see today.

- There were organisms fixed to the ocean floor, such as algae photosynthesising in the dim filtered light, and numerous sponges, filter-feeding organic material and microplankton raining down from above. Most of the mobile creatures were dominated by forms that moved around on the ocean floor, probably eating mats of algae and microbes or processing sediment on the ocean floor for organic material. Some creatures lived in the sediment. Compared to today, there was not the diversity of creatures swimming in the water column.

- If it were not for the exceptional preservation of the soft-bodied Burgess Shale animals, our picture of the Cambrian world would have been very different—one that would appear impoverished, with only shelly, hard-parted creatures represented.

- Like today, the dominant life-form in the Burgess Shale, in sheer numbers of species, were the arthropods. There were a variety of trilobite as well as early ancestors of the crustaceans.

- By the time of the Burgess Shale at 510 million years ago, an entire suite of arthropods is present, with most of the major arthropod subgroups represented—not only here, but in all of the other Burgess-type sites around the world.

- But there are also some odd arthropod-like creatures associated with our Burgess arthropods. One is the *Tyrannosaurus rex* of the Cambrian oceans: *Anomalocaris*, some specimens of which from China are up to 6 feet long. The giant limbs in front of this creatures were used to capture and hold its prey.

- On the underside of its head is a strange squared-ring mouthpart full of sharp teeth, probably designed to crunch arthropods or other prey. It had well-developed eyes, and its body was flanked with flexible lobes, which would have made it a strong swimmer.

- A recent discovery of a new anomalocarid has shed more light on the relationships among the arthropods of the Burgess world. This particular find doesn't come from the Cambrian, though; it was found in Morocco in Ordovician rocks, 30 million years after the Burgess Shale.

- The creature was enormous, about 6.5 feet (2 meters) in length, but rather than the fierce predator from the Cambrian, *Aegirocassis benmoulae* appeared to have its front appendages modified for filter feeding, probably swimming through the ocean filtering microplankton.

- But what is significant for our understanding of the evolution of the arthropods is the nature of the swimming lobes in this fossil. All previous anomalocarids were assumed to have a single set of flaps per segment for swimming, but *Aegirocassis* possessed 2. The upper flaps were equivalent to the upper limb branch (called the exopodite) of modern arthropods, while lower flaps are the equivalent of the lower walking limb branch (called the endopodite) of modern arthropods.

- An examination of the Cambrian anomalocarids has shown that these, too, had paired flaps but had been overlooked. The reason they were found in *Aegirocassis* is the nature of the preservation, which is less flattened than the Cambrian forms.

- Before this came to light, the anomalocarids were an anomaly in our understanding of arthropod evolution and did not quite fit comfortably in the general arthropod story.

- The discovery that they possessed 2 flaps on their segments and not one and that those 2 flaps were separate and not branched put them on a stem leading to what some call the Euarthropoda, or true arthropods. They represent a stage before the fusion of exopodite and endopodite into the modern arthropod biramous limb we see today.

- In other words, anomalocarids are more basal than the trilobites and today's arthropods. As such, the more we find out about their morphology, the more hints we get regarding the evolution of arthropods—in this particular case, an insight into the typical Euarthropodan biramous limb.

- We are still a long way from a complete understanding of the evolution of the early arthropods, but new discoveries, particularly of beautifully preserved Burgess-type material, are certainly helping to clear the fog that has surrounded the origins of this most important phylum.

The Trilobites

- Dr. Robert Hazen, a senior scientist at the Carnegie Institution of Washington, has one of the most wonderful trilobite collections in the world. In 2007, he donated the collection to the Smithsonian's National Museum of Natural History's Department of Paleobiology.

- The trilobites first appeared about 521 million years ago, about 21 million years before the abundant arthropod and trilobite fauna we find in the Burgess Shale. They were a very diverse group, with 20,000 species, so far, described.

- Trilobites, meaning 3 lobes, have quite an association with the number 3. Their bodies are divided into 3 segments, and they have 3 major longitudinal lobes. The living Paleozoic trilobite was provided with jointed legs, gills, and antennae like many other arthropods. In general, though, it is only the hardened calcium carbonate of the outer shells that get preserved. Like all arthropods, trilobites had to molt to grow.

- We even have some clues as to how at least some trilobites would mate. There are a number of instances where trilobites of one species, a monospecific assemblage, have been found preserved in large clusters that appear to represent a life assemblage. A remarkable discovery in Portugal of some large trilobites was published in 2009 by The Geological Society of America. Researchers from Portugal and Spain describe incredible clusters and long lines of trilobites. They suggest that this may be equivalent to a mass spawning like we see in their closest relatives, the horseshoe crabs, of today.

- Trilobites are probably some of the first animals to gaze upon the world they lived in, as they possessed fairly sophisticated eyes. Some trilobite even had extremely enlarged eyes.

- Their adaptability is seen in the variety of habitats they lived in; they have been found in virtually all marine environments, from the shallow to the deep ocean. They were not restricted to the ocean floor, either. Some trilobites with extremely enlarged eyes are thought to have been pelagic floaters. Some larger forms were likely active swimmers.

- Their diversity is also seen in the different ways in which they fed. Important clues come from a structure on the underside of the animal called the hypostome situated by the trilobite's mouth. There is a whole range of different types of hypostome, probably reflecting the different types of food these creatures were scavenging or hunting.

- The trilobites would also develop a whole range of spines and complicated structures, some of which are pretty difficult to interpret. Perhaps some of the most bizarre are those that have been extracted from the Anti-Atlas Mountains of Morocco. For example, *Walliserops*, a trilobite that comes from the Lower-Middle Devonian period, is not only provided with various long curved spines and processes, but at the front end is a long 3-pronged trident.

- The trilobites would suffer major extinctions at the end of the Ordovician and then again at the end of the Devonian. They would recover a little through the Carboniferous but never regain the diversity they attained in the Cambrian and Ordovician periods.

- By the end of the Permian, their numbers had dwindled to 5 genera. At the final devastation that rocked the biosphere 252 million years ago, the trilobites were in a very precarious position, with a restricted distribution in shallow marine environments that would be hit hard during the extinction. Even with the sad passing of the trilobites, arthropods would continue to spread and diversify.

Making a Break for the Land

- Arthropods were probably some of the first animals to make the break from the ocean to the land. Their external exoskeletons could act a bit like a spacesuit on land, and many groups of aquatic arthropods had already evolved limbs, placing them a step ahead of our group, the vertebrates. We would have to turn limbs into fins through transitional forms.

- Some of the oldest obligate terrestrially adapted creatures on land were probably related to the myriapods, the group that contains millipedes, centipedes, and the gigantic Late Paleozoic arthropleurids. The oldest terrestrial animal fossil, *Pneumodesmus newmani*, is a species of millipede-like myriapod dating to the Late Silurian of Scotland, about 428 million years ago. These initial invaders of the land were probably feasting on plant litter, just like their modern equivalents do today, and were likely an important factor in the development of soils.

- These detritus feeders were rapidly followed by a wave of other arthropods, the Chelicerata, the group that includes mites, scorpions, and spiders.

- Another major group of arthropods, the crustaceans, would also get in on the terrestrial act. Examples today include wood lice, or pill bugs, and coconut crabs.

- Another important group of arthropods are the hexapods, from whom we gain the incredibly diverse insects. Research into the paleontology of insects and other arthropods is continuing to flesh out the dynamic history of this important group of animals.

Questions to consider:

1. Why are arthropods not the dominant large creatures on Earth today?
2. Could our world function without arthropods?

Suggested Reading:

Conway-Morris, *The Crucible of Creation*.

Fortey, *Trilobite*.

Lecture 12 Transcript: Arthropod Rule on Planet Earth

Arthropods are everywhere—flying and floating in the atmosphere, crawling on the tops of mountains and in the deep dark ocean, in baking deserts and at the poles, and living with you in your home. Everywhere you look, you will find our friends the arthropods. The exoskeleton of this diverse group of animals is composed of chitin, which, in some members of the phylum like crabs and millipedes, may be mineralized with calcium carbonate and, due to their external suit of armor, all arthropods have to molt in order to grow.

Today they are represented by four major groups: the Chelicerata, which includes the spiders and scorpions and the horseshoe crabs; the Myriapoda, that includes centipedes and millipedes; the Crustacea, the tasty group, including the crabs, lobsters, shrimp, but also the tiny ostracods that secrete a bivalve shell; and perhaps the largest group that includes both the most beautiful and the most annoying arthropods, the Hexapoda to which the insects belong, a group that has received special attention at the Smithsonian in recent years.

There is one major group that is missing, though, a group of arthropods that would be extremely important in Earth's oceans for over 250 million years, but would be wiped out geologically overnight in a calamity that almost wiped out all of life on Earth. These are the trilobites.

The Department of Paleobiology at the Smithsonian's National Museum of Natural History has had a long and important association with the study of fossil arthropods. In this lecture, we will examine some of the past history and collections, and some of the current research that is being undertaken on this extremely important group of animals. So in this lecture let's ask: What are the origins of the arthropods? And how would a former Secretary of the Smithsonian help change our perception of arthropods after life's big bang? We'll look at a collection of fossils at the Smithsonian started by an 8-year-old and the picture they paint of the arthropods in the Paleozoic seas, and consider which arthropods would make a break for the land.

The last common ancestor of the arthropods is out there, somewhere, in rocks over half a billion years old. By removing all the recent modifications in arthropod design, you can figure out the basic characteristics of this time-distant creature, or ur-arthropod as it's called, after the cradle of human urbanization, the ancient Sumerian city of Ur in Iraq.

The ur-arthropod is imagined as a segmented, bilaterally symmetrical, highly appendaged creature with each segment covered by its own armor plate or sclerite. Each undifferentiated segment would be provided with a pair of biramous—that's branched—limbs with a mouth positioned under the body at the head end. The head would have eyes, often compound, and probably one or more pairs of antennae. Given the nature of some of the very early arthropods, it would most probably feed by processing sediment for organics with quite complicated mouthparts.

Some of the earliest arthropod-looking creatures in the fossil record come from the Ediacara Hills just north of Adelaide in Australia. The fossils are found in a geological formation called the Rawnsley Quartzite, which were originally sands deposited in shallow tidal waters around 555 million years ago during the Late Precambrian-Ediacaran period.

One of the members of this diverse and sometimes strange fauna of the Ediacaran Hills is a worm-like fossil called Spriggina that we met in an earlier lecture. It is named for Reginald Sprigg who discovered the Ediacaran fauna back in 1946. Spriggina was around 1 to 2 inches long and appears to be segmented, supplied with rows of plates along its back. In addition, unlike many other creatures in the Ediacaran fauna, it has an obvious head or cephalon not unlike the head shields we find in trilobites, which are definitive arthropods, occurring later during the Cambrian period.

If it does represent an early arthropod then, Spriggina is a problematic fossil. For a start, one of the key arthropod features of arthropods—jointed legs—have not been found on any specimens thus far. In addition, although the creature appears to be symmetrical, Spriggina actually has a special form of symmetry called glide symmetry, where the segments running down the center line of the creature are imbricated, forming a kind of a step-like pattern. Another contender for the arthropod ancestor from this time is Parvancorina, meaning small anchor, which has been compared to trilobites. As you can see, this is a fairly simple creature with a shield-like body and blunt head—if, indeed, that is the head end of the animal. It is bilaterally symmetrical but no segments or limbs have been found.

So, perhaps in order to try and tie down our ur-arthropod, we need to go forward in order to look back. In order to do that, we have to look at collections from the Burgess Shale in Canada that are held at the National Museum of Natural History. They were collected in the early 1900s by one Charles Doolittle Walcott, Secretary of the Smithsonian Institution. The Burgess Shale is located in the Rocky Mountains of British Columbia, close to the town of Field and the border with Alberta.

It was discovered by Walcott at the end of his 1909 field season and named for nearby Mount Burgess. He would return a number of times and amass a wonderful collection of fossils. Walcott's fossil quarry is now part of the Yoho National Park. You can't just wander up and visit, but you can book an organized tour. This is a very important site for biologists and paleontologists and has been recognized as such by UNESCO.

The Burgess shale was originally a dark mud deposited around 505 to 510 million years ago on the edge of an underwater feature called the Cathedral Escarpment, a massive 650 feet, 200 meter, submarine cliff that ran along the edge of North America that has now been transformed into limestone. Today, that limestone marks the extent of one end of Walcott's original quarry. North America at that time was arranged in an east-west orientation along the equator.

The fossils that Walcott found were originally living at the base of this cliff in calm waters, well below the churning waves but probably just within the photic zone—that's the depth to which sunlight can penetrate through water—so the environment would have been somewhat gloomy but lit well enough to allow various forms of algae to photosynthesize. Occasionally, sediments from the escarpment would flow down the side of the cliff in an event called a turbidity flow, not unlike an underwater avalanche. These sediments covered the ocean floor and buried, very rapidly, the creatures living there—not just once, but in repeated events, perhaps initiated by earthquakes. In total, Walcott would recover over 65,000 specimens that he would faithfully record, extract, and return to the Smithsonian Institution, forming one of the most important collections of Cambrian fossils in the world.

Although its full importance was not really recognized until the 1970s, Walcott's discovery sheds light on an incredible ecosystem, a world that had just recently come through the Cambrian Explosion that would see the relatively simple animals from the Ediacaran diversify into all the body plans of animals we see today. Everything from a worm to an elephant to a ladybug traces its origins back to the Cambrian Explosion. And the Burgess Shale records the world just after that spectacular event.

There were organisms fixed to the ocean floor, algae like Margaretia photosynthesizing in the dim, filtered light, and numerous sponges such as *Vauxia gracilenta* filter-feeding organic material and microplankton raining down from above. Most of the mobile creatures were dominated by forms that moved around on the ocean floor, animals like the mollusk precursor Wiwaxia and the trilobite-like Naraoia, probably eating mats of algae and microbes or processing sediment on the ocean floor for organic material. Some creatures lived in the sediment. Ottoia, a priapulid worm, lived in U-shaped burrows from which it extended its proboscis in search of prey. In particular, Ottoia had a preference for the hyoliths, which were possibly some sort of shelled mollusk, fragments of which have been found in Ottoia's gut.

Compared to today, there was not the diversity of creatures swimming in the water column. Notable exceptions, though, were creatures like Marrella, or the lace crab, which were the first fossils that Walcott found and also some of the most common, with 15,000 individuals recovered from Walcott's quarry. There was also one of the few members of our particular family tree, Pikaia, an early chordate only about as long as your little pinkie. Another swimmer from the Burgess shale is everybody's favorite weirdo, Opabinia. It had 5 eyes and a long flexible proboscis with grasping spines. Opabinia could swim using its lobes, but probably spent most of its time on the seabed probing the soft sediment for worms.

If it were not for the exceptional preservation of soft-bodied Burgess Shale animals, our picture of the Cambrian world would be very different, one that would appear impoverished, with only shelly, hard-parted creatures represented. Like today, the dominant life-form in the Burgess Shale in sheer numbers of species were the arthropods. There were a variety of trilobites; for example, Olenoides, a 4-inch-long trilobite who, in this beautiful specimen, still shows a pair of curling antennae.

As we noted, Marrella, the lace crab, probably swam in large shoals but shared the ocean with a number of other arthropods, like Sidneya, a predator about 2 to 5 centimeters long that probably dined on crustaceans, trilobites, and hyoliths. There were early ancestors of the crustaceans such as Canadaspis and the lobster-like Waptia, who possibly spent most of its time walking around on the ocean floor on its jointed legs.

And then there is the 2-inch-long, blind Leanchoilia with its prominent frontal appendages and long, whip-like extensions. These may have helped it locate food and kept Leanchoilia informed about its surroundings. It is thought that this species might be a member of the Chelicerata, a major arthropod group that includes present day scorpions, mites, and spiders.

The principal point here is that by the time of the Burgess Shale at 510 million years ago, an entire suite of arthropods is present, with most of the major arthropod subgroups represented. Not only here but in all of the other Burgess-type sites around the world, such as Sirius Passet in Greenland and Chenjiang in China, the latter of which is even older than the Burgess Shale at about 525 million years old.

But there are also some odd arthropod-like creatures associated with our Burgess arthropods. We've already met one of them, Opabinia, but another is the T-rex of the Cambrian oceans: Anomalocaris. Some specimens from China are up to 6 feet long. The giant limbs in front of this creature were used to capture and hold its prey. On the underside of its head is a strange square-ringed mouthpart full of sharp teeth, probably designed to crunch arthropods or other prey. It had well developed eyes and its body was flanked with flexible lobes, which would have made it a probably fairly able swimmer.

A recent discovery of a new anomalocarid has shed a bit more light on the relationships among the arthropods of the Burgess world. This particular find, though, doesn't come from the Cambrian; it was found in the Ordovician rocks of Morocco, 30 million years after the Burgess Shale. The creature was enormous, about 6.5 feet—that's 2 meters—in length, but rather than a fierce predator from the Cambrian, *Aegirocassis benmoulae* appears to have its front appendages adapted for filter feeding, probably swimming through the ocean filtering microplankton.

But what is significant for our understanding of the evolution of the arthropods is the nature of the swimming lobes in this fossil. All previous anomalocarids were assumed to have a single set of flaps per segment for swimming, but Aegirocassis possessed 2. The upper flaps were equivalent to the upper limb branch, called the exopodite, of modern arthropods while lower flaps are the equivalent of the lower walking limb branch, called the endopodite, of modern arthropods, as well. An examination of the Cambrian anomalocarids has shown that these too had paired flaps, but had been overlooked. The reason they were found in Aegirocassis is the nature of the preservation, which is less flattened than those of the Cambrian forms.

I guess you might be thinking that all this fuss about flaps is, well, just a bit of a fuss about flaps. Actually, no—it could be vital. Before this came to light, the anomalocarids were a bit of a problem, an anomaly in our understanding of arthropod evolution, and did not quite fit comfortably with the general arthropod story. The discovery that they possessed 2 flaps on their segments and not one, and that those 2 flaps were separate and not branched, put them on a stem leading to what some call the Euarthropods, or true arthropods. They represent a stage before the fusion of the exopodite and the endopodite into the modern arthropod biramous limb that we see today. Or put another way, anomalocarids are more basal than the trilobites and today's arthropods. As such, the more we find out about their morphology, the more hints we get regarding the evolution of the arthropods, in this particular case an insight into the typical euarthropodan biramous limb.

As you can see, we are still a long way off a complete understanding of the evolution of the early arthropods, but new discoveries, particularly of beautifully preserved Burgess-type material, is certainly helping to clear the fog that surrounded the origins of this most important phylum. Right now, though, I would like to focus on just one of the components of the Burgess Shale fauna that would diversify into one of the most successful groups that spread through the oceans of the Paleozoic: the trilobites.

Dr. Robert Hazen is a senior scientist at the Carnegie Institution of Washington. He is a mineralogist and astrobiologist, and has a keen interest in the role minerals have in the origin of life. We explored his ideas in an earlier lecture. He also just happens to have one of the most wonderful trilobite collections in the world that he has been collecting since he was 8 years old, and in 2007 he donated the collection to the Department of Paleobiology. Fifty of these beautiful fossils are currently on display at the National Museum of Natural History.

So who were the trilobites? They first appeared about 521 million years ago—that's about 21 million years before the abundant arthropod and trilobite fauna we find in the Burgess Shale. They were a very diverse group with 20,000 species—so far—described. This was a highly successful group of animals.

Trilobites, meaning 3 lobes, have quite an association with the number 3. Their bodies are divided into 3 segments. From the anterior to the posterior: the cephalon, or head; thorax—body; and the pygidium—the tail. The name trilobite, though, refers to the 3 major longitudinal lobes: the left pleural lobe, the axial lobe, and the right pleural lobe. The living Paleozoic trilobite was provided with jointed legs, gills, and antennae, like other arthropods. In general, though, it is only the hardened calcium carbonate, or calcite, of the outer shells that gets preserved. Like all arthropods, trilobites had to molt to grow.

They had a life cycle that started with a tiny larva that was a fraction of a millimeter across called a protaspid that probably had a planktonic lifestyle. As it matured, it would start to differentiate into a head and tail, and then expand by adding segments to the thorax until they reached a point where no new segments were added. After the addition of segments stopped, the creature just grew in a general size increase after each molt.

We even have some clues as to how at least some trilobites would mate. There are a number of instances where trilobites of one species, what we call a monospecific assemblage, have been found preserved in large clusters that appear to represent a life assemblage. In other words, they have not been transported and concentrated together after death.

A remarkable discovery in Portugal of some large trilobites around 27 inches long—that's 70 centimeters—was published in 2009 by the Geological Society of America. Researchers from Portugal and Spain, including Juan Gutierrez-Marco, describe incredible clusters and long lines of trilobites. They suggest that this may be equivalent of a mass spawning like we see in the horseshoe crabs of today.

Trilobites are probably some of the first animals to gaze upon each other and the world that they lived in, as they possessed fairly sophisticated eyes. Like the rest of the trilobite carapace, the lenses in the compound eyes of trilobites were composed of calcite, but in this case composed of a single crystal of calcite arranged so that light could pass right down the center of the crystal.

Some trilobite eyes went to extremes, though. Consider this—this is *Erbenochile erbeni*, described by Dr. Richard Fortey of London's Natural History Museum from Devonian rocks in Morocco. It has remarkable tower-like eyes with straight sides that gave this trilobite probably very good vision over a long distance, both in front and behind the animal. However, notice the lip at the top. Dr. Fortey suggests that this is a sunshade. This species probably lived in shallow, well-lit waters, so may have needed to shield its eyes from the Sun. The trilobite was doing that with this lip, placing the lenses in shadow from the Sun above.

Their adaptability is seen in the variety of habitats they lived in. They have been found in virtually all marine environments, from the shallow to the deep ocean. They were not restricted to the ocean floor, either. Some trilobites with extremely enlarged eyes are thought to have been pelagic—that means they floated. For example, Carolinites who probably floated in the upper reaches of the Ordovician oceans, what we call an epipelagic life-habit, preying upon microplankton and looking out for predators with those enormous eyes that gave them near 360° vision.

Some larger forms like Parabarrandia, about 1 foot long, were likely active swimmers, their streamlined bodies cutting through the water efficiently while being propelled by strong swimming legs, as has been suggested by the large muscle attachments found on this animal. Their diversity is also seen in the different ways in which they fed. Important clues come from a structure on the underside of the animal called the hypostome situated by the trilobite's mouth. In many trilobites, the hypostome is fixed to the animal and acted as a surface against which food could be processed. Unfortunately, we know very little about trilobite mouthparts.

The forms with such a fixed hypostome, though, were probably scavengers and predators, using the structure to help rip and shred material before passing it into the gullet. There is a whole range of different types of hypostome, probably reflecting the different types of food these creatures were scavenging or hunting for. Other trilobites have a detached hypostome that in life was attached probably to a membrane. It acted very much like a scoop, which it used to funnel sediment into its mouth. These sediment-processing forms are probably responsible for some of the multiple trilobite trails called *Cruziana*, the animal plowing through a particular organic-rich sediment multiple times and processing the sediment as it went.

A group of trilobites with a very particular and very specialized mode of feeding is the trinucleoids. They're generally found in rocks that were originally soft and soupy, and have been interpreted as using their limbs to pump sediment-laden water through pits around the edge of their cephalon, the head shield, filtering any organic yummies they could find. This particular posture has been supported by trace fossils that match a trinucleid trilobite filtering food in this manner.

The trilobites would have also developed a whole range of spines and complicated structures, some of which are pretty difficult to interpret. Perhaps some of the most bizarre are those that have been extracted from the Anti-Atlas Mountains of Morocco. For example, I present Walliserops, a truly strange form and another trilobite that has intrigued Dr. Richard Fortey. It comes from the lower Middle Devonian period.

Not only is it provided with various long curved spines and processes, but at the front end is a long, three-pronged trident. This structure could not be retracted so it couldn't be used to spear prey, and in some species the arrangement of the trident would have meant that the trilobite would have had to raise its head in order to move forward to prevent the trident burying itself in the sediment. A display structure perhaps, like the antlers on a moose?

But, as the phrase goes, all good things must pass. The trilobites would suffer major extinctions at the end of the Ordovician and then again at the end of the Devonian. They would recover a little through the Carboniferous but never regain the diversity they attained in the Cambrian and Ordovician periods. By the end of the Permian, their numbers had dwindled to 5 genera. At the final devastation that rocked the biosphere 252 million years ago, the trilobites were in a very precarious position with a restricted distribution in shallow marine environments that would be hit hard during the mass extinction.

Even with the sad passing of trilobites, arthropods would continue to spread and diversify. But there was another vast area of opportunity that they were primed to exploit. The land was waiting and arthropods would be first out of the blocks. Examine any arthropod and you can clearly see why they were probably some of the first animals to make the break from the ocean to the land. Their external exoskeletons could act a bit like a spacesuit on land, and many groups of aquatic arthropods had already evolved limbs, placing them a step ahead of our group, the vertebrates, who would have to turn fins into limbs through transitional forms like the famous Tiktaalik.

Some of the oldest obligate—terrestrially adapted—creatures on land were probably related to the myriapods, the group that contains millipedes, centipedes, and the gigantic Late Paleozoic arthropleurids. The oldest terrestrial animal fossils, *Pneumodesmus newmani*, is a species of millipede-like myriapod dating to the late Silurian of Scotland some 428 million years ago. These initial invaders of the land were probably feasting on plant litter just like their modern equivalents do today, and were likely an important factor in the development of soils.

These detritus feeders were rapidly followed by a wave of other arthropods: the Chelicerata, the group that includes mites, scorpions, and spiders. Some of the first predators in these low-lying Paleozoic forests of plants, that probably would have been no higher than me or you, were the trigonotarbids, an arachnid but not yet a true spider. These animals couldn't spin silk and didn't have the complex feeding apparatus of modern spiders. Instead, it is thought that they ambushed their prey and kind of, well, vomited digestive enzymes on them, sucking up the predigested food through a filter plate in their mouthparts.

Another major group of arthropods, the crustaceans, would also get in on the terrestrial act. Examples today include wood lice—the pill bugs—and the fantastic coconut crabs, which hold the current record of being the largest land-living arthropod, growing up to 3 feet 3 inches from leg to leg.

Of course, we're missing an important group of arthropods so far, the Hexapods, and from them we get the incredibly diverse insects. Some of the first insects are found along with those early arachnids in a deposit called the Rhynie Chert from Scotland, a window into the Early Devonian landscape at 410 million years ago. The fossils of plants, fungi, lichens, and a variety of arthropods are well preserved in chert that was being deposited around hot springs and geysers, instantly preserving the flora and fauna in fantastic detail.

The early insect *Rhyniella praecursor*, found in the Rhynie Chert, belongs to a primitive group of flightless insects called springtails. They, like millipedes, were probably detritus feeders, but the wider insects would diversify into a dizzying array of creatures with an astounding variety of lifestyles. Dr. Conrad Labandeira at the Smithsonian's National Museum of Natural History thinks that the group diversified significantly earlier than this though, because of several specializations they already have, particularly the distinctive jumping device at the end of their abdomens. Springtails likely originated in freshwater during the Late Silurian.

Conrad's research group at the Smithsonian, and colleagues throughout the globe, have been investigating the remarkable history of insects and how they have been interacting with other elements of the biosphere, adding richness and depth to ecosystems through time. His research has shed light into the origins of the most diverse groups of insects, the Holometabola—that is, insects that go through egg, larva, pupa, and adult stages, a very useful life cycle as larva and adult inhabit different areas of the ecosystem and typically don't compete for food.

You can see the earliest 311-million-year-old larval representative of the Holometabola in this remarkable fossil from Mazon Creek from northeastern Illinois, U.S.A., an exceptional deposit of terrestrial and brackish environments. This is *Srokalarva berthei*—it was preserved in an iron carbonate nodule and mineralized with calcite, pyrite, and other minerals. The creature resembled a caterpillar and fed on the leafy tissues of plants, evidence of which comes from damage found to seed plant fossils in the same deposit.

In this series we will also investigate how insects have adapted and coevolved with flowers, and how they could respond in the post-Permian extinction world during the Triassic. Research into the paleontology of insects and other arthropods is continuing to flesh out the dynamic history of this important group of animals. Even with the sad passing of the trilobites, arthropods have continued to spread and diversify, and can probably still be regarded as the animal kings of planet Earth, no matter what we might think.

Lecture 13

Devonian Death and the Spread of Forests

Alfred Sherwood Romer was a U.S. paleontologist whose research points to a drop in diversity in the fossil record of early tetrapods—vertebrates that have 4 limbs—from 360 to 345 million years ago during the first 15 million years of the Carboniferous period. Prior to the gap, during the Late Devonian, there was an expanding population of early tetrapods. This gap has been a matter of great debate. One explanation is that diversity had crashed at the end of the Devonian period. In this lecture, you will consider what happened at the boundary between the Carboniferous and the Devonian.

The Late Devonian Earth

- The climate at the start of the Devonian was generally warm and dry, with the situation getting more tropical and sometimes rainy as the Devonian continents started to move toward the equator. During the Late Devonian, however, there are indications of successive advances and retreats of glacial ice at the poles.

- Glacial indicators—such as glacial sediments called till, grooves on bedrock, and dropstones (which form as rocks are released from floating icebergs and fall into the sediment beneath)—have been found in various locations in Africa and South America, at this time part of Gondwana. This suggests the presence of ice sheets at various times in the southern polar regions, with perhaps a number of expansions and retreats of ice over the South Pole. This period of glaciation in Gondwana lasted until the mid-Permian.

- An important feature of the later Devonian oceans is the development of black shales in many ocean basins around the world. The black color of the shale is in part caused by the presence of the mineral pyrite that is finely dispersed throughout the rock, but also due to its high organic content. These sediments often give off a very characteristic rotten-egg stench, the marker of hydrogen sulfide, and bacteria that like to live and respire in such low-oxygen conditions. This hydrogen sulfide is also responsible for the high pyrite content of these rocks.

- Black shales are common in the oceans from about the Middle Devonian, but prior to this, extensive reefs were very common in the shallow oceans that surrounded the still-fragmented continents of the Devonian world. An example of a spectacular Devonian reef can be found in the Canning Basin of Australia. This reef developed during the Middle to Late Devonian, when a shallow tropical sea covered this area of Australia. The reefs were constructed by calcareous algae, corals, and spongelike encrusting creatures called stromatoporoids. Reefs were much more common in the Devonian than they are today and supported a thriving community of invertebrates.

- But the Devonian had seen considerable innovations on land, too. The Devonian boasts the first tetrapods that were starting to tentatively explore the land.

- But one of the most striking features of the Late Devonian world was the spread of green along coastlines of the continents, perhaps extending inland in more favorable settings. Plants were expanding their colonization of the land.

- Plants had made it to land earlier, during the Silurian about 430 million years ago, but plants would remain pretty small, inauspicious, and tied to open sources of water in those times.

- Even by the time we get into the Early Devonian, the landscape was still dominated by small wetland-dwelling plants. But by the early Middle Devonian, plants had risen off the ground with the evolution of horsetaillike forms and the beginnings of the fern lineage. But it was not until the late Middle Devonian that the real revolution occurred, with the evolution of seeds, true roots, wood, and multiple origins of leaves. This dramatically changed the reproductive biology and stature of land plants.

- It is important to note that each time we see an innovation in plants, there is an associated increase in the diversity of herbivores. Dr. Conrad Labandeira, a paleoecologist in the Smithsonian's National Museum of Natural History's Department of Paleobiology, notes that herbivory (eating plants) in arthropods developed just 20 million years after the first land plants had evolved during the Silurian. Herbivory was an important development that became a major driving force in the processing of live plant tissue into organic carbon.

- Innovations in the plants permitted tall trees to spread across the landscape and into highland and more inland areas, finally breaking ties with standing water. This was the start of the greening of the Earth beyond the coastlines and the first forests.

Crisis in the Late Devonian

- The Late Devonian was a time of change, not only in the biosphere, but also in the state of the oceans and atmosphere. Oceanic anoxia was present in some areas, as demonstrated by the presence of black shales in many Late Devonian strata. There is also evidence of global cooling with the advance of glaciers and associated sea-level changes, and it is possible that these changes would stress the biosphere over a period of around 20 to 25 million years, producing a series of about 8 to 10 extinction pulses.

- There would be 2 particularly intense spikes of extinction at 372 million years ago called the Kellwasser event, lasting about 2 million years, and another at the end of the Devonian period around 358 million years ago called the Hangenberg event, lasting about 1 million years.

- The earlier Kellwasser event would mostly affect marine species and in particular the beautiful Devonian reef systems. Many invertebrate groups that lived in and around those reefs would be severely impacted. For example, the number of trilobite families, each of which represented numerous species, would be reduced from 9 to 5.

- Reef systems following this Kellwasser event would tend to be dominated by those spongelike stromatoporoids and microbially constructed, laminated structures called stromatolites. Corals, which had played such an important part in younger reef systems, would be decimated. Overall, tropical warmwater forms were hit the hardest in this extinction.

- It is the final extinction pulse, the Hangenberg event, that marks the boundary between the Devonian and the Carboniferous periods, in which invertebrates and many of the surviving reefs are hit again. It would also affect both the marine and freshwater environments. It is estimated that around 44% of the higher-level vertebrate groups are removed.

- In total, around 19% of families and about 50% of genera would go extinct, but the decimation was probably more severe in the oceans, with perhaps around 22% of families dying. It is possible that around 79 to 87% of all species in the ocean went into extinction. This extinction is referred to as the Devonian mass extinction event.

The Trigger of the Crisis

- What could cause all of these changes at the end of the Devonian? The Devonian extinction is recognized as one of the big 5 mass extinctions that have occurred during the last half a billion years on Earth.

- All of these extinctions, with the exception of the first one at the end of the Ordovician, have been associated with large volcanic events that produced extensive flood basalts. It is well known that such intense volcanic episodes can have varied effects on climate, including global cooling and ozone destruction but also global warming.

- There are at least 2 glacial episodes about the same time as the Kellwasser and Hangenberg events. A cooling scenario for extinction is supported by a decline in the number of warm-tropical–adapted species and a spread of cooler-water–adapted species toward the equator.

- Other culprits for the Late Devonian extinctions have been suggested. It is known that there were at least 2 impact events in the later Devonian. A large enough impact could have serious and sudden consequences for the biosphere.

- The idea that something we equate today with a healthy biosphere—namely, plants—could cause a mass extinction is kind of counterintuitive, but this is just the scenario suggested by researchers such as Thomas Algeo at the University of Cincinnati.

- Consider these features of the Devonian extinction: Both the Kellwasser and Hangenberg events are associated with the development of black shales in shallow marine settings and are associated with an increase in ^{34}S isotope, an isotope of sulfur. These anomalies are regarded as good indicators of anoxia in ocean water. There is also an associated drawdown in atmospheric carbon dioxide and an increase in sediment delivery to the oceans. Algeo suggests that these effects were caused by plants colonizing new habitats.

- As we move through the Devonian, the depth and complexity of root systems increase. This would cause a short-term increase in global weathering as plants colonizing new areas started to break up rocks with their root systems and increased the production of sediment. This sediment, delivered to the oceans by rivers and streams, would muddy the water, making it cloudier.

- This weathering also produces lots of calcium and magnesium carbonates, using a lot of carbon dioxide in the process. This effectively pumps carbon dioxide out of the atmosphere and into the soil. In addition, increased burial of organic plant material helps lock organic carbon away from oxygen and prevents carbon dioxide from being produced by the oxidation of organics.

- Through these mechanisms, it is possible that levels of atmospheric carbon dioxide would be reduced, decreasing the greenhouse effect and initiating a period of global cooling, perhaps allowing the advance of glaciers and the fall of sea level.

- Weathering would also increase rates of nutrient flux and organic delivery to the oceans, causing vast algal blooms in the upper well-lit/oxygenated levels of the ocean. The organic material produced in such blooms would sink, rapidly use up available oxygen in bottom waters, and help generate black anoxic shales and the associated positive ^{34}S anomalies.

- Evidence has been recovered that suggests that oxygen-poor hydrogen sulfide–charged bottom waters would spread through the shallow Devonian oceans, poisoning creatures living on the ocean floor. It is possible that hydrogen sulfide–charged water could have risen to the surface, causing problems for other creatures higher in the water column. In addition, release of hydrogen sulfide into the atmosphere would cause associated problems for animals on land, not the least of which is assisting destruction of the ozone layer.

- For the Hangenberg event at the end of the Devonian, Algeo suggested that there would be a further proliferation of plants following the development of seeds, a major innovation in plant reproduction. This resulted in a global flora spreading to highland areas with an associated peat accumulation and further drawdown of atmospheric carbon dioxide, locking it away into the geosphere in the form of coal. Advances and retreats in ice continued to stress the biosphere.

- This new diversification of seed plants would likewise be accompanied with increased weathering, nutrient delivery to the ocean, algal blooms, and all the problems associated with hydrogen sulfide. In this scenario, the Late Devonian kill zone was a combination of effects, including glacially driven cooling and sea-level fall as well as sick oceans full of poisonous hydrogen sulfide—all potentially triggered by an unlikely source: the plants.

Consequences of the Extinctions to Our Modern Biosphere

- The Devonian witnessed a resetting of the biosphere and the whole character of reefs would change with a significant loss of the old Paleozoic reef-building organisms.

- There would also be an important change for the vertebrates. The Devonian did support a wide variety of fish, many of which are extinct today. The Devonian extinction would bring our more modern-looking fish to the fore. Following the Devonian, we see the rise and dominance of the sharks, the rays, and the bony fish. The tetrapods that had evolved from lobe-finned fish would also be put through the sieve of extinction.

- So, a global catastrophe might be the cause of Romer's gap—a catastrophe that the world was still recovering from in the first 15 million years of the Carboniferous period. But this catastrophe was also a pivotal event, as are all extinction events, clearing ecological space and providing opportunities for the evolution of the new life that was to follow.

Questions to consider:

1. How is it possible that life itself could act as a trigger for a mass extinction?
2. How would the Devonian determine how you put your gloves on?

Suggested Reading:

Beerling, *The Emerald Planet.*

Levin and King Jr, *The Earth Through Time*, chaps. 11–12.

Lecture 13 Transcript

Devonian Death and the Spread of Forests

Welcome back. Let me tell you about a man named Alfred Sherwood Romer, son of a newspaper man born in 1894 in White Plains New York. A man who as a boy was bitten by his pet fox terrier and from such random happenstances the whole course of his life would change.

Unfortunately for Al and presumably the dog, the little terrier was found to be rabid and Al was whisked away to the Pasteur Institute in New York for treatment The young boy would stay with his aunts in Brooklyn while he received a long series of painful injections in his stomach. As a break from being jabbed Al would wander around the halls of the American Museum of Natural History where he caught another bug, the fossil bug, and this would stay with him for the rest of his life

He would start his studies at Amherst college but would have a 2-year break when he decided to serve with the US Air Force during WWI, initially as a private driving an ammunition truck but ultimately rising to the rank of 2nd Lieutenant where he was in charge of 500 French women who were employed sewing covers on to the wings of planes. Al's career would take a number of such interesting twists and turns.

eventually though he would complete graduate studies in zoology and in 1921gained a doctorate from Columbia University. He joined the Department of Geology and Paleontology at The University of Chicago as an associate professor but ultimately was appointed professor of biology at Harvard in 1934.

He was a committed teacher and researcher, and by the time of his death in 1973, he had amassed not only a fine body of research but also a host of awards, medals and honorary degrees for his contributions to science. Awards that apparently never went to his head, for he was known for being an easy going, likable guy. His research would focus on the evolution of the vertebrates and his textbook, *Vertebrate Paleontology*, would be become the touchstone for a more orderly classification of the vertebrates.

The part of Romer's research I would like to consider here relates to a proposed drop in diversity in the fossil record of early tetrapods—that is vertebrates that have four limbs—from about 360 million years ago to 345 million years ago during the first 15 million years of the Carboniferous Period. This low diversity part of the fossil record is called Romer's Gap.

So why do we have a gap? Prior to the gap during the Late Devonian, there was an expanding population of early tetrapods. Tetrapods, like *Ichthyostega*, a five-foot-long amphibian, and its smaller relative, *Acanthostega*. These were just two of the new tetrapods that had evolved from transitional fishapod forms like *Tiktaalik* that were making the break for the land.

The gap has been a matter of great debate, though. It is not utterly devoid of tetrapods; for example, fossils of the six-and-a-half foot long, streamlined *Crassigyrunus* have been found in Scotland. But in general, the recovery rate of fossils during this 15 million-year time interval is quite diminished.

Some claim we have just not looked in the right places to find the fossils. Others have suggested that the creatures existed, but due to some geochemical reason, it was just not a good time to form fossils.

But of course, we have a third option—that diversity had crashed at the end of the Devonian period; a drop in diversity that has subsequently also been noted to occur in another important group of animals—the arthropods.

So what had happened at the boundary between the Carboniferous and the Devonian? To answer that we need to address the following questions. What was the world of the Late Devonian like? What was the crisis in the late Devonian? What triggered this crisis? Should the finger of blame be pointed at plants? What would be the consequences of these extinctions to our modern biosphere?

This is the Late Devonian Earth around 370 million years ago. The continent of Gondwana dominated the Southern Hemisphere, with Siberia in the north. Laurussia, which had formed by the earlier collision of Baltica consisting of much of Eastern Europe, and Laurentia, composed of much of North America, was positioned between Siberia and Gondwana.

The collision of these continental blocks had created a significant mountain chain, the much-diminished remnants of which now can be seen in Appalachia, the Scottish Highlands, and Scandinavia. But soon after their formation in the late Devonian these Caledonian and Acadian Mountains must have been an imposing feature, not unlike the Himalaya Mountains today.

In general, we are looking at the amalgamation of continental areas which will continue through the Paleozoic, ending with the formation of the supercontinent of Pangaea about 70 million years later during the Permian.

The climate at the start of the Devonian was generally warm and dry with the situation getting more tropical and sometimes rainy, as the Devonian continents started to move toward the equator. During the late Devonian, however, there are indications of successive advances and retreats of glacial ice at the poles.

Glacial sediments called till and grooves on bedrock caused as a glacier drags rocks over a rocky surface, are good indicators of the presence of land ice Dropstones have also been identified. These form as rocks are released from floating icebergs and fall into the sediment beneath. These glacial indicators have been found in various locations in Africa and South America at this time part of Gondwana. This suggests the presence of ice sheets at various times in the southern polar regions with perhaps a number of expansions and retreats of ice over the South Pole This period of glaciation in Gondwana lasted until the mid-Permian.

An important feature of the later Devonian oceans is the development of black shales in many ocean basins around the world A good example from North America is the Chattanooga Shale today found in Kentucky, Missouri, and Tennessee.

During the late Devonian, this area was west of mountains that were rising under the influence of a subduction zone that had developed along the eastern edge of Laurasia. As the mountains were eroded, coarse sediments transported by rivers coursing to the west were deposited in the Kaskaskia Sea close to shoreline The finer components of this sediment load were carried further to the east and deposited as fine-grained, muddy sediment now turned to shale.

The black color of this shale is in part caused by the presence of the mineral pyrite that is finely dispersed throughout the rock, but also due to its high organic content These sediments originally had the consistency of the black muck that you might find at the bottom of a stagnant pond during hot summers where organic material is preserved under water that contains very little dissolved oxygen.

Such sediments often give off a very characteristic rotten egg stench the marker of hydrogen sulfide and bacteria that like to live and respire in such low oxygen conditions This hydrogen sulfide is also is responsible for the high pyrite content of these rocks. It would also explain why no benthic fossils—fossils of creatures that lived on the ocean floor—are found in the Chattanooga Shale All the available oxygen had been used up by oxidizing the organic material.

Black shales are common in the oceans from about the Middle Devonian, but prior to this, extensive reefs were very common in the shallow oceans that surrounded the still-fragmented continents of the Devonian world.

An example of a really spectacular Devonian reef can be found in the Canning Basin of Australia. Here is a reef that today rises 50–100 meters above a dry plain that is beautifully preserved just as if the ocean had been drained away, leaving the reef behind This reef developed during the middle to late Devonian when a shallow tropical sea covered this area of Australia.

The reefs were constructed by calcareous algae, corals and sponge-like encrusting creatures called stromatoporoids. Reefs were far more common in the Devonian than they are today and supported a thriving community of invertebrates like these nautiloids and trilobites.

Armored placoderm fish were top dog in the water. Some forms like Dunkleosteus reached around 33 feet in length. Spiny Acanthodians could be found swimming around in large shoals avoiding a variety of sharks There were also representatives of modern bony fish that had evolved just before the Devonian that would diversify into a number of forms.

Armored Placoderm fish were top dog in the water. Some forms like Dunkleosteus here reached around 33 feet in length. Spiny Acanthodians could be found swimming around in large shoals avoiding a variety of sharks There were also representatives of modern bony fish that had evolved just before the Devonian that would diversify into a number of forms.

But the Devonian had seen considerable innovations on land too. As we have already seen the Devonian boasts the first tetrapods—vertebrates with four limbs—that were really starting to tentatively explore the land.

But if you were a space traveler flying over the late Devonian world, one of the most striking features you would have noted was the spread of green along coastlines of the continents; perhaps extending inland in more favorable settings. Plants were expanding their colonization of the land

Plants had made it to land earlier during the Silurian with forms like *Cooksonia* about 430 million years ago, but plants would remain pretty small and inconspicuous and tied to open sources of water in those times.

Even by the time we get into the early Devonian the landscape was still dominated by small wetland-dwelling plants—a thin ribbon of green along sources of open water One of the largest life forms on land was the massive fungus-like organism, Prototaxites, an enormous vegetative structure standing some 26 feet tall. Interestingly, some specimens of Prototaxites show evidence of sizable arthropod borings too.

But by the early Middle Devonian plants had risen a little further off the ground with the evolution of horsetail-like forms and the beginnings of the fern lineage. But it was not till the late Middle Devonian that the real revolution occurred with the evolution of seeds, true roots, wood and multiple origins of leaves This dramatically changed the reproductive biology and stature of land plants.

Seeds were a particularly important development. Seedless plants require moist conditions so that sperm can travel to the egg on the gametophyte. This is the strategy used by plants like ferns.

In seed bearing plants spores are not released but are retained on the spore-bearing plant where they grow into male and female forms of the gamete-bearing generation. In the case of flowerless seed plants gymnosperms, these are the male and female cones that you can see on modern conifers.

This plant revolution would see the evolution of a very special plant about 385 million years ago called Archaeopteris This was a conifer-like tree with the first true wood and one of the earliest plants that bore true seeds.

As an aside here, it is important to note that each time we see an innovation in plants there is an associated increase in the diversity of herbivores Dr. Conrad La Bandera a paleontologist in the Department of paleobiology notes that herbivory—that's eating plants—in arthropods developed just 20 million years after the first land plants had evolved during the Silurian. Conrad's research that explores the co-association of plants and the insects that feed on them highlights how important it is to consider the wider implications of developments in the Earth's system. Herbivory, for example, was an important development that became a major driving force in the processing of live plant tissue into organic carbon.

Conrad recovered remarkable fossils of Middle Devonian liverworts from New York State that show several types of herbivory by microarthropods. In these images, you can see various types of damage on liverworts caused by various feeding strategies including a possible anti-herbivore counter-defense mounted by the plant host in the form of terpenoid-laden oil cells that appear dark in these images.

These innovations in the plants, though, permitted tall trees to spread across the landscape and into the highland and more inland areas, finally breaking ties with standing water. This was the start of the greening of the Earth beyond the coastlines and the first forests.

As you can see the late Devonian was a time of change not only in the biosphere but also in the state of the oceans and atmosphere too. Oceanic anoxia was present in some areas as demonstrated by the presence of black shales in many late Devonian strata. There is also evidence of global cooling with the advance of glaciers and associated sea level changes. And it is possible that these changes would stress the biosphere over a period of around 20–25 million years, producing a series of about 8–10 extinction pulses.

There would be two particularly intense spikes of extinction at 372 million years ago called the Kellwasser event lasting about two million years, and another right at the end of the Devonian Period around 358 million years ago called the Hangenberg event, lasting about one million years.

The earlier Kellwasser event would mostly affect marine species, and in particular, those beautiful Devonian reef systems Many invertebrate groups that lived in and around those reefs would be severely impacted. For example, the number of trilobite families each of which represented numerous species would be reduced from nine to five.

Trilobites show an interesting morphological change through this interval too. Many evolutionary lineages record a reduction in the size of eyes up to this event with some losing their eyes entirely. It's been speculated that this might represent increased turbidity in the oceans. In addition, many exhibit an increase in the width of brims—the rim-like flange that surrounds the trilobite head shields, the cephalon. These are thought to play a role in respiration and perhaps their change may be a response to falling oxygen levels in the world's oceans.

The jawless agnathan fish would also suffer through this event with most of these ancient fish disappearing in the Devonian, as would many of the free swimming cephalopods.

Reef systems following this Kellwasser event would tend to be dominated by those sponge-like stromatoporoids and microbially constructed, laminated structures called stromatolites. Corals that had played such an important part in younger reef systems would be decimated. Overall, tropical warm-water forms were hit the hardest in this extinction event.

It is the final extinction pulse—the Hangenberg event—that marks the boundary between the Devonian and the Carboniferous periods, in which invertebrates and many of the surviving reefs would be hit yet again It would also affect both the marine and freshwater environments It is estimated that around 44% of the higher-level vertebrate groups are removed including all placoderms and many of the lobe-finned fish.

In total around 19% of families and about 50% of genera would go extinct, but the decimation was probably more severe in the oceans with perhaps around 22% of families biting the dust. Although you have to be careful about estimating how many species became extinct due to problems in sampling at this taxonomic level, it is possible that around 79–87% of all species in the ocean went into extinction.

This extinction is referred to as the Devonian Mass Extinction event. However, due to the protracted nature of the diversity decline occurring over 25 million years with a series of pulses, it is perhaps better to call this event the Devonian Biodiversity Crisis.

So what could cause all these changes at the end of the Devonian and potentially be identified as the smoking gun? The Devonian extinction is recognized as one of the Big Five mass extinctions that have occurred during the last half a billion years on Earth. All of these extinctions, with the exception of the first one at the end of the Ordovician, have been associated with large volcanic events that produced extensive flood basalts, the Devonian included.

A flood basalt is a vast outpouring of dark, basaltic lava that often forms these stepped landscapes like you can see here along the Columbia River in Washington State. Each of these steps is formed by a lava flow These volcanic events are often referred to as traps from the Swedish word *trappa* meaning "step."

For the Devonian, there is a large traps-volcanic area in Siberia called the Vilyuy Traps. Interestingly these strata are located close to another even larger Traps volcanic deposit that erupted much later at 251 million years ago that may have been the trigger for one of the largest mass extinction events that the planet has ever seen—the Permian mass extinction that we meet in a later lecture.

The actual cause of the Vilyuy volcanic episode is somewhat unclear, but it is possible that a large plume of material from the mantle was impacting the base of the crust of the Siberian continent. This would have caused melting of the lithosphere, rifting and extensive volcanic activity.

It is well known that such intense volcanic episodes can have varied effects on climate. These effects include global cooling and ozone destruction as a result of fine ash and sulfate aerosols suspended in the atmosphere, but also global warming caused by the release of carbon dioxide. Significantly, two major pulses of Vilyuy volcanic episode at 364.4 and 376.6 million years ago are close in time to the Kellwasser and Hangenberg events.

So if these volcanic episodes are extinction triggers based in fire, how about one based on ice? There are at least two glacial episodes about the same time as the Kellwasser and Hangenberg events. A cooling scenario for extinction is supported by a decline in the number of warm tropical adapted species, and a spread of cooler water adapted species toward the equator.

As glaciers expanded and retreated across the land, there would be corresponding changes in sea level too. Lowest sea level would occur when more water was locked on land during times of maximum development of ice.

These sea-level changes can be seen recorded in Devonian reef systems as an erosion surface that we call an unconformity, which demonstrates the exposure of reefs above water before sea levels rose again allowing marine sediments to be deposited once more.

This fluctuating availability of reef ecosystems accompanied by cooling water temperatures would obviously have had an effect on the abundant and biodiverse ecosystem that thrived around them. Additionally, if we enter a glacial phase with exposure and erosion of continental shelves, we may also increase both the nutrient content and cloudiness of the oceans around the continents. Remember those small trilobite eyes? These are not good conditions for corals which prefer warm, clear and nutrient poor waters.

Other culprits for the Late Devonian extinctions have been suggested, though, as it is known that there were at least two impact events in the later Devonian One impact the Alamo event occurred in southeastern Nevada. Although the precise structure of the impact is not clear, deformed and shattered rocks—called impact breccia—tell us that something big hit this area in the Devonian It is estimated to have deformed or destroyed over 1000 cubic kilometers of rock It probably impacted in the ocean as well and consequently would have generated a large tsunami.

A large enough impact could have serious and sudden consequences for the biosphere, but unfortunately for this hypothesis, the Alamo impact is dated occurring some 3.5 million years before the Kellwasser event

There is possibly another large impact structure though around 75 miles in diameter in West Australia the Woodleigh Impact. This impact structure is only known from subsurface material and remote gravity surveys. The precise timing of the impact although close to the end of the Devonian remains uncertain.

So perhaps we should look to the Vilyuy Traps as the trigger for extinction then You could imagine pulses of volcanic activity causing global cooling and the advance of ice from glaciers, followed by a build-up of carbon dioxide from the eruptions, driving the climate in the opposite direction, melting the glacial ice, and causing sea levels to rise again. But is there another finger on the trigger of this environmental gun? Could the cause be plants?

The idea that something we equate today with a healthy biosphere, namely plants, could cause a mass extinction is kind of counterintuitive isn't it? This, though, is just the scenario suggested by researchers such as Thomas Algeo at the University of Cincinnati.

Consider these features of the Devonian extinction. Both the Kellwasser and Hangenberg events are associated with the development of black shales in shallow marine settings and are associated with an increase in super 34 isotope; these anomalies as regarded as good indicators of anoxia in ocean water There is also an associated drawdown in atmospheric carbon dioxide, and an increase in sediment delivery to the oceans. He suggests that these effects were caused by plants colonizing new habitats. Let me explain how.

As we move through the Devonian, the depth and complexity of root systems increases. This would cause a short-term increase in global weathering as plants colonizing new areas started to break up rocks with their root systems and increased the production of sediment This sediment delivered to the oceans by rivers and streams would muddy the water making it more cloudy. Perhaps this is a reason for the reduction in trilobite eye size that we noted earlier.

Secondly, this weathering also produces lots of calcium and magnesium carbonates using a lot of CO_2 in the process. This effectively pumps carbon dioxide out of the atmosphere and into the soil. In addition, increased burial of organic plant material helps lock organic carbon away from oxygen and prevents carbon dioxide from being produced by the oxidation of organics.

Through these mechanisms it is possible that levels of atmospheric carbon dioxide would be reduced decreasing the greenhouse effect and initiating a period of global cooling, perhaps allowing the advance of glaciers and sea level to fall.

Thirdly, weathering would also increase rates of nutrient flux and organic delivery to the oceans causing vast algal blooms in the upper well-lit oxygenated levels of the ocean The organic material produced in such blooms would sink, rapidly use up available oxygen in bottom waters and help generate black anoxic shales and those associated positive sulfur anomalies

Evidence has been recovered that suggests that oxygen-poor hydrogen sulfide-charged bottom waters would spread through the shallow Devonian oceans, poisoning creatures living on the ocean floor It is possible that hydrogen sulfide-charged water could have risen to the surface causing problems for other creatures in the water column, In addition, release of hydrogen sulfide into the atmosphere would cause associated problems for animals on land, not the least of which is assisting destruction of the ozone layer.

And finally, the Hangenberg event at the end of the Devonian Algeo suggested that there would be a further proliferation of plants following the development of seeds a major innovation in plant reproduction This resulted in a global flora spreading to highland areas with an associated peat accumulation and further drawdown of carbon dioxide locking it away into the geosphere in the form of coal as you can see here in the vast Devonian coal reserves in China. Advances and retreats in ice possibly controlled by Earth's tilt, wobble, and other orbital factors continued to stress the biosphere.

This new diversification of seed plants would likewise be accompanied with increased weathering, nutrient delivery to the ocean, algal blooms and all the problems associated with hydrogen sulfide that we just described. In this scenario, the Late Devonian kill zone was a combination of effects including glacially driven cooling and sea level fall, and sick oceans full of poisonous hydrogen sulfide, all potentially triggered by an unlikely source—our friends the plants.

The Devonian witnessed a resetting of the biosphere, and the whole character of reefs would change with a significant loss of old Paleozoic reef-building organisms.

There would also be an important change for the vertebrates The Devonian is often called the Age of Fishes, and it's true that the Devonian does support a wide variety of fish, many of which are extinct today The Devonian extinction though would bring our more modern-looking fish to the fore.

Following the Devonian, we see the rise and dominance of the sharks, the rays, and the bony fish, a crown they have held in the Earth's oceans ever since.

The tetrapods that had evolved from lobe-finned fish like Tiktaalik would also be put through the sieve of extinction. For example, prior to the final extinction at the end of the Devonian many tetrapods had variable numbers of digits. Ichthyostega, for example, had seven digits on each limb.

Following the extinction, the rules that govern tetrapod body plans had stabilized with five digits being the ground-plan and typical of all tetrapods including us today.

So we may have an answer for what caused Romer's gap a global catastrophe A catastrophe that the world was still recovering from in the first 15 million years of the Carboniferous Period. But this catastrophe was also a pivotal event as are all extinction events clearing ecological space and providing opportunities for the evolution of the new life that was to follow.

Lecture 14

Life's Greatest Crisis: The Permian

Although most volcanic and seismic activity occurs at plate boundaries, not all of it does. Mantle plumes, which are thought to develop at the core-mantle boundary, might cause volcanic activity away from plate margins. It is possible that the development of a plume below Siberia triggered a cascade of events that would bring all of life to its knees in the greatest extinction this planet has ever seen. In this lecture, you will learn about the Permian extinction.

Before Catastrophe

- The Permian period was an interval that experienced a gradual warming as the continents came together to form Pangaea, the supercontinent that Alfred Wegener had proposed in the early 1900s as part of the hypothesis of continental drift.

- It was a world with oceans containing beautiful coral reefs, above which would swim numerous creatures. On land, you would see plants colonizing lowland areas.

- In the later Permian, terrestrial vertebrates would diversify into a variety of forms, including archosaurs (ancestors of the dinosaurs), herbivorous reptiles, large predators, and cynodonts (who would evolve into the Triassic and were ancestral to the first true mammals).

- And all of this beautiful world collapses—fundamentally, almost completely. By the end of the Permian, it is possible that up to 95% of species had been driven into extinction in life's greatest crisis in the last 540 million years.

- Paleontologist Dr. Douglas Erwin in the Department of Paleobiology at the Smithsonian's National Museum of Natural History has long wrestled with this most catastrophic event in life's history. The end of the Permian saw the sweeping away of many of the main players of what you could call an old Paleozoic world, a world full of trilobites, corals, and brachiopods. As Dr. Erwin has suggested, the Permian extinction has fundamentally shaped the biosphere we live in today.

Plumes as the Trigger

- Actually, we now think that 2 plumes could potentially be involved in this event, causing 2 extinction events. The first occurred around 260 million years ago, about 8 million years before the second, more catastrophic event at the end of the Permian.

- This so-called Guadalupian extinction occurred in both marine and terrestrial environments. Marine extinctions were varied among locations and the types of taxa impacted. Brachiopods and corals record severe losses, as did important groups of microfossils.

- Recent research in the Karoo region of South Africa led by Michael Day of the University of the Witwatersrand in Johannesburg, and including Dr. Erwin, has shown that around 74 to 80% of terrestrial species became extinct.

- The plume event that is thought to be related to this extinction occurs in China and produced a volcanic outpouring called the Emeishan Traps. This volcanic event is dwarfed, though, by the next plume to ascend from the depths. This would develop under Siberia at the end of the Permian, generating volcanism known as the Siberian Traps.

- Due to its scale, the Permian extinction receives a lot of research attention. As a result, our understanding of this event changes rapidly. One of the recent updates regards the timing of Siberian Traps volcanic activity. It is still known to be coincident with the extinction event, but it would appear that the duration of the event was shorter and sharper than previously thought.

- It is not the vast amount of lava that would be the most significant problem during these events. Creatures could easily migrate out of the affected area. Rather, the greatest problem are the gases that it produced.

- Research at the Carnegie Institution of Washington by Benjamin Black and Linda Elkins-Tanton investigated inclusions of gas trapped in the lava from the Siberian basalts. In effect, these so-called melt inclusions provide a sample of the gases that were actually being produced during the event. The Carnegie research team had a particular interest in the amount of sulfur, chlorine, and fluorine gases that were being produced.

- From this, they estimated that between 6300 and 7800 gigatons of sulfur, between 3400 and 8700 gigatons of chlorine, and between 7100 and 13700 gigatons of fluorine were released to the atmosphere. If these gases reached the upper atmosphere, they could cause significant environmental impact. Add to this the vast amounts of carbon dioxide also produced by the traps volcanism and you have a deadly cocktail.

- But perhaps an even more significant product of the traps were the vast amounts of the greenhouse gas carbon dioxide that were produced.

- A rise in carbon dioxide would have inevitably led to a rise in global temperatures due to greenhouse warming. It is thought that this could have helped destabilize certain gas-rich deposits in the oceans. These deposits form in offshore continental margins, where it is cold even in tropical areas. The cold temperatures and high pressures cause ice to complex with microbially generated methane, forming what is called a methane clathrate or gas clathrate.

- Clathrates are very sensitive to temperature changes. It's proposed that the traps-induced warming raised ocean temperatures to the point where the clathrates started to destabilize and release their methane, an even more effective greenhouse gas than carbon dioxide. And although methane oxidizes quite readily in the atmosphere, what it produces, when it does oxidize, is more carbon dioxide.

- Today, there is enough methane in clathrates to total twice the amount of carbon to be found in all known fossil fuels on Earth. There were probably plenty of clathrates in the Permian, too. So, not only do we have an initial global warming from the traps-produced carbon dioxide, but gas clathrates may have also contributed to the situation, perhaps raising global temperatures by the end of the Permian by as much as 10° Celsius.

- At the moment, the most likely triggers of the end-Permian crisis are the Siberian Traps and the consequences they would have on global climate. But what were the killing mechanisms?

- On land, we have a signal that suggests the collapse of the flora—a signal that is recorded in the style of the river systems transitioning from meandering "plant stabilized" ones to the more disorganized braided systems, suggesting a sudden decimation of plants on land. This collapse of land flora may also explain the lack of coal in the Early Triassic.

- It has been suggested by some that extinction in the oceans lagged somewhat behind the initial extinction on land. The reason for this is that water bodies act as temperature buffers and respond more slowly. More recent evidence, though, suggests that extinction was more or less simultaneous in both realms. Whenever it hit, the catastrophe was probably even greater in the marine realm, with decreasing oxygen levels and increasing carbon dioxide levels.

An Acid Nightmare

- As carbon dioxide dissolved in the oceans, they started to become more acidic. This development of acidic oceans matches the patterns of organisms that were hit hardest during the extinction. Creatures that secrete calcium carbonate shells—brachiopods, echinoderms, and corals—were severely affected. This is especially significant when considering that corals, and the organisms that lived on, in, and around them, constituted an entire complex reef ecosystem that soon would also collapse.

- A team led by Dr. Matthew Clarkson at the University of Edinburgh in Scotland has been studying this phenomenon by analyzing carbonate rocks in the United Arab Emirates (UAE), which, at the time of the extinction, laid in a huge embayment encircled by the eastern edge of Pangaea called the Tethys Ocean.

- The team studied changes in the ratio of boron isotopes that are known to vary with pH. Their analysis suggests a decrease of around 0.6 to 0.7 pH units during the extinction. This represents a catastrophic change for ocean chemistry.

- The boron acidity signal occurs about 50,000 years after the initiation of the extinction event. They propose that the Permian-Triassic extinction was a 2-phase event. The first pulse, over 50,000 years, was fairly slow, as carbon dioxide was added incrementally to the Earth's oceans and atmosphere by the Siberian Traps.

- It is possible that the oceans, which are pretty alkaline, were able to buffer the carbon dioxide being dissolved in the seawater until a critical point was reached—when a second pulse of carbon dioxide was released over a period of about 10,000 years.

- They suggest that this pulse released about 24,000 gigatons of carbon dioxide at a rate of about 2.4 gigatons per year. Such a massive flush of carbon dioxide into the Earth system would have overcome the buffering potential of the oceans, seawater would be driven into an acidic range, and the reef ecosystems crashed.

- Acid rain on land would likely have caused problems for the terrestrial ecosystem. Evidence for this has not been forthcoming, although most agree that it is likely to have occurred. However, a team of scientists, including Mark Sephton of Imperial College London, have found some intriguing evidence that might act as a proxy for the acidification on soils on land.

- By investigating a section of soil in the Southern Alps of northern Italy, the team found a pattern that pointed toward a series of pulses of acidity on land and suggested that these pulses were related to individual episodes of volcanic activity from the Siberian Traps and associated acid rain events. This is still pretty new research, but this study could be some of the first concrete evidence that we have an acid planet both in the oceans and on land during the Permian-Triassic transition.

The Role of Microbes

- The microbial world is often overlooked, but microbes are a fundamental part of the biosphere. Their presence can be both essential and catastrophic to the health of the biosphere. Given that they can reproduce and spread very rapidly, they can respond very rapidly to environmental changes and opportunities, as may be the case for microbial life at the end of the Permian.

- In a recent paper, a team including Massachusetts Institute of Technology researchers Daniel Rothman and Gregory Fournier suggested a potential source of methane in addition to clathrate methane that may have come from a microbe called *Methanosarcina* that belongs to an ancient group of microbes belonging to the Archaea.

- Their research into the genome of this microbe points to a change that occurred at about the time of the Permian-Triassic extinction. This change would have permitted *Methanosarcina* to produce significant quantities of methane and, as a consequence, more global warming.

- In this scenario, the Siberian Traps still act as a trigger—or, more accurately, a fertilizer. One of the elements that the traps would release into the environment is nickel, which is a vital "fertilizing" element for these particular microbes and would have permitted their rapid blooming and the production of large quantities of methane.

- There is another type of microbe that would likely have had serious impacts on the biosphere, too. The Permian oceans were becoming oxygen depleted, or anoxic. Certain microbes have a preference for living in such anoxic conditions and gain energy from stripping oxygen off sulfate molecules. In doing so, these sulfate-reducing bacteria produce a dangerous by-product: hydrogen sulfide.

- Hydrogen sulfide is highly toxic to the metabolism of most organisms and would have had a serious impact on oceanic ecosystems. This is not mere conjecture, though, as biomarkers of sulfate-reducing bacteria have been found in Chinese sections that cover the Permian-Triassic boundary.

- Significantly, these sections were originally deposited under shallow-water conditions. The placement of these deposits indicate that oxygen-poor, hydrogen-sulfide–rich water encroached and extended from deeper ocean water well into the shallow well-lit areas of the oceans—just where the Permian coral reefs grew.

- It would appear that during the Late Permian, in addition to increasing temperatures, increasing oceanic acidity, and reduced oxygen levels, the Permian reef systems were being poisoned, too.

The World That Remained

- What was left was a decimated world, baking in a hot Sun with slow, sluggish, poisoned oceans. The Permian extinction provides an insight into a biosphere that ceased to function. There wasn't any single cause why the extinction occurred, even though there might have been an initial trigger in the plume that rose from the core and impacted the base of the lithosphere in Siberia.

- In dealing with a complex interacting system like the Earth, a catastrophic event like the Permian-Triassic extinction has to be viewed as a cascade of events that rapidly spread and expanded until it impacted almost every corner and level of the Permian biosphere.

- It often takes hundreds of thousands of years for the biosphere to start to recover from such events, but things would be different for the Permian-

Triassic extinction, as 5 million years into the Triassic, the Earth was still a pretty devastated planet.

Questions to consider:

1. Are the triggers of the Permian extinction absent in the modern world?
2. In which ways do extinctions progress—in a linear manner like a line of dominoes or as a cascade of events?

Suggested Reading:

Erwin, *Extinction*.

Plummer, Carlson, and Hammersley, *Physical Geology*, chaps. 4 and 17.

Lecture 14 Transcript

Life's Greatest Crisis: The Permian

The advent of Plate Tectonics was like a veil being removed from in front of our eyes. It explained so much about the way the world was like it was. For example, the interaction of plates specifically at their boundaries accounted for the distribution of earthquakes, the location of mountain ranges along the edges and in the interior of continents and for the distribution and types of volcanoes across the globe. In short, it deserved its title The Grand Unifying Theory of Geology—it explained everything!

Well actually not quite everything. There were some troublesome points. For example, although most volcanic and seismic activity occurred at plate boundaries not all of it did. This was a wrinkle in the theory of everything geological. A wrinkle that can be summed up in a particularly beautiful series of islands in the Pacific Hawai'i.

Hawai'i didn't fit the existing model. It occurs in the middle of the Pacific plate, not at a plate boundary and as we know Hawai'i is very volcanically active. How do we account for Hawai'i? This was a question that geophysicist John Tuzo Wilson would try to answer.

Professor Wilson would propose a workaround to explain the anomalies like Hawai'i. He suggested that the Islands were created as the northwesterly motion of the Pacific Plate moved over a fixed hotspot in the Earth's mantle.

As the plate moved over this hot spot a new volcano was generated, this neatly explaining how the islands to the northwest of the big island of Hawai'i get progressively older, smaller and eroded as they are removed from the source of new material over the hot spot.

Through the pioneering work of Wilson and other researchers like Geophysicist W. J. Morgan we now have a working hypothesis of how these hot spots form; a hypothesis that involves deep Earth structures called mantle plumes.

Mantle plumes are thought to develop at the core–mantle boundary at around 1800 miles deep. This was identified as the place where plumes might develop due to the dramatic change in temperature that occurs there. Temperatures in the core are around 2000°C higher than the mantle. It is thought that there is little if any transfer of material between the core and the mantle so heat across the interface must be transferred by conduction.

Plumes are postulated to form in a layer called the D layer at the bottom of the mantle as the rocks there heat up and become buoyant It is important to note that we are not generating magma the pressures are probably too high to allow significant melting to occur instead the rocks of the mantle move slowly over millions of years like slow, flowing taffy.

According to the plume hypothesis, this produces a cylindrical column of hot mantle rocks moving away from the core and towards the Earth's crust. As the plume moves up through the mantle, it starts to develop a bulbous head. This head is formed as material moves faster through the conduit than the head of the plume moves through the mantle. These differences in speed cause the characteristic mushroom structure seen in plume modeling experiments.

Eventually, the plume impacts the base of the crust, flattens out, and spreads out over 1000's of kilometers. At these higher levels, pressures are much less than at the core–mantle boundary. This allows these hot rocks to undergo decompression melting and start to generate what is called a partial melt. In a partial melt, only the minerals with the lower melting points will actually melt, producing a compositionally distinctive magma. When you partially melt mantle rocks called peridotite the magma you produce is basaltic in composition.

Under this hypothesis, the Hawai'ian islands are the surface representation of a mantle plume that has been and still is impacting the base of the Pacific Plate. The partial melting of the peridotite generates the basaltic magma that erupts as basaltic lava, forming the Hawai'ian chain. Interesting stuff right? Yes, but also sobering as it is possible that the development of just such a plume below Siberia would trigger a cascade of events that would bring all of life to its knees in the greatest extinction this planet has ever seen.

So in this episode let's ask: What was the world like before the hammer dropped? How did plumes trigger catastrophe? Was the Permian Earth turning into an acid nightmare? What role did microbes play in catastrophe? and consider the world that remained as a new era dawned.

The Permian Period was an interval that experienced a gradual warming as the continents came together to form Pangaea, the supercontinent that Alfred Wegener had proposed in the early 1900s as part of this hypothesis of continental drift.

It was a world with oceans containing beautiful coral reefs above which would swim numerous creatures like these nautiloids, relatives of squids and the modern nautilus.

Other aquatic organisms included numerous groups of fish and sharks like the bizarre helicoprion with its buzz-saw teeth that has no modern dental analog.

On land, you would see plants colonizing lowland areas. In northern Pangaea, there was a diverse flora of horsetails, ferns, cycads, a variety of major seed-fern lineages and conifers, some of which had a modern cast to them by now.

By contrast in southern Pangaea, the flora of the north was mostly absent. Instead, the flora was dominated by glossopteris seed ferns, such as glossopteris with their mostly tongue-shaped leaves.

In the later Permian terrestrial vertebrates would diversify into a variety of forms including archosaurs, ancestors of the dinosaurs; herbivorous reptiles like dicynodonts in a variety of forms; large Predators like the gorgonopsians; and cynodonts who would evolve into the Triassic and were ancestral to the first true mammals.

And all of this beautiful world collapses fundamentally almost totally! This is not just a pruning of the tree of life. The tree of life would be reduced to a few wandering branches, a poor shadow of its former self. By the end of the Permian, it is possible that up to 95% of species had been driven to extinction in life's greatest crisis in the last 540 million years.

Paleontologist Dr. Doug Erwin in the Department of Paleobiology at the Smithsonian's National Museum of Natural History has long wrestled with this most catastrophic event in life's history.

Doug visited my current hometown Vancouver for a conference in 2014 that was hosted by the Geological Society of America. He was one of many fascinating presenters at a session entitled Mass Extinctions: Volcanism, impacts and Catastrophic Environmental Changes. One of the many insights he provided, this one stuck in my mind as it really brings home to me the extent of the event.

The end-Permian saw the sweeping away of many of the main players of what you could call an old Paleozoic world a world full of trilobites, corals, and brachiopods. As Doug put it, "The composition of a tidal pool today is a reflection of who lost and who won during the event 252.4 million years ago. In short, the Permian extinction has fundamentally shaped the biosphere we live in today.

Actually, we now think that two plumes could potentially be involved in this event causing two and not one extinction event. The first occurred around 260 million years ago, about 8 million years before the second more catastrophic event at the end of the Permian. This first, so-called Guadalupian extinction occurred in both marine and terrestrial environments. Marine extinctions were varied among locations and the types of taxa impacted. Brachiopods and corals record severe losses as did important groups of microfossils like fusulinid foraminifera.

Recent research in the Karoo region of South Africa led by Michael Day of the University of the Witwatersrand in Johannesburg and including Doug Erwin of the Department of Paleobiology has shown that around 74–80% of terrestrial species become extinct This includes the impressive dinocephalia, a group of large, heavily built, therapsid mammal-like reptiles the most famous of which is probably the wonderfully named Moschops, a herbivore around 9 feet long.

The plume event, thought to be related to this extinction, occurs in China and produced a volcanic outpouring called the Emeishan Traps. This volcanic event is dwarfed, though, by the next plume to ascend from the depths. This would develop under Siberia at the end of the Permian, generating volcanism known as the Siberian Traps.

The Siberian Traps volcanic deposits cover two million square kilometers today, an area comparable to that covered by Western Europe. The original coverage at the end of the Permian was about 7 million square kilometers.

Due to its scale, the Permian extinction rightfully receives a lot of research attention. As a result, our understanding of this event changes rapidly. One of the recent updates regards the timing of Siberian Traps volcanic activity.

It is still known to be coincident with the extinction event, but it would appear that the duration of the event was shorter and sharper than previously thought.

Research at the Massachusetts Institute of Technology by Sam Bowring and Richard Shrock suggests that the traps activity was around 10 times faster than previously thought taking place in just 60,000 years. That is an instant in geological time.

As we have mentioned in earlier lectures, it is not the vast amount of lava that would be the most significant problem during these events. Creatures could easily migrate out of the affected area. Rather the greatest problem is the gases that it produced. Research at the Carnegie Institute of Washington by Benjamin Black and Elkins Tanton investigated inclusions of gas trapped in the lava from the Siberian basalts. In effect, these melt inclusions provide a sample of the gases that were actually being produced during the event. The Carnegie research team had a particular interest in the amount of sulfur, chlorine and fluorine gases that were being produced.

From this, they estimated that between 6300–7800 gigatons of sulfur, 3400–8700 gigatons of chlorine and 7100–13700 gigatons of fluorine were released to the atmosphere. A gigaton is equal to one billion tons.

If these gases reached the upper atmosphere, they could cause significant environmental impact. Add to this the vast amounts of carbon dioxide also produced and the traps volcanism have produced a deadly cocktail.

Both sulfur dioxide and carbon dioxide can be dissolved in rainwater forming highly acidic rain that could impact plants and soils on land. Some think that the acid rain fell from the skies with a pH of around about 2, close to the acidity of undiluted lemon juice. This could have been falling in the Northern Hemisphere at this time. The pH of rain would likely return to normal levels after each volcanic episode, but the regularity of these events would have meant a regular occurrence of acidic lemon juice from the skies when it occurred.

The pulsing activity of the volcanic traps would also inject chlorine and fluorine into the atmosphere. These halogen gases are known to have serious impact on the ozone layer. Ben Black at MIT has estimated that ozone levels could have decreased by a conservative 5% to an extreme 65% particular at the poles. Such ozone depletion would expose animals on land and in shallow levels in the oceans to dangerous levels of ultra-violet radiation. But perhaps an even more significant product of the traps were the vast amounts of the greenhouse gas, carbon dioxide that was produced.

The problem was compounded by the rocks that the traps were erupted through; rocks rich in calcium carbonate—limestones! When you heat limestone, you guessed it you produce lots of carbon dioxide. There were other rocks causing problems too as there were also significant deposits of coal in the same area. As the hot magma and lava combusted the coal, it generated even more carbon dioxide and additional toxic gases. Even if not combusted the heating of coal could also produce significant quantities of methane, another important greenhouse gas.

A rise in carbon dioxide would have inevitably led to a rise in global temperatures due to greenhouse effect. It is thought that this could have helped to destabilize certain gas-rich deposits in the oceans. These deposits form in offshore continental margins where it is cold even in tropical areas. The cold temperatures and high pressures cause ice to complex with microbially generated methane, forming what is called a methane- or gas clathrate.

Clathrates are very sensitive to temperature changes. It's proposed that the traps-induced warming raised ocean temperatures to the point where the clathrates started to destabilize and release their methane. Methane is an even more effective greenhouse gas than carbon dioxide, and although methane oxidizes quite readily in the atmosphere what it produces, when it does is even more carbon dioxide.

Today there is enough methane in clathrates to total twice the amount of carbon to be found in all known fossil fuels on Earth. There were probably plenty of clathrates in the Permian too. So not only do we have an initial global warming from the traps-produced carbon dioxide, but gas clathrates may have also contributed to the situation perhaps rising global temperatures by the end of the Permian by as much as 10°C. You can see methane bubbling out of muddy sediments on the sea floor today. Can you imagine vast clouds of methane that might have been released from the Permian oceans?

In Doug Erwin's excellent book, *Extinction*, he calls the chapter dealing with causes of the extinction "A Cacophony of Causes," a very appropriate title. And apart from the Siberian Traps Doug lists a whole bunch of tipping mechanisms that have been proposed as possible causes of the end-Permian crisis. Everything from drifting continents, rises and falls of sea levels, cosmic radiation, impacts from space and my personal favorite interaction of our planet with dark matter.

It would appear, though, that at the moment the most likely triggers are the Siberian Traps and the consequences they would have on global climate. But what were the killing mechanisms?

On land? We have a signal that suggests the collapse of the flora a signal that is recorded in the style of the river systems transitioning from meandering plant stabilized to the more disorganized braided systems suggesting a sudden decimation of plants on land. This collapse of land flora may also explain the lack of coal in the early Triassic the so-called Coal Gap

And why do we see this collapse? Here are some ideas. First, the Siberian Traps were likely a dirty eruption. The coal we mentioned ignited by magma would burn and produce a lot of fine particulate material which could have blocked sunlight. In addition, sulfate aerosols produced by the eruption may have also contributed to global dimming. As the skies started to dim behind the haze photosynthesis started to diminish.

Even after the ash was washed from the skies the carbon dioxide and sulfate aerosols would likely fall as acid rain and poison the soils. Probably more significantly though would be the carbon dioxide that would initiate global warming and drive the already dry Permian world into drought. Plants and the rest of the terrestrial life that it supported collapsed catastrophically.

But what of the oceans? It's been suggested by some that extinction in the oceans lagged somewhat behind the initial extinction on land The reason for this is that water bodies act as temperature buffers and respond more slowly. More recent evidence, though, suggests that extinction was more or less simultaneous in both realms. Whenever it hit, the catastrophe was probably even greater in the marine realm What happened here?

First, the oceans slowly became sluggish as a warm layer capped the oceans and reduced vertical mixing and circulation. This means, the oceans' ability to circulate oxygen from the atmosphere to the ocean was seriously impacted.

Microplankton would still have thrived in the warm oxygen-rich surface waters. When they died and sank, though, they decayed and used up those dwindling oxygen supplies in the deeper water, driving the oceans into a state we call anoxia.

It has been suggested that an even more significant effect than decreasing oxygen levels though was an increase in carbon dioxide. Carbon dioxide is more soluble in water than oxygen and thanks to the Siberian Traps there was a lot of extra carbon dioxide in the atmosphere that was starting to dissolve in the Permian oceans. Elevated carbon dioxide in the oceans may have led to elevated levels of carbon dioxide in the bloodstreams of marine-dwelling organisms, a condition known as hypercapnia.

Terrestrial organisms are actually better adapted to cope with increased levels of carbon dioxide in their blood when compared to marine organisms. For us landlubbers, carbon dioxide is removed relatively slowly from our bodies. We have to use complex organs like our lungs to facilitate gas exchange. As a result, carbon dioxide tends to hang around in our bodies longer than for marine creatures. This is why we have evolved a higher tolerance for elevated levels of CO_2.

In aquatic species, carbon dioxide is more easily flushed from the body—gas exchange is a much simpler process in an aquatic environment. This means that even moderate levels of hypercapnia in poorly carbon dioxide adapted marine species could cause various metabolic problems such as reduced protein production and decreased fertility rates.

But there would be another consequence of increased levels of carbon dioxide, for as carbon dioxide dissolved in the oceans, they started to become more acidic. This development of acidic oceans matches the patterns of organisms that were hit hardest during the extinction. Creatures that secrete calcium carbonate shells brachiopods, echinoderms and corals were severely affected. This is especially significant when considering that coral and the organisms that lived on, in, and around them constituted an entire complex reef ecosystem that would collapse as well.

A team led by Dr. Mathew Clarkson at the University of Edinburgh, Scotland has been studying this phenomenon by analyzing carbonate rocks in the United Arab Emirates. At the time of the extinction, the UAE lay in a huge embayment encircled by the eastern edge of Pangaea called the Tethys Ocean. They studied changes in the ratio of boron isotopes that are known to vary with pH. Their analysis suggests a decrease of around 0.6–0.7 pH units during the extinction. This might not sound like much, but that represents a catastrophic change for ocean chemistry.

The boron acidity signal occurs about 50,000 years after the initiation of the extinction event. They propose that the Permian-Triassic extinction was a two-phase event. The first pulse over 50,000 years was fairly slow as carbon dioxide was added incrementally to the Earth's oceans and atmosphere by the Siberian Traps. It is possible that the oceans, which are pretty alkaline, were able to buffer the carbon dioxide being dissolved in the sea water until a critical point was reached when a second pulse of CO_2 was released over a period of about 10,000 years.

They suggest that this pulse released about 24,000 gigatons of carbon dioxide at a rate of about 2.4 gigatons per year. Such a massive flush of carbon dioxide into the Earth system would have overcome the buffering potential of the oceans, seawater would be driven into an acidic range, and the reef ecosystems crashed.

We have already described how acid rain on land would likely have caused problems for the terrestrial ecosystem. Evidence for this though has not been all that forthcoming although most agree that it is likely to have occurred. However, a team of scientists including Mark Sephton of Imperial College in London have found some intriguing evidence that might act as a proxy for the acidification on soils on land.

To explain what that proxy is we need to look at what acid rain does to soils. Acid rain increases the solubility of plant toxic metals ions like aluminum. It will also leach calcium, magnesium, and other nutrients from the soil profile. In this way, acidity in soil causes problems for root systems in their development and in their ability to take up nutrients causing the plants above ground to wither and die.

Acid rain percolating into the soil profile would also seriously impact vital microbial and fungal diversity that occurs in soils. Microbes in particular show very low diversity in acidic soil conditions and fungi don't thrive too well in pH's lower than 4.5.

Fungi and bacteria in healthy soils help in the breakdown of complex plant molecules like cellulose and lignins, wood and bark, and animal chitin, exoskeletons, and so on, as well as fungal molecules into simpler organic molecules that can be used by other life forms. This breakdown is facilitated by enzymes that fungi, bacteria, and other microorganisms secrete into the soil. With elevated levels of acidity, these enzymes are severely inhibited or destroyed.

An important step in this enzymatic decomposition of lignin is the production of a substance called vanillin. Vanillin undergoes further enzymatic alteration and is oxidized to vanillic acid. This part of the process, though, is very pH sensitive with optimal chemical change at pH greater than 5. As such, the relative proportions of vanillin to vanillic acid may give us an idea of the degree of soil acidity.

Finding old soils or paleosols is difficult though especially as it is likely that much plant material in soils was eroded and washed away from the fossil record. Fortunately, this soil washout can be picked up in marine sediments by the presence of things like filaments of soil fungi and also of a common organic component of soil profiles called Polycyclic Aromatic Hydrocarbons.

The team investigated a section in the Southern Alps of Northern Italy that contained evidence of these soil wash-out markers and analyzed the ratio of vanillic acid to vanillin. What they found was a pattern that pointed towards a series of pulses of acidity on land. The team suggested that these pulses were related to individual episodes of volcanic activity from the Siberian Traps and associated acid rain events.

This is still pretty new research, and there are a lot of wrinkles that need to be sorted out, but this study could be some of the first concrete evidence that we have an acid planet both in the oceans and on land during the Permian-Triassic transition.

It is possible that microbes may have played an important role in the Permian extinction, too. The microbial world is often overlooked, though, it's not showy. It doesn't work well in simulations in TV. A reconstruction of an ancient microbe floating in the ocean really doesn't really have the same appeal as seeing a Tyrannosaurus Rex hunting down a Triceratops.

Nevertheless, microbes are a fundamental part of the biosphere. Their presence can be both essential and catastrophic to the health of the biosphere. Given that they can reproduce and spread very rapidly, they can respond very rapidly to environmental changes and opportunities as may be the case for microbial life at the end of the Permian.

In a recent paper, a team including MIT researchers Daniel Rothman and Gregory Fournier suggested a potential source of methane in addition to the clathrate methane we described earlier. This methane they suggest may have come from a microbe called methanosarcina that belongs to an ancient group of microbes belonging to the Archaea.

Their research into the genome of this microbe points to a change that occurred at about the time of the Permian–Triassic extinction. This change would have permitted methanosarcina to produce significant quantities of methane and as a consequence more global warming.

In this scenario, the Siberian Traps still act as a trigger or more accurately a fertilizer. One of the elements that the traps would release into the environment is Nickel. Nickel is a vital fertilizing element for these particular microbes and would have permitted their rapid blooming and the production of large quantities of methane.

There is another type of microbe that would likely have had serious impacts on the biosphere as well. As we have already discussed the Permian oceans were becoming oxygen depleted, anoxic. Certain microbes have a preference for living in such anoxic conditions and gain energy from stripping oxygen off sulfate molecules. In doing so, these sulfate-reducing bacteria produce a dangerous byproduct hydrogen sulfide.

Hydrogen sulfide is highly toxic to the metabolism of most organisms and would have had a serious impact on oceanic ecosystems. This is not mere conjecture, though, as biomarkers of sulfate-reducing bacteria have been found in Chinese sections that cover the Permian-Triassic boundary.

Significantly these sections were originally deposited under shallow-water conditions. The placement of these deposits indicates that oxygen-poor hydrogen sulfide-rich water encroached and extended from deeper ocean water well into the shallow well-lit areas of the oceans just where that Permian coral reefs grew.

It would appear that during the late Permian in addition to increasing temperatures, increasing oceanic acidity, and reduced oxygen levels the Permian reef systems were being poisoned too.

What was left was a decimated world baking in a hot sun with slow, sluggish, poisoned oceans. The Permian extinction provides an insight into a biosphere which ceased to function. There wasn't any single cause why the extinction occurred even though there might have been an initial trigger in the plume that rose from the core and impacted the base of the lithosphere in Siberia.

It is becoming increasingly obvious that they only way to understand our world is by considering it as a system composed of integrated interacting parts. Although we like to partition the world up into convenient packets geology, climatology, oceanography and even economy, in trying to understand any of these human-defined divisions of reality, we cannot truly understand them if we view them as individual stand-alone disciplines.

A simplistic discipline-driven view of the world will obviously be limited and miss the bigger picture. Even more significantly a failure to appreciate that the impacts on one system will generate impacts on other systems is also very dangerous. The Permian extinction was not a simple cause and effect paradigm like dominoes falling in a row. In dealing with a complex interacting system like the Earth, a catastrophic event like the Permian-Triassic extinction has to be viewed as a cascade of events that rapidly spread and expanded until it impacted almost every corner and level of the Permian biosphere.

It often takes hundreds of thousands of years for the biosphere to start to recover from such events, but things would be different for the Permian-Triassic extinction as 5 million years into the Triassic the Earth was still a pretty devastated planet. We will investigate why in the next episode.

Lecture 15 Life's Slow Recovery after the Permian

In this lecture, you will examine the world of the Early Triassic just after the end-Permian extinction and attempt to track the biosphere's recovery. This lecture will address several questions: What was the world like in the Early Triassic? What was left of life following the greatest of all extinctions? What was driving the impoverished Early Triassic, and why did it last so long? When did the Earth start to recover? Was there more doom at the end of the Triassic?

The Early Triassic

- During the Early Triassic, our planet would have looked very different. A giant landmass, Pangaea, dominated one side of a planet surrounded by the vast Panthalassic Ocean. The Early Triassic climate was harsh on Pangaea. Hot, arid deserts covered most of the interior of the supercontinent.

- It is possible that this was the hottest, most arid time in more than half a billion years. In fact, there is likely no ice at the poles. Indeed, it is possible that the poles might have been relatively temperate places, where forests and a more diverse biota of plants, fungi, and animals could survive.

- This is the inverse of what we have today, because in the Early Triassic, the temperate polar regions were more habitable than the equatorial regions.

- Analysis of oxygen isotopes taken from conodonts, the toothlike mouthparts of an early chordate, paint a disturbing picture of the temperature change during the Early Triassic.

- Research in equatorial deposits from southern China, spanning the period of the extinction at the end of the Permian and continuing into the Early Triassic, show a rapid warming in the oceans from 21 to 36° Celsius. The warming peaks at around 252.1 million years ago, the end of the Permian.

- There is a cooling following this, and then a second rise in temperatures occurring around 250.7 million years ago, a period of the Early Triassic called the Early Olenekian when temperatures in the water column rose again to about 38° Celsius and perhaps even exceeded 40° Celsius at the surface. At such marine surface temperatures, organisms such as corals cannot survive.

What Life Remained

- The life-forms that would limp across the Permian-Triassic boundary boundary would inherit a truly impoverished world. In general, they are referred to as disaster taxa.

- Dr. Conrad Labandeira is a paleoecologist at the Smithsonian's National Museum of Natural History's Department of Paleobiology with a particular interest in plant-insect interactions over geological time. His research shows the dramatic changes that occurred in insect faunas across the extinction boundary, an event that he describes as being one of the most profound in the evolutionary history of insects.

- Dr. Labandeira asserts that the Permian extinction divided the history of insects into 2 evolutionary faunas. Many Paleozoic lineages became extinct; in fact, we lose most insect species. This is a linked plant-insect ecological event that would have profound effects on the associations between plants and insects.

- But the situation on land was just half of the story. Life in the oceans had been decimated, too. The oceans, just like the continents, were dominated by high-abundance but low-diversity faunas. The rich invertebrate assemblages of the pre-extinction Permian world were reduced to just a few species, such as the brachiopod *Lingula* and the bivalve *Claraia*—classic disaster taxa found in the oceans at this time.

- This is also a world with no corals. In fact, just like the coal gap on land, there is a similar reef gap in the oceans, with coral reefs not returning to the planet until about 9 million years after the start of the Triassic period.

- The only reefal structures at this time would be stromatolites, columns of cemented sediment created by laminated mats of microbes. Stromatolites had been pretty well absent from most environments since the Ordovician, 190 million years earlier, but probably make an appearance in the Triassic due to the decimation of all the grazing creatures that would usually keep them in check.

- There is also a geographical pattern to the impoverishment of the taxa during the Early Triassic in both the marine and terrestrial realms. This pattern is particularly noticeable during the peak temperature rise during the Olenekian. At this time, a disturbing gap in fossils occurs at the equator—it appears that the equator had become a dead zone, with most life absent at low latitudes.

- In the oceans, fish, marine reptiles, and corals are missing. Life in these zones tend to be invertebrates of limited diversity and stromatolites. The situation is the same on land. The majority of the impoverished fauna that survived the extinction at the end of the Permian retreated toward the more hospitable poles.

- Contrast that with the distribution of life today, where a latitudinal band north and south of the equator shows overwhelmingly the greatest biodiversity on the planet—a biodiversity that typically decreases as you move toward the poles.

Driving the Impoverished Early Triassic

- It appears that the increase in lethal temperatures during the Early Triassic, especially at equatorial latitudes, pushed organisms beyond their thermal tolerances and were responsible for driving this period of extremely low biodiversity. But isn't this just what we should expect? After all, we have just come through the largest mass extinction ever, at the end of the Permian.

- While it would obviously take time for the Earth to cool down and recover, in most extinctions, the biosphere is well on the way to recovery within hundreds of thousands of years. For the Permian extinction, however, things were still pretty awful up to 5 to 7 million years into the Triassic. In fact, the only reason why another mass extinction is not registered at this time is that there is very little life on the planet left to go extinct.

- It is important to note, though, that it was not just rising temperature that was the problem. Increased temperature had caused a whole cascade of related problems during the Permian extinction—problems that persisited well into the Triassic. These difficulties included reduced oxygen conditions in various parts of the oceans and the possibility of an associated rise in hydrogen sulfide, due to the proliferation of certain sulfur-loving bacteria that like to live in these oxygen-poor conditions.

- Increases in carbon dioxide levels would also mean more dissolved dioxide levels in the oceans, making seawater acidic. This would ensure that creatures that secrete thick calcium carbonate shells and skeletons, such as corals, would have a long wait before they could make a comeback.

- At the moment, it is still unclear what was maintaining these environmental conditions way beyond the Permian-Triassic. It is possible that the Siberian Traps, one proposed smoking gun for the end-Permian extinction, was still active and releasing carbon dioxide. This delay may be responsible for the continued warming we see at the end of the Olenekian. Perhaps this warming also helped destabilize more methane hydrates, further warming the planet. At the moment, though, evidence for this, or another source or warming, has yet to be found.

The Path to Recovery

- Eventually, though, through the Middle and Late Triassic, in both the continental and marine realms, the numbers and diversities of lineages started to increase significantly. In addition, there is evidence of more complex ecological associations developing.

- For example, by the time we get into the Middle Triassic, we start to see long-term reef development initiated again. In addition, there is a quantum leap in plant-insect activity commencing during the Middle Triassic and expanding, especially throughout the Late Triassic. Innovative interactions between different life-forms were starting to increase, and food webs—an important engine in the evolution of new species and increasing biodiversity—were becoming more complex.

- The Middle Triassic also sees the spread of mollusks, such as bivalves (clams) and gastropods (snails). In addition, the Middle Triassic to Early Jurassic interval would see the expansion of terrestrial and freshwater vertebrate faunas. An early example is the rise of archosaurs, which, by the mid-Triassic, had started to replace the mammal-like therapsid reptiles.

- Archosaurs included rauisuchia, which resembled crocodiles on long legs that ranged in size from 4 to 6 meters. But another archosaur had emerged—not yet quite as imposing and impressive as they would ultimately become, but showing great signs of promise. The dinosaurs had arrived. Dinosaurs did not dominate the Triassic but were a part of rapidly diversifying vertebrate fauna.

- It is likely that some of the earliest mammals had evolved by the Late Triassic from some of those mammal-like therapsid reptiles that had been so common in the Permian and Early Triassic.

The End of the Triassic

- But as encouraging as this explosion of life in the later Triassic may sound, there would be another setback before the flourishing of the dinosaur world we recognize today. At the end of the Triassic, at around 201.3 million years ago, an event called the Triassic-Jurassic mass extinction occurred—the fourth extinction in the big 5 mass extinctions.

- In the oceans, the conodonts that had been such an important part of the Paleozoic fauna would be extinguished. Reef systems would suffer once again, and ammonites, brachiopods, and bivalves all would suffer significant extinctions. In total, it is estimated that 22% of all marine families, 53% of all genera, and an estimated 76 to 84% of all species would be driven into extinction.

- The Triassic-Jurassic extinction is a difficult extinction to tie down, as was formerly the case with the Permian-Triassic extinction, because there are relatively few sections of sediments that cover the interval of the extinction. In fact, some paleobiologists suggest that no extinction took place.

- It is possible that falling sea levels could have caused the crisis. This would have reduced the area of shallow, warm seas and restricted the spread of reefs. But the reason for these sea-level changes is uncertain at the moment.

- We do know that something very significant was happening in the Earth system at this time, though. Pangaea, a familiar geographical feature of the Triassic period, was really starting to fragment, with a series of rifts opening between the Americas and Africa and Europe. Evidence of this rifting can be seen in sediments, mostly Late Triassic to Early Jurassic in age, located in eastern North America, called the Newark Supergroup. Many of these sediments were deposited in lakes that developed along the line of the rift.

- The rift that would eventually widen to form the Atlantic Ocean was a center of igneous activity. Huge volumes of hot magma were being intruded into this area of Pangean crust at the end of the Triassic period about 200 million years ago. This igneous activity was part of what is known as the Central Atlantic magmatic province (CAMP). Not only were intrusive bodies of igneous rock that never got to the surface being produced, but so were vast outpourings of lava found today in northwestern Africa, southwestern Europe, and North America.

- The CAMP eruptions are concurrent with the extinction producing a volume of magma that would cover an area of about 11 million square kilometers. The CAMP eruptions are the first murmurings of the splitting apart of Pangaea.

- Samuel Bowring and Robert Schrock at the Massachusetts Institute of Technology have radiometrically dated the thickest deposits of lava, found in the High Atlas Mountains of Morocco, and have concluded that all this material was erupted in a fury of activity lasting, incredibly, only 40,000 years—a rapid shock to the Earth system.

- Such vast volcanic activity could have caused the effects of the Permian-Triassic extinction: global cooling due to the release of sulfur dioxide and aerosols, followed by intense warming as carbon dioxide levels started to rise in the atmosphere. Such warming may have also caused the destabilization and dissociation of gas hydrates in sediments on the ocean floor, thereby releasing methane and causing even more global warming.

- Currently, CAMP and possibly sea-level changes are the best explanations for the Triassic-Jurassic extinction. However, there has been a suggestion that the decrease in diversity was caused more by a decrease in speciation than by an increase in extinction. This decrease in speciation is a kind of slowing down of the engine of biodiversity, with normal rates of extinction not being met by a similar rate of new species evolving.

- The world following the Triassic would see a glorious new ecosystem dawn. The Jurassic is sometimes considered the golden age of the dinosaurs, but to many paleontologists, it was a golden age of life—a life that had been proven through earlier, very trying times in Earth's history.

Questions to consider:

1. Should we expect generalists or specialists to survive mass extinctions?
2. Does luck have a role to play in who survives a mass extinction event?

Suggested Reading:

Levin and King Jr, *The Earth Through Time*, chaps. 13–14.

Sues and Fraser, *Triassic Life on Land*.

Lecture 15 Transcript

Life's Slow Recovery after the Permian

The Permian-Triassic mass extinction event is probably the greatest catastrophe the biosphere faced since complex life evolved about 600 million years ago. A cascading series of events probably triggered by extensive volcanic activity in Siberia that would see the development of a runaway global warming and extinction across vast swathes of the biosphere. Roughly 90% with sometimes estimates as high as 96% of species go into extinction in a geological snap of the fingers.

We have already charted those difficult times in this series but what about the aftermath? It's easy to get distracted by the fireworks of the big event and forget about the world that would follow the disaster.

Doug Irwin of the Smithsonian's Department of Paleobiology makes an interesting point. Does the cause of these extinction events really matter? Perhaps generating a general theory based on the causes of mass extinctions is as he put it misplaced. Perhaps we should be more focused on how ecosystems fell apart as this may have great implications for biodiversity losses occurring today. Although the trigger of an ecosystem collapse may vary, perhaps they all fall apart in a similar way. To gain insight into this let's look at the world of the early Triassic just after the end-Permian bomb was dropped and see if we can chart the biospheres recovery.

So in this episode let's ask: What was the world like in the Early Triassic? What was left of life following the greatest of all extinctions? What was driving the impoverished Early Triassic and why did it last so long? When did the Earth start to recover? And what was there that could possibly give us even more doom at the end of the Triassic?

During the Early Triassic, our planet would have looked very, very different. A giant land mass Pangaea dominated one side of a planet surrounded by the vast Panthalassic Ocean.

The Early Triassic climate was harsh on Pangaea. Hot, arid deserts covered most of the interior of the supercontinent. Evidence of this is found in many of the red desert sandstones we find that occasionally contain evidence of fossil sand dunes.

Further proxies of hot conditions come from evaporite deposits. These can form by the evaporation of inland bodies of water and the precipitation of various salts as the water continues to concentrate and dwindle. You can see some modern examples here in Deep Springs Lake California and along the margins of the Dead Sea in Israel.

It is possible that this was the hottest and most arid time in over half a billion years. In fact, there is likely no ice at the poles. Indeed, it is possible that the poles might have been relatively temperate places where forests and a more diverse biota of plants, fungi, and animals could survive. This is the inverse of what we have today, as in the early Triassic the temperate, polar regions were more habitable than the equatorial regions.

Analysis of oxygen isotopes taken from conodonts—the tooth like mouthparts of an early chordate—paint a disturbing picture of the temperature change during the early Triassic.

Research in equatorial deposits from southern China spanning the period of the extinction at the end of the Permian and continuing into the early Triassic show a rapid warming in the oceans from 21 to 36°C. The warming peaks at around 252.1 million years ago, the end of the Permian.

There is a cooling following this and then a second rise in temperatures occurring around 250.7 million years ago, a period of the Early Triassic called the early Olenekian when temperatures in the water column rose again to about 38°C and perhaps even exceeded 40°C at the surface. At such marine surface temperatures, organisms such as corals cannot survive.

The life forms that would limp across the Permian-Triassic boundary would inherit a truly impoverished world. In general, they are referred to as disaster taxa. One of the poster kids for these disaster taxa is a strange shovel faced animal called *Lystrosaurus*. *Lystrosaurus* was a member of a group of reptiles called the dicynodonts—that means "two dog-like teeth"—that are named for their paired tusks They belong to a wider family of reptiles called the therapsids—or colloquially mammal-like reptiles—that are important in our evolution of our own group, the mammals.

Although it evolved during the Permian Lystrosaurus was most abundant during the early Triassic between 251–247 million years ago where it made up over 90% of the terrestrial vertebrate species on Earth. The animal probably cropped vegetation on the surface but may have been able to dig for material, if times got tough, with those powerful front limbs. It is this adaptability that might have allowed the *Lystrosaurus* to struggle through the difficult Early Triassic times. On the whole, it would be the generalists like this stubby reptile and not the specialists that would make it through the troubled Early Triassic.

And what of the plant life that *Lystrosaurus* may have dined upon? Forests were not common in the desert planet, and where plants were present, they would be dominated by smaller herbaceous forms like neo calamites, horsetails, and the lycopod *Pleuromeia*.

This may also explain why we see that transition in the early Triassic from the meandering river systems in the Permian with banks stabilized by plants to more chaotic braided systems that followed afterward.

This is a signal that plant communities have retreated and along with them their ability to stabilize the banks of river channels. At the same time, as we discussed in the previous lecture, there is an increased delivery of sediment to the oceans that contain soil-derived biomarkers. Soils were being stripped and washed away.

This reduction in flora may also explain the coal gap in the Early Triassic that suggests a lack of wetland plant communities and peat swamps. Coals would not appear in equatorial regions until about 15 million years after the Permian-Triassic extinction.

Dr. Conrad La Bandera is a paleoecologist in the department of Paleobiology with a particular interest in plant/insect interactions over geological time. His research shows the dramatic changes that occurred in insect faunas across the extinction boundary, an event that he describes as being one of the most profound in the evolutionary history of insects.

As you can see from this illustration in a paper Conrad published in 2005, the Mother of all Extinctions as his colleague Doug Erwin calls it divides the history of insects into two evolutionary faunas. Many Paleozoic lineages became extinct. In fact, we lose most insect species. This is a linked plant-insect ecological event that would have profound effects on the associations between plants and insects. For example, pre-extinction Permian herbivorous insects favored consumption of plants from the outside, while post-extinction Triassic species consumed their plant hosts from within. You can see that on these drawings made by Conrad of fossils from South Africa.

Of course, the situation on land was just half of the story. Life in the oceans had been decimated too. The oceans, just like the continents, were dominated by high-abundance but low-diversity faunas. The rich invertebrate assemblages of the pre-extinction Permian world were reduced to just a few species such as the brachiopod *Lingula* and the bivalve *Claraia*, classic disaster taxa' found in the oceans at this time.

This is also a world with no corals. In fact, just like the coal gap on land there is a similar reef gap in the oceans with coral reefs not returning to the planet until about nine million years after the start of the Triassic Period

The only reefal structures at this time would be stromatolites, those columns of cemented sediment created by laminated mats of microbes. Stromatolites had been pretty well absent from most environments since the Ordovician, 190 million years earlier, but probably make an appearance in the Triassic due to the decimation of all the grazing creatures that would usually keep them in check.

There is also a geographical pattern to the impoverishment of the taxa during the Early Triassic in both the marine and terrestrial realms. This pattern is particularly noticeable during that peak temperature rise during the Olenekian. At this time a disturbing gap in fossils occurs at the equator; it appears that the equator had become a dead zone.

This is not due to a lack of the right aged rocks or because of conditions that were not conducive to fossilization. Simply put, most life was just absent at low latitudes. In the oceans fish, marine reptiles and corals are missing. Lie in these zones tends to be invertebrates of limited diversity and of course, stromatolites.

The situation is the same on land. The majority of the impoverished fauna that survived the extinction at the end of the Permian retreated toward the more hospitable poles.

Contrast that with the distribution of life today where a latitudinal band north and south of the equator shows overwhelmingly the greatest biodiversity on the planet. A biodiversity that typically decreases as you move toward the poles. By contrast, life on planet Triassic had not only been impoverished, but that pattern of remaining diversity had flipped.

Another feature of many but not all of the taxa found during these times is their size. They're generally very small. This is thought to be related to a phenomena called the Lilliput Effect—named for the tiny people Gulliver visited on his travels. The Lilliput Effect is seen to occur after crises in the biosphere and manifests in smaller adult size and also in an increased juvenile mortality rate. Together these demographic features result in a fossil record characterized by smaller individuals. This probably also explains why the trace fossils only record the presence of small organisms since there were no large species that were burrowing very deeply at this time.

It would appear that the increase in lethal temperatures during the Early Triassic especially at equatorial latitudes is pushing organisms beyond their thermal tolerances. For many plants that threshold temperature is about 35°C with few being able to survive over 40°C.

For animals, temperatures over 45°C are very serious physiologically with marine animals having a much-reduced tolerance to rising temperature—about 35°C is their limit. This is due to increased risk of hypoxemia—the lack of oxygen in the blood—as rising temperature will reduce the amount of dissolved oxygen available in water.

This also explains why active marine creatures like fish and cephalopods are absent or rare at the equator; the more active you are, the more active your metabolism; and the more active your metabolism, the greater your demand for oxygen. With reduced potential prey the number of marine reptiles would be reduced too.

So it would appear that elevated surface temperatures were driving this period of extreme low biodiversity. But isn't this just what we should expect? After all, we have just come through the largest mass extinction ever, at the end of the Permian. Surely it would take time for the Earth to cool down and recover?

That's true, but in most extinctions, the biosphere is well on the way to recovery within hundreds of thousands of years. For the Permian extinction, however, things were still pretty awful up to 5–7 million years into the Triassic. In fact, the only reason why another mass extinction is not registered at this time is that there was pretty not much life left to go into extinction.

It is important to note, though, that it was not just rising temperature that was the problem. Increased temperature had caused a whole cascade of related problems during the Permian extinction; problems that persisted well into the Triassic. These difficulties included reduced oxygen conditions—what we call dysoxic or anoxic—in various parts of the oceans and the possibility of associated rises in hydrogen sulfide, due to the proliferation of certain sulfur loving bacteria that like to live in these oxygen-poor conditions.

Increases in carbon dioxide levels would also mean more dissolved Carbon dioxide in the oceans making seawater acidic. This would ensure that creatures such as corals that secrete thick carbonate shells and skeletons would have a long wait before they could make their comeback. Bivalves from this period like

Claraia that still produced calcium carbonate hard parts could only manage to secrete shells that were relatively thin.

Dr. Richard Bambach, a research associate in the Department of Paleobiology, was an author in a 2007 paper, published in *Earth and Planetary Science Letters*. This paper examined the selective nature that the extinction had on marine invertebrates, and it concluded that this was best explained—but not exclusively explained—by elevated levels of carbon dioxide, or hypercapnia He also notes that early Triassic land vertebrates display features such as respiratory turbinals, structures that increase the surface area over which air is processed in preparation for the lungs and also display burrowing behaviors that may have helped them survive a hot, carbon dioxide–poisoned world.

So what was maintaining these environmental conditions way beyond the Permian-Triassic? At the moment it is still unclear.

It is possible that the Siberian Traps, one proposed smoking gun for the end-Permian extinction, was still active and releasing carbon dioxide. This delay may be responsible for the continued warming that we see at the end of the Olenekian. Perhaps this warming also helped destabilize more methane hydrates, further warming the planet. At the moment, though, evidence for this or another source or warming has yet to be found.

Eventually, though through the Middle and Late Triassic in both the continental and marine realms the numbers and diversities of lineages started to increase significantly.

In addition, there is evidence of more complex ecological associations developing too. For example, by the time we see the Middle Triassic, we start to see long-term reef development initiated again.

The principal reef builders of the Pre-Permian extinction events, the rugose and tabulate coral were gone though as were many of the invertebrates and vertebrates associated with them.

Although there is evidence of some minor reef formation about 1.5 million years after the Permian-Triassic mostly by sponges and serpulids—those are worms that secrete tubes out of calcium carbonate—it is not really until around five million years after the extinction that stony corals really start to make a comeback. It is during the Middle Triassic that we see the first scleractinian corals at about 240 million years ago. The corals are the main reef-forming metazoans today.

Another example from Triassic comes from terrestrial plant/insect interactions. There is a quantum leap in plant/insect activity commencing during the Middle Triassic and expanding especially throughout the Late Triassic. Dr. Conrad La Bandera has been investigating how insects were initiating these associations with plants, starting around this time. For example, these are the scars and eggs left by a female insect as she used her sword-like ovipositor to lay eggs in a fossil plant stem from the Middle to Late Triassic of Kyrgyzstan.

Conrad has also found a whole spectrum of new plant/insect, plant/mite and fungal associations from the Late Triassic of the Karoo Basin in South Africa. For example, insects chewing holes through leaves, munching along leaf margins, and mining within leaf tissues. Not only does this show that plant and insect biodiversity was improving, but also that innovative interactions between different life forms were starting to increase, and food webs were becoming more complex; an important engine in the evolution of new species and increasing biodiversity.

The Middle Triassic also sees the spread of mollusks like bivalves—clams—and gastropods—snails. Prior to the Permian Extinction, the shells you would have commonly found washed up on a beach would likely have been brachiopods, sometimes called lamp shells. Brachiopods may superficially resemble clams, but they belong to a very different group of animals than the mollusks. They have a lophophore—that's a ring of ciliated tentacles to capture material suspended in the water column. You can see a lophophore here on a related animal called a bryozoan.

By the time we get into the Middle Triassic the bivalves are recovering and diversifying but the brachiopods although they are still with us today will never dominate the ocean floor the way they did in the Paleozoic. The sea shells of the Mesozoic and the Cenozoic will be dominated by mollusks.

The Middle Triassic to Early Jurassic interval would also see the expansion of terrestrial and freshwater vertebrate faunas. An early example is in the rise of archosaurs, which by the mid-Triassic had started to replace the mammal-like therapsid reptiles such as Lystrosaurus. Archosaurs included *Rauisuchid*. These were scary animals resembling a crocodile on long legs that ranged in size from about 4–6 meters. But another archosaur had emerged not yet quite as imposing and impressive as they would ultimately become but showing great signs of promise. The dinosaurs had arrived.

Dinosaurs did not dominate the Triassic but were a part of rapidly diversifying vertebrate fauna. Probably one of the best known of these Triassic dinosaurs is Coelophysis a slender bipedal carnivore about 3 meters long. Coelophysis is very common in the southwestern United States but has been found in other regions of the Triassic world too.

It was probably an agile runner with its forward facing eyes providing stereoscopic vision with good depth perception. 1000 specimens were discovered at the Whitaker Quarry in Ghost Ranch, New Mexico. This had led to speculation that Coelophysis may have roamed the Triassic in large flocks. This is difficult to prove however as it is also possible that these creatures had gathered at this location to drink before they were overcome by a flash flood event. Nevertheless, it does show what a common and successful creature this little dinosaur was.

The archosaurs would also move the vertebrates into the atmosphere during the later Triassic. Early gliding creatures like Sharovipteryx are found in the mid-Triassic of Central Asia, but it was the pterosaurs in the late Triassic like Eudimorphodon that would develop truly powered flapping flight.

And in a way it is from this burst of diversification following the Permian Extinction and the barren, hot Earth of the early Triassic that our story would start.

It is likely that some of the earliest mammals had evolved by the Late Triassic from some of those mammal-like therapsid reptiles that had been so common in the Permian and early Triassic.

It's kind of difficult to distinguishing between the last of the therapsids that were starting to look increasingly like mammals and true mammals, but Late Triassic animals like Megazostrodon are certainly starting to look very mammalian.

As encouraging as all this explosion of life in the later Triassic may sound though there would be another setback before the flourishing of the world, we are all familiar with from movies like Jurassic Park. For at the end of the Triassic at around 201.3 million years ago it would appear that the wheels are coming off the wagon again; this time in an event called the Triassic–Jurassic Mass Extinction; extinction number 4 in the Big 5 mass extinctions.

In the oceans the conodonts that had been such an important part of the Paleozoic fauna would be extinguished. Reef systems would suffer once again, and ammonites, brachiopods, and bivalves all would suffer significant extinctions. In total, it is estimated that 22% of all marine families, 53% of genera, and an estimated 76–84% of all species may have been driven into extinction.

On land, many plant groups were extinguished as Dr. Peter Wagner from the Department of Paleobiology has noted in his studies of the floras of Greenland. The plants Peter was studying do not just show a simple decrease in taxa but also demonstrate decreases in the number of common taxa, a change that was attributed to increased carbon dioxide and an associated global warming.

For tetrapods on land, the story is pretty grim too. Many of the archosaurs, the remaining therapsids, and many large amphibians became extinct. So what went wrong this time?

In truth, the Triassic-Jurassic extinction is a difficult extinction to tie down as was formerly the case with the Permian-Triassic extinction because there are so few sections of sediments that cover the extinction interval. In fact, some paleobiologists as we will see later suggest, in fact, that no extinction took place.

It is possible that falling sea levels could have caused the crisis; this would have reduced the area of shallow warm seas and restricted the spread of reefs. But the reason for these sea-level changes? Uncertain at the moment.

We do know that something very significant was happening in the Earth system at this time, though. Pangaea, a familiar geographical feature of the Triassic Period, was really starting to fragment with a series of rifts opening between the Americas, and Africa and Europe. Evidence of this rifting can be seen in sediments mostly Late Triassic to Early Jurassic in age located in eastern North America, called the Newark Supergroup. These sediments are very thick around 6 km at Newark in New Jersey and consist of a package of mostly terrestrial sediments deposited in fault-created down-dropped valleys that we call half grabens.

Many of these sediments were deposited in lakes that developed along the line of the rift. For a modern rift analogy think of the East African rift system today. This is a rift that is currently developing under East Africa that may at some point see the development of new ocean floor and the separation of this part of Africa from the Mainland.

The East Africa Rift is also associated with volcanic activity, for example, Mount Kilimanjaro and Mount Nyiragongo. In the same way, the rift that would eventually widen to form the Atlantic Ocean was also a center of igneous activity.

Dramatic evidence of all this activity can be seen if you take a boat down the Hudson River. This large cliff is the New Jersey Palisades Sill, originally a basaltic magma that was injected into the crust. The Palisades Sill is just one example of the huge volumes of hot magma that were being intruded into this area of Pangaean crust at the end of the Triassic Period about 200 million years ago.

This igneous activity was all part of what is known as the Central Atlantic Magmatic Province or CAMP. Not only were intrusive bodies of igneous rock that never got to the surface—like the Palisades Sill—being produced, but also vast outpourings of lava found today in NW Africa, SW Europe, as well as North America.

The CAMP eruptions are concurrent with the extinction producing a volume of magma that would cover an area of about 11 million square kilometers. The thickest deposits of lava today are found in the High Atlas Mountains of Morocco, where flows can reach as much as 300 meters thick. The CAMP eruptions are the first murmurings of the splitting apart of Pangaea.

Research by Sam Bowring and Robert Shrock of MIT have radiometrically dated these deposits and have concluded that all this material was erupted in a fury of activity lasting incredibly only 40,000 years, a rapid shock to the Earth system.

Such vast volcanic activity could have caused the effects we have previously described for the Permian-Triassic; extinction global cooling due to the release of sulfur dioxide and aerosols, followed by intense warming as Carbon dioxide levels started to rise in the atmosphere. Such warming may have also caused the destabilization and dissociation of gas hydrates in sediments on the ocean floor, therefore releasing methane and causing even more global warming.

A team of scientists including researchers at the University of Southampton in the United Kingdom and the Geological Survey of Canada has found evidence from sediments now exposed on Haida Gwaii, the Queen Charlotte Islands, of this warming. At the time of deposition, these islands were located in the northeastern Panthalassic Ocean.

They found chemical fossils of photosynthesizing brown-pigmented, green sulfur bacteria. Such bacteria only live in anoxic conditions, producing hydrogen sulfide as a waste product of their respiration.

This demonstrates that is this area that the upper ocean suffered severe oxygen depletion and hydrogen sulfide poisoning. The global warming caused by the CAMP carbon dioxide release may have acted as the trigger that would slow down oceanic circulation, restricting the ventilation of the oceans with oxygen and allowing these bacteria to thrive. The oceans just as they had at the end of the Permian may have become sick again.

There have been alternate triggers suggested for the extinction. For example, the eye of Quebec in Canada is a pronounced geomorphic structure occupied today by Lake Manicouagan. This is actually the remains of an ancient impact crater the 6th largest on Earth. It formed when a 5-kilometer diameter asteroid impacted this area during the late Triassic.

The crater was originally about 100 km wide but has now been eroded down to where it is only about 72 km wide. At its central point is Mount Babel. Mount Babel is interpreted as being formed due to the post-impact uplift of the ground, as it rebounded after the asteroid hit. Could this be the smoking gun of the end-Triassic extinction event? Unfortunately for the impact hypothesis recent dating of the Manicouagan crater places it some 12 million years before the extinction occurred.

So that leaves us with CAMP and possibly sea level changes as currently the best explanation for the Triassic-Jurassic extinction. However, there has been a suggestion that the decreasing diversity was caused more by a decrease in speciation than by an increase in extinction. This decrease in speciation is a kind of slowing down of the engine of biodiversity with normal rates of extinction not being met by a similar rate of new species evolving. As you can see from this figure in Conrad La Bandera's 2005 paper about family-level insect extinction, there is no significant downturn across the proposed extinction boundary

Whatever this event actually represents the world following the Triassic would see a glorious new ecosystem dawn. Dr. Bambach in a paper that he co-authored with Andrew Bush of the University of Connecticut notes that from the Jurassic there was as he puts it a sustained taxonomic radiation. The Jurassic is sometimes called the golden age of the dinosaurs, but to many paleontologists, it was a golden age of life; a life that had been proven though earlier very trying times in Earth's history.

Lecture 16

Dinosaur Interpretations and *Spinosaurus*

New fossils are continually pushing paleobiological research forward, and our insights into ancient creatures are only going to become clearer as new discoveries are made. In this lecture, you will discover how dinosaurs become fossils and how they are found. You will also learn about the 2 broad groups of dinosaurs. In addition, you will learn who first discovered the *Spinosaurus*, why it was puzzling, why it was a "lost dinosaur" for many years, and why it is special.

Finding a Dinosaur Fossil

- There are many factors that have to come together for a living organism to become a fossil. One of the most important is, in most but not all cases, to cover the carcass of the organism with sediments as soon as possible. This effectively gets it out of the way of scavengers and, preferably, into conditions where oxygen levels may be sufficiently low to help slow decay. This increases the chances that the processes of fossilization may occur and preserve some of the organism.

- The problem is that dinosaurs, along with many terrestrial vertebrate creatures, tended to live in environments that are net areas of erosion rather than deposition of sediment. Although the landscapes might have had lakes and rivers, most of the land was exposed to erosion.

- This explains why the majority of fossils that are found are from aquatic environments and why the majority of dinosaur fossils are from rocks that were deposited in sediments in, or close to, rivers and lakes. So, for a dinosaur to maximize its chance of becoming a fossil, it really needs to be close to one of these environments when it dies. One notable exception is if dinosaurs gets caught in volcanic ashfall or mudflows deposits.

- To maximize your chance of finding a dinosaur, you need to satisfy 2 basic things:
    - It is vital to look in sedimentary rocks of the right age. The dinosaurs' reign is sandwiched between the largest extinction in Earth's history, at 251 million years ago, and the extinction that wiped them from the planet, probably due to a massive meteor strike centered on the Yucatán Peninsula at 66 million years ago. Dinosaurs probably evolved around 230 million years ago, during the Late Triassic, making their time on Earth about 164 million years.

    - We also need rocks of the right type. These rocks need to be terrestrial and not marine, and they most likely need to be rocks deposited in or near rivers or lakes, although there are notable exceptions, such as volcanic deposits.

Broad Groups of Dinosaurs

- The earliest dinosaurs, in the Late Triassic, were bipedal, but they were not the terrifying meat-eating giants that would evolve later in the Mesozoic. Dinosaurs would evolve and diversify throughout the Mesozoic, producing a variety of forms and, in general, demonstrating an increase in size through their evolutionary history. Dinosaurs are only very rarely found to evolve into smaller sizes, and the average weight of a Mesozoic dinosaur was about 100 kilograms.

- Dinosaurs can be split into 2 broad groups and are generally differentiated on the basis of the structure of the pelvis.
    - In the group of dinosaurs called Saurischia, also called the lizard-hipped dinosaurs, the pubis bone points forward. This group of dinosaurs includes the 2-legged theropod dinosaurs (such as *Tyrannosaurus rex*) and the sauropod dinosaurs (such as *Diplodocus*).

- The Ornithischia, or bird-hipped dinosaurs, have a pubis that points backward. These forms tend to be more common in the Cretaceous and include dinosaurs such as *Triceratops*.

- Many different types of dinosaurs would evolve within these 2 broad groupings, with only one member surviving to the present day: the avian dinosaurs, or birds.

Spinosaurus

- The first dinosaur fossil discovered in Malaysia was uncovered in 2012 by a team from the University of Malaya in Malaysia and Waseda University and Kumamoto University in Japan. The fossil was recovered from Pahang, but the exact location of the site was kept a secret to deter illegal fossil hunting.

- They discovered a single dark-colored tooth that was quite distinctive, with 2 sharp edges on the front and back called carinas that exhibit serrations.

These are typical features of carnivorous theropod dinosaurs, such as *Tyrannosaurus*.

- But the particular theropod tooth that was found in Malaysia had very specific ridges running down its length and micro-ornamentation on its surface. The tooth was also quite conical in shape. These features are indicative of a particular type of theropod dinosaur: a spinosaur.

- The first spinosaur to be revealed to the scientific community was named and scientifically described by a famous German paleontologist named Ernst Stromer von Reichenbach. It was found by Richard Markgraf, an Austrian fossil collector living close to Cairo, Egypt.

- In 1912, Markgraf uncovered a most remarkable fossil contained in rocks dating to the Upper Cretaceous, about 100 million years in age, near el-Bahariya, about 230 miles (370 kilometers) from Cairo. He shipped this to Stromer, who was in Germany at the time, and by 1915, Stromer had described and officially named the fossil *Spinosaurus aegypticus*, the Egyptian spine lizard.

- Even though it was an incomplete skeleton, it was certainly unlike any other large theropod dinosaur the world had seen before. It had a unique long, narrow, crocodile-like jaw. Its teeth were conical, not blade-like, and rising off its back vertebrae were enormous spines, which might have supported a large sail, perhaps for use in display or thermoregulation.

- The spinosaur was only one of at least 2 other large theropod dinosaurs, including *Bahariasaurus* and *Carcharodontosaurus*, that were around 40 feet (12 meters) long. This just didn't make sense: How could such an ecosystem support so many large apex predators living in such close proximity? This became known as Stromer's riddle.

- The story of Stromer's remarkable dinosaur takes a rather sad turn during World War II. The spinosaur remains, along with many other specimens from his expeditions, were proudly displayed at the Bavarian State Collection for Paleontology and Geology in Munich, southern Germany. On April 24

1944, the Royal Air Force targeted Munich for a nighttime bombing raid. Unfortunately, one of the buildings hit was the museum where Stromer's fossils were held—all reduced to dust.

- All that remained of Stromer's spinosaur were some of his notes and sketches. Even the photographic records of the spinosaur were lost, only turning up in 1995 in a collection of Stromer's records donated to the museum by his son.

- The holotype—the originally described specimen—was lost to science, but other spinosaurid theropods started to be discovered around the world. Spinosaurs are now classified as the family Spinosauridae that consists of 2 subfamilies: the Baryonchinae and the Spinosaurinae, the latter of which includes Stromer's *Spinosaurus aegyptiacus*.

- The family Spinosauridae is now known to be a wide-ranging group of dinosaurs with specimens found in Africa, Europe, South America, Asia, and Australia. They first appear during the Late Jurassic (about 155 million years ago), then start to dwindle in numbers between 93 and 100 million years ago, and are last known around 85 million years ago in the mid–Late Cretaceous.

- They are still puzzling. They all have elongated crocodile-like skulls with conical teeth and spines along their back that likely supported a sail. The crocodile-like skull suggests that part of the diet of this group of dinosaurs consisted of fish. This opens up an interesting possibility: This could be the only known predatory dinosaur that spent at least some of its life in water.

- It is important to address a common misconception. There were plenty of fully aquatic reptiles that lived at the same time as the dinosaurs, but they were not dinosaurs. Dinosaurs are a group of very specific animals with particular diagnostic features, such as the structure of their hips. Like modern whales, though, all of the marine reptiles had evolved from older animals, not dinosaurs, that once lived on land.

- How do we go about testing this hypothesis of an aquatic—or, at least, semiaquatic—dinosaur? The spinosaur's skull and teeth do resemble fish-eating crocodiles, and partly digested fish scales have been found fossilized in the stomach contents of some specimens. However, non-semiaquatic creatures, such as bears and wolves, eat fish, too.

- A paper released in 2010 by a team including Romain Amiot at the University of Lyon in France suggested that oxygen isotopes might be used to determine how much of an aquatic lifestyle the Spinosauridae may have exhibited.

- Oxygen occurs in both light oxygen-16 and heavy oxygen-18 forms. Land-dwelling creatures lose a lot of water through breathing and evaporation, and it is the light form of water that contains the oxygen-16 that gets evaporated. As a result, it is the heavy oxygen-18 that is concentrated in tooth enamel.

- Aquatic animals lose less water than terrestrial creatures and, as such, have less oxygen-18 in their teeth. Aquatic creatures also drink more and are constantly flushing water through their bodies, keeping the oxygen-18 levels low.

- The researchers collected samples from 133 Cretaceous specimens of various species, including spinosaurs, other dinosaurs, and crocodiles, and found that the oxygen-16/oxygen-18 ratio for spinosaurs was more similar to crocodiles than to other dinosaurs. Were the spinosaurs aquatic, then?

- At the time of this paper, there were relatively few, or well-preserved, complete skeletons that could be examined to see if the rest of the spinosaur skeleton demonstrated any aquatic adaptations.

- For his Ph.D. research at University College Dublin in Ireland, Nizar Ibrahim was studying all of the fauna in the Kem Kem beds in Morocco. These are Late Cretaceous sediments deposited between 100 to 94 million years ago. From a fossil found in Erfoud, Morocco, that is the same species as Stromer's spinosaur—*Spinosaurus aegypticus*—Ibrahim, along with Samir

Zouhri from the University of Casablanca and David Martill of the University of Portsmouth in the United Kingdom, was able to deduce that the environment that *Spinosaurus* lived in was a lush plain over which rivers meandered about 100 million years ago, during a short interval between the transition from the Early to Late Cretaceous.

- Ibrahim was able to make a 3-dimensional reconstruction of the spinosaur that suggested that an adult *Spinosaurus aegypticus* would have been about 50 feet (15 meters) long, which would have made it larger than *Tyrannosaurus rex* at 40.5 feet (12.3 meters) long. As a result, *Spinosaurus aegypticus* would have been the largest carnivorous dinosaur that had been discovered at that time.

- Overall the model presented—with a small pelvis; somewhat stumpy paddling back legs; and long, narrow jaws—somewhat resembles the ancestors of the whales, carnivorous terrestrial creatures who had also adopted a semiaquatic lifestyle. If this creature was truly adapted for life in the water, then this is the second reason why Spinosauridae are so special: Not only would they contain the largest carnivorous dinosaurs that ever lived, they would also be the only truly aquatically adapted dinosaurs known.

- If this is the case, it also helps us solve Stromer's riddle: How could the giant *Spinosaurus* live alongside other giant theropod dinosaurs? If *Spinosaurus* is semiaquatic, they would be living in a different environment and mostly preying on aquatic rather than terrestrial organisms.

- *Spinosaurus* was a large creature, probably because it was preying on other large creatures. If we follow the semiaquatic model, the fossils of very large turtles, 8-feet-long lungfish, 13-feet-long coelacanth fish, and 25-feet-long sawfish found in the same sedimentary deposits may have formed some of the prey items for this animal.

- This interpretation of spinosaurs paints this dinosaur as a creature that was in transition between the terrestrial and aquatic environments, a kind of transitional form. But it is important to remember that any model is a hypothesis, and this one is certainly being highly debated in the circles of vertebrate paleontology.

Questions to consider:

1. Given the success of the dinosaurs, why are their fossils not more common?
2. Are the anatomical and ecological questions about *Spinosaurus* now answered?

Suggested Reading:

Lanham, *The Bone Hunters*.

Pim, *Dinosaurs*.

Lecture 16 Transcript: Dinosaur Interpretations and *Spinosaurus*

I love the ocean, but I'm not really a beach person. Spending days, just lying on the sand never really appealed to me. I more of an "OK that was nice—what's next?" kind of a guy. So when some friends suggested I might want to join them on a cruise ship, I was a little wary. Yes, I would be on the ocean but being stuck on a ship sounded to me not too different to being stuck on a beach for 2 weeks, but this time with no way to escape.

Much to my surprise, I loved it. Being at sea really helps me relax While on the ship I attended some guest lectures given by various people about some of the locations we would be visiting.

Now I love talking about science you might have noticed, so the idea of being at sea on a beautiful ship and chatting all day about science sounded great. When my sabbatical came up in 2015, I got the opportunity traveling as a scientific lecturer around Southeastern Asia. It was a great experience, and there was plenty of material to lecture about. Southeastern Asia is like a geological laboratory. It is one of the most volcanically and seismically active places on the planet and also one of the most wonderfully biodiverse places on Earth too.

I really wanted to present a talk on dinosaurs but relate it to the locations we would be visiting The lion's share of the ports of call though would be in Indonesia. Indonesia has rocks from the Mesozoic Era, the time in which we find dinosaurs, but by 2015 no dinosaurs fossil had been found although ichthyosaurs a marine reptile—not a dinosaur from the Middle Triassic Period had been found.

The problem is dinosaur fossils are not really as common as you might expect from all the books, articles and TV shows you might have seen. One does not come up to an outcrop remove a little loose sand and voilà the perfectly preserved head of a T. rex.

There are many factors that have to come together for a living organism to become a fossil. One of the most important is in most, but not all cases, to cover the carcass of the organism with sediments as soon as possible. This effectively gets it out of the way of scavengers and preferably into conditions where oxygen levels may be sufficiently low to help slow decay This increases the chances that the processes of fossilization may occur and preserve some of the organism.

The problem that dinosaurs had along with many terrestrial vertebrate creatures is that they tended to live in environments that are net areas of erosion rather than deposition of sediment. Have a look any landscape Although you may see lakes and rivers most of the land is exposed to erosion. Creatures that die on land will most likely be exposed to scavengers the rest of their remains strewn about and rotting in our oxygen-rich atmosphere.

This explains why A, the majority of fossils that are found are from aquatic environments and B, why the majority of dinosaur fossils are from rocks that were deposited in sediments in or close to rivers and lakes. So for a dinosaur to maximize its chance of becoming a fossil it really needs to be close to one of these environments when it dies. One notable exception is if dinosaurs get caught in volcanic ash-fall or mudflows. These are lahar deposits similar to what happened to humans in Pompeii and Herculaneum in AD79 as has been proposed for some spectacular dinosaur and bird fossils found in the Early Cretaceous Lujiatan deposits of Northeast China.

So to maximize your chance of finding a dinosaur you need to satisfy two basic things. One, it is vital to look in sedimentary rocks of the right age. The dinosaurs' reign is sandwiched between the largest extinction in Earth's history at 251 million years ago and the extinction that wiped them out from the planet probably due to a massive meteor strike centered on the Yucatan peninsula at 66 million years ago.

Dinosaurs probably evolved around 230 million years ago during the Late Triassic making their time on Earth some 164 million years. For comparison, we *Homo sapiens* have been around for the last 200,000 years with our genus Homo perhaps around 2.5 million years old. In terms of longevity, dinosaurs were a very successful group of creatures.

Secondly, we also need to find rocks of the right type too. These rocks need to be terrestrial and not marine, and they most likely need to be rocks deposited in or near rivers or lakes although there are notable exceptions such as the volcanic deposits we mentioned earlier.

But getting back to my desire to write a lecture about dinosaurs while in Southeast Asia. As I noted there had been fossil marine reptiles of the right age found in shallow marine deposits from Indonesia but what I needed was sedimentary rocks of terrestrial origin.

So how could I bring a dinosaur lecture into the schedule? I could, of course, just throw one in but how much better if I could make it relevant to the areas we would be visiting.

There were rocks relatively close in Thailand that fit the bill, sedimentary rocks belonging to the Khorat Group are known to be terrestrial in origin They range from about the middle Triassic to the Cretaceous that's the right time for dinosaurs and were deposited mostly in coastal settings and alluvial—that's river flood plains—so the right type of rocks too.

They have a distinctive red color common to many rocks of this type caused by the presence of oxidized iron minerals and are often called red beds, in effect, these rocks are rusty. You can see similar but older rocks in this image from the Orkney Islands of Scotland. Fossil plant pollen when present in the Thailand sections indicated that the environment was dominated by gymnosperm woodland in a seasonally dry subtropical climate.

This group of sediments also contains horizons that have produced dinosaur fossils. A unit called the Sao Khua Formation had produced dinosaurs of Early Cretaceous age, including phuwiangosaurus, a dinosaur belonging to a group we call the titanosaurian sauropods that contain some of the largest animals to have walked the Earth such as Argentinosaurus. Estimates vary, but it may have weighed about 90 tons and been around 100 feet, that's 30 meters, long.

Fragmentary remains of the Carnivorous dinosaur Siamotyrannus, the Siamese tyrant have also been found in Thailand that possibly belongs to a broad group called the carnosaurs. These included various fearsome bipedal dinosaurs from the Jurassic and Cretaceous Periods such as Giganotosaurus.

Many of these finds have been fragmentary but still are significant in providing a broader picture of the global distribution of dinosaurs during the Mesozoic Era.

But I still had a problem. The cruise I was on would not be visiting any ports in Thailand—darn. Fortunately, my research turned up something useful. Recently, fragments of dinosaurs had started to turn up in Malaysia, and that was a country we would visit.

In 2014, paleontologists at the University of Malaya revealed that they had found a tooth of a Cretaceous ornithischian dinosaur an order of beaked herbivorous dinosaurs that includes one of the kids' favorites, Triceratops. Subsequently, there have also been reports of possible Iguanodon fossils too.

But it wasn't these discoveries that caught my eye. In fact, the fossil of greatest interest to me was also the first dinosaur fossil discovered in Malaysia. It was uncovered in 2012 by a team from the University of Malaya in Malaysia and Waseda University and Kumamoto University in Japan The fossil in question was recovered from Pa-hang, but the exact location of the site was kept a secret due to illegal fossil hunting.

What had they discovered? A single, dark, colored tooth just under an inch long and around 0.3 inches wide. The tooth is quite distinctive with two sharp edges on the front and back called carinae that exhibit serrations.

These are typical features of carnivorous theropod dinosaurs like Tyrannosaurus and Allosaurus.

But before we move on, I think I should provide you with a little dinosaur primer, so I can show you where very broadly this dinosaur fits in the rest of the dinosaur family picture.

The earliest dinosaurs were bipedal like Eoraptor, who lived in the late Triassic about 231 million years ago in Argentina. As you can see, Eoraptor—at about 1m long—might have been able to give your knees a bit of a nibble but we have a long way to go before the terrifying meat-eating giants that evolve in the later Mesozoic.

Dinosaurs would evolve and diversify throughout the Mesozoic producing a variety of forms and, in general, demonstrating an increase in size through their evolutionary history, an observation called Copes Rule. Dinosaurs are only very rarely seen to evolve in to smaller sizes, and the average weight of Mesozoic dinosaur was about 100 kg.

Dinosaurs can be split into two broad groups and are generally differentiated on the basis of the structure of the pelvis. So we can get a better feel of this, let's look at the human pelvis for comparison. The bones we should be keeping in mind are pubis, ischium, and ilium.

So to those dinosaur groups. Firstly, the Saurischia also called the lizard-hipped dinosaurs. In this dinosaur, the pubis bone points forward. This group of dinosaurs includes the two-legged theropod dinosaurs like T-rex and the sauropod dinosaurs like Diplodocus.

The second group of dinosaurs is referred to as the Ornithischia or bird-hipped dinosaurs. These dinosaurs have a pubis that points backward. These forms tend to be more common in the Cretaceous and include dinosaurs such as Edmontosaurus and Triceratops. Just to clear up some potential confusion here before we move on, this group of dinosaurs is called bird-hipped dinosaurs as the pelvis looks a bit like a bird's. But birds, in fact, probably evolved from the lizard-hipped saurischian dinosaurs, specifically from the bipedal theropods during the Jurassic period.

As you can see from this figure, many different types of dinosaurs would evolve within these broad groupings with only one member surviving to the present day, the avian dinosaurs, or birds if you would prefer. And it is to this group the theropods that our particular dinosaur belongs.

But the particular theropod tooth that was found in Malaysia had very specific ridge pattern running down its length and micro-ornamentation on its surface. The tooth is also quite conical in shape. These features are indicative of a particular type of theropod dinosaur. This particular dinosaur is a paleontological star; this is a tooth from a spinosaur.

So, for the rest of this episode, let's look at the following: Who was the first to discover the *Spinosaurus* and why was it puzzling? Why was this a lost dinosaur for so many years? Why are spinosaurs so special, and how do we solve what I'll be calling Stromer's Riddle?

The first spinosaur to be revealed to the scientific community has an interesting history in its own right. It was named and scientifically described by a famous German paleontologist with the grand name of Ernst Stromer von Reichenbach.

Stromer had an aristocratic background dating back to the 13th century that included courtiers, lawyers, architects, and scientists. In fact, he had the title Freiherr, kind of the equivalent of baron. He was apparently no fan of roughing it in the field but even so he would complete a number of expeditions to Egypt in less than ideal conditions. It is amazing how a love of fossils will drive you out of your comfort zone.

His first paleontological love was mammals, not dinosaurs. So in November 1910 he embarked on an expedition to find evidence for early mammals from the Late Cretaceous. By 1911, following a brief time in quarantine on his ship, Cleopatra, due to a cholera outbreak, he was in the Egyptian desert with his friend Richard Markgraf, an Austrian, who had fallen in love with Egypt and its western desert and who was living close to Cairo.

It was a tense time politically in Egypt and Stromer, as a German had some difficulty in obtaining permits to explore the area from the British and French

Authorities. Eventually, with permits secured, they started to hunt for fossils and, over the course of a number of months, although not having too much luck finding early mammals they uncovered the remains of turtles, crocodiles, marine reptiles and, of course, dinosaurs.

Stromer had to return to Germany in 1912, leaving Markgraf to continue to explore. Later that year Markgraf started to uncover a most remarkable fossil contained in rocks dating to the Upper Cretaceous, about 100 million years in age near El Bahariya about 230 miles away from Cairo. He shipped this to Stromer in Germany who by 1915 had described and officially named the fossil *Spinosaurus aegypticus* the Egyptian Spine Lizard.

Even though it was an incomplete skeleton, it was certainly unlike any other large theropod dinosaur the world had seen before. It had a unique long, around 15-foot narrow, very crocodile-like jaw. Its teeth were conical, not blade-like and rising off its back vertebrae were enormous spines some as much as 5 feet in height. Stromer first speculated that these might support a large, fatty hump like you might see in Bison or zebu cattle, but this hypothesis was later rejected in favor of a large sail perhaps used in display or thermoregulation.

What was even more strange to Stromer though, was the fact that the Spinosaur was only one out of at least two other large theropod dinosaurs, including Bahariasaurus and Carcharodontosaurus—that were around 40 feet long. This just didn't make sense, how could such an ecosystem support so many apex predators—top dogs—living in such close proximity. This became known as Stromer's Riddle.

The story of Stromer's remarkable dinosaur takes a rather sad turn though during World War II. The spinosaur remains along with many of the other specimens from his expeditions were proudly displayed at the Bavarian State Collection for Paleontology and Geology in Munich, southern Germany. These fossils had made Stromer really quite famous.

This did not impress, though, one Karl Beurlen, head of the paleontology collections and supporter of the Nazi party. Stromer had openly criticized the Nazis, and as a result, Beurlen refused to move Stromer's collections to safety during the war, even though most works of art and important scientific collections had been moved to caves or salt mines for protection.

The inevitable happened. On the evening of April 24th, 1944 the Royal Air Force targeted Munich for a night-time bombing raid. Unfortunately, one of the buildings hit was the museum where Stromer's fossils were held, all reduced to dust in one evening.

In addition to this scientific tragedy, Stromers's dislike for the Nazi regime also caused him personal tragedy. Although his fame and aristocratic background protected him, his three sons were sent to the front-line Two were killed and the other son was captured by the Soviets and not returned to Germany until after the war. All that remained of Stromer's spinosaur were some of his notes and sketches, Even the photographic records of the spinosaur were lost only turning up in 1995 in a collection of Stromer's records donated to the museum by his only surviving son, Wolfgang.

So the holotype the originally described specimen was lost to science but other spinosaurid theropods started to be discovered around the world. Spinosaurs are now classified as the family spinosauridae that consists of two subfamilies, the Baryonchinae and the Spinosaurinae, the latter of which includes Stromer's *Spinosaurus aegypticus*.

The family spinosauridae is now known to be a wide-ranging group of dinosaurs with specimens found in Africa, Europe, South America, Asia, and Australia. They first appear during the Late Jurassic about 155 million years ago, then start to dwindle in numbers between 93 and 100 million years ago, and are last known around 85 million years ago in the mid-Late Cretaceous.

They are still puzzling, though. They all have elongated crocodile-like skulls, and conical teeth, and spines along their back that might have, as we said, supported a sail The crocodile-like skull suggests that part of the diet of the group of these dinosaurs consisted of fish. This opens up an interesting possibility—that this could be the only known predatory dinosaur that spent at least some of its life in water.

As a quick aside here, it is important to address a common misconception. There were plenty of fully aquatic reptiles that lived at the same time as the dinosaurs. However, these were not dinosaurs. Dinosaurs are a group of very specific animals with particular diagnostic features like the structure of their hips that we mentioned earlier. Like modern whales, though, all of the marine reptiles had evolved from older animals, not dinosaurs that once lived on land. Let's have a quick look at some of the main groups that were swimming through the Mesozoic oceans.

Ichthyosaurs were streamlined marine reptiles, many of which closely resembled modern dolphins in today's oceans. They are often regarded as a classic example of convergent evolution. Basically unrelated creatures evolve similar features as the result of living in similar environments and interacting with that environment in a similar way. In the case of these two creatures, as high-speed marine predators, although Ichthyosaurs probably had cephalopods like ammonites and belemnites on the menu in addition to fish. Plesiosaurs that contained many of the long-necked marine reptiles like the Elasmosaurs.

But by the Cretaceous, a new marine reptile had evolved which, by the end of the period, were certainly the kings of the oceans. Some of the large members of the group were around 14 meters long. These are the wonderfully scary Mosasaurs. A member of that group, Tylosaurus, on loan from the National Museum of Natural History, has been my companion through this series. Just look at those teeth and these extra pterygoid teeth that are on the palate here—you certainly would not want to meet this guy on a Cretaceous scuba diving vacation.

So we know we have aquatic marine reptiles, how do go about testing this hypothesis of an aquatic or at least, semi-aquatic dinosaur? It is true that the spinosaur's skull and teeth do resemble fish-eating crocodiles like the gharial. In addition partly digested fish scales have been found fossilized in the stomach contents of some specimens. However, nonsemiaquatic creatures like bears and wolves eat fish too.

A paper released in 2010 by a team including Romain Amiot at the University of Lyon, suggested that oxygen isotopes might be used to determine how much of an aquatic lifestyle the Spinosauridae may have exhibited.

Oxygen occurs in two forms, light oxygen 16 with 8 protons and 8 neutrons, and heavy oxygen 18 with 8 protons, and 10 neutrons. Land-dwelling creatures lose a lot of water through breathing and evaporation, and it is the light form of water that contains the oxygen 16 that gets evaporated. As a result, it is the heavy oxygen 18 that is concentrated in tooth enamel.

Aquatic animals lose less water than terrestrial creatures, and as such have less oxygen 18 in their teeth, Aquatic creatures also drink more and are constantly flushing water through their bodies keeping the oxygen 18 levels low.

The researchers collected samples from 133 Cretaceous specimens of various species including spinosaurs, other dinosaurs, and crocodiles and found that the oxygen 16/oxygen 18 ratio for spinosaurs were more similar to crocodiles than to other dinosaurs. Were the spinosaurs aquatic then? At the time of this paper, there were relatively few or well preserved complete skeletons that could be examined to see if the rest of the spinosaur skeleton demonstrated any aquatic adaptations.

Our story now moves to Nizar Ibrahim, currently a postdoctoral researcher at The University of Chicago. For his Ph.D. research at University College Dublin in Ireland, he was studying all of the fauna in the Kem Kem Beds in Morocco. These are Late Cretaceous sediments deposited between 100–94 million years ago This was quite the undertaking, but Nazir was hoping to paint a picture of the entire environment as it existed back then in northwestern Africa in an attempt bring a lost world back to life.

This part of the *Spinosaurus* story starts with Nizar Ibrahim's chance encounter with a Bedouin fossil hunter during 2008 in the town of Erfoud, in Morocco.

Ibrahim was shown a collection of dinosaur bones including one that looked like a hand. The bones were associated with a very distinctive purple sediment with yellow streaks.

Cut now to 2009, where we find Ibrahim visiting the Natural History Museum of Milan, in Italy. Here he was shown a partial skeleton of an exciting fossil a fossil that is obviously the same species as Stromer's Spinosaur—*Spinosaurus aegypticus*. This in its own right would be exciting, especially as it appeared to be more complete than Stromer's specimen, but there was something else On examining some of the rock still clinging to the bones Ibrahim realized that it was exactly the same type of sedimentary matrix as he had seen in Erfoud a year earlier. This was part of the same specimen.

This was great news not, only because a more complete picture of the anatomy of *Spinosaurus* could be gained, but there was also the possibility of putting this creature in the context of the environment in which it lived. This is why illegal fossil hunting is such a problem to paleontologists. Without the context of the rocks where a specimen was found, the value of any fossil is greatly diminished. A fossil found in location can tell us when it lived, but so much more too. It can tell us about the environment of the creature that he inhabited and other creatures that shared that environment, including details of plant life, climate, and geography; vital information to bring life to these ancient bones.

So all Ibrahim had to do was locate the fossil hunter he met in 2008 and find out from him where he found it. Unfortunately, Ibrahim did not know his name and all he could remember about him was that he wore white clothes and had a mustache This, unfortunately, is a profile that fits many men in Morocco.

This is why he along with Samir Zouhri from the University of Casablanca and Dave Martill of the University of Portsmouth in the U.K. found themselves in a cafe in Erfoud at the end of a fruitless search for the fossil hunter when a man in white with a mustache walked by. It was him.

Apparently, he took a little convincing, but eventually, he led the scientists to the extraction site of the fossil.

From it, they were able to deduce that the environment that *Spinosaurus* lived in was a lush plain over which rivers meandered about 100 million years ago during a short interval between the transition from the Early to the Late Cretaceous.

Ibrahim was able to scan all the bones from this new find but also combine them on a computer scaling some for size with other specimens in other collections.

From this, he made a reconstruction of the spinosaur that suggested that an adult *Spinosaurus aegypticus* would have been about 50 feet long, which made it larger than T-rex at around about 40.5 feet long. As a result, *Spinosaurus aegypticus* would have been the largest carnivorous dinosaur that had been discovered at that time.

This particular reconstruction also suggests that *Spinosaurus* had a very odd body plan when compared to other theropod dinosaurs, with a barrel-shaped torso like modern whales and dolphins and a rather small pelvis and short stumpy hind limbs with flat feet. When modeled it was found that in this reconstruction the dinosaurs' center of gravity would be well forward making it unlikely that it could walk in an effective bipedal manner like the rest of its theropod cousins and unlike many of the earlier reconstructions that showed it as a fully bipedal animal. It was suggested that it may have employed its front legs to help it walk in a quadrupedal manner on land, similar to Late Cretaceous duckbill dinosaurs.

Under this reconstruction, those rather short hind limbs could be interpreted to have acted as paddles in water a bit like modern otters.

Its bones were also extremely dense, a feature shared by many modern semiaquatic mammals such as the hippopotamus, where one of the main problems is not floating but combatting natural buoyancy.

In this model of *Spinosaurus*, the tail is long and contains bones that are rather loosely connected Theropods tend to have very stiff tails that they used as a counterbalance when walking. Under this particular hypothesis, the long flexible tail of *Spinosaurus* may have been able to bend laterally like those of crocodiles perhaps propelling it through water.

And then there is that marvelous long skull and its narrow jaws armed with conical teeth. The skull resembles that of fish-eating crocodiles like the gharial. In the dinosaur's snout, you can find pits resembling the foramina found in crocodiles. In living crocodiles, these foramina house the pressure sensors that are used to detect the movement of prey in water. Is it possible the *Spinosaurus* had this useful skill too?

Overall, the model presented with a small pelvis and somewhat stumpy paddling back legs, and long narrow jaws somewhat resembles the ancestors of the whales—carnivorous terrestrial creatures who had also adopted a semiaquatic lifestyle. Could this be a creature that was truly adapted for life in the water? If this is the case, then this is the second reason why Spinosauridae is so special. Not only would they contain the largest carnivorous dinosaurs that ever lived, they would also be the only known, truly aquatically adapted dinosaurs.

If this is the case, it also helps us solve Stromer's Riddle. How could the giant *Spinosaurus* live alongside other giant theropod dinosaurs? If *Spinosaurus* is semi-aquatic, it's easy. They would be living in different environments and mostly preying upon aquatic rather than terrestrial organisms. In technical terms, ecological niche partitioning had occurred, thus allowing these giants to live close to one another.

Spinosaurus was a large creature, probably because it was preying on other large creatures. If we follow the semi-aquatic model, the fossils of very large turtles, eight-feet-long lungfish, thirteen-feet-long coelacanth fish and twenty-five-feet-long sawfish found in the same sedimentary deposits may have formed some of the prey items for this animal.

If this interpretation of *Spinosaurus* is correct, what a fantastic sight this beast would have been, stalking perhaps heron-like at the side of a river dipping its long snout in to the water to feel the vibrations of fish as they swam by. With other individuals actively swimming through the water like crocodiles, with that magnificent sail perhaps brightly colored, defining its territory or advertising for a mate.

This interpretation of spinosaurs paints this dinosaur as a creature that was in transition between the terrestrial and aquatic environments, a kind of transitional form. A word of caution, though. It is important to remember that any model is a hypothesis, and this one is certainly being highly debated in the circles of vertebrate paleontology. Some researchers are not yet convinced about this new anatomical reconstruction of *Spinosaurus*, or some of the interpretations that are being drawn from it.

This investigation has shown, though, how new fossils are continually pushing paleobiological research forward even if pushing forward means generating more debate and alternate hypotheses. The paleontological discipline of paleobiology is alive and well, and our insights into ancient creatures like this are only going to get clearer as new discoveries are made, and more flesh is placed back onto old bones.

Lecture 17

Whales: Throwing Away Legs for the Sea

In this lecture, you will learn about the fantastic evolutionary journey of the group of mammals known as whales and the incredible diversity that can result from natural selection in an instant of geological time. This lecture will address several questions: Why would creatures that evolved on land move back into the ocean? Which group of mammals would start the whales along a path to the ocean? And how can we explain the wonderful paleontological treasure at Cerro Ballena?

From Land Back to Ocean

- Since we started to consider animals scientifically, comparing them to other creatures in the animal kingdom, it was abundantly obvious that whales, or cetaceans, were mammals that probably had ancestors that lived on land. We see features such as the bones in flippers that very closely resemble the limbs of land mammals, a vestigial hind limb, a vertical movement of the spine when swimming that shares more in common with a mammal running than a fish swimming, and the fact that whales need to breathe air—all of which suggested a land animal link.

- But why make the move back into the water? Vertebrates had to overcome several difficulties in leaving the oceans and adopting a life on land, including how to obtain oxygen from air rather than water, develop a more robust skeleton that would make up for the buoyancy effects that would no longer be enjoyed on land, accommodate hearing in a gas environment rather than a liquid one, and develop limbs rather than fins for getting around.

- Despite all the challenges vertebrates faced, and adaptations they evolved in making the break for land, some of them returned to the water—some so completely that they can no longer survive on land. But why?

- It is important to understand that evolution does not have a goal in mind. Creatures will be selected for in particular environments, or as environments change, based on certain physical and biological shifts through time. The characteristics, or mutations, that organisms possess are not produced because they "choose" to evolve things like flippers or echolocation. These features evolve through natural selection over long stretches of time.

- It is important to understand this while we consider the evolution of cetaceans from their land-based ancestors. The "proto-whales" did not consciously move back into the oceans, forcing their own evolution in some way. Rather, certain forms would have a selective advantage, allowing them to inhabit increasingly more aquatic environments in a step-by-step manner over millions of years. This story is now all the more fascinating, as many wonderful fossils have been found in recent years that have turned mere speculation about the evolution of cetaceans into a true, well-documented family history.

Whale Evolution

- Genetically, we know that whales fall into a group of mammals called the even-toed ungulates, or the artiodactyls. The Artiodactyla include familiar modern animals, such as pigs, camels, giraffes, and deer, but the closest land-living relatives of the whales today is the hippopotamus.

- This is not suggesting that Whales evolved from hippos; rather, they share a common ancestor somewhere in the biosphere's deep past. Who is the best paleontological candidate for this common ancestor? To answer this question, we need to roll back the clock to around 54 million years ago, about 12 million years after the death of the dinosaurs, during the Early Eocene and in a region of the planet defined by the Tethys Sea.

- Back then, the Earth was much warmer than it is today and climatically more homogeneous, with less difference in temperature between the equator and the poles. The Early Eocene environment had just come off the heels of the Paleocene-Eocene thermal maximum, which was the warmest period of the Cenozoic, just 2 million years earlier.

- The Eocene world had started to look a little more like the planet we know today, with familiar-looking continents and a widening Atlantic Ocean. Australia was still connected to Antarctica, and India was starting to collide with Asia, building the Tibetan plateau and the Himalayas.

- And on India, a small herbivorous deerlike creature about the size of a raccoon could be found living along the edge of rivers and lakes in an area not far above sea level but that is now elevated high in the Himalayas. This creature, about 2 feet long, is *Indohyus*, a member of a sister group of the cetaceans called the raoellids. Most paleontologists now agree that it was from animals like this that the lineage that would give rise to modern whales would evolve, at about 54 million years ago.

- Why do we presume a cetacean relationship to this little beast? Although it doesn't look like a whale, there are similarities. For example, the auditory bulla, the bones that surround the inner ear, are very distinctive—adapted for hearing underwater and only shared by this group and the cetaceans. In addition, the bones of *Indohyus* are thickened, not unlike modern hippos, an adaptation that semiaquatic animals possess to help them overcome buoyancy effects and allow them to stay underwater.

- If *Indohyus*, or a creature very much like it, gave rise to the whales, who is the first true member of the whale lineage, the cetaceans? *Pakicetus* is, so far, the oldest member of the cetaceans for which we have fossil evidence. We are now around 50 million years before present, and the small herbivorous "deer" from which whales evolved has grown to about 6.6 feet long, and this creature has also developed a taste for meat.

- *Pakicetus* lived on the shores of the Tethys Sea in what is now northern Pakistan. It lived in a freshwater floodplain environment but was likely a poor swimmer. It probably waded through shallow water, ambushing animals that came to the water's edge to drink.

- This creature still doesn't resemble modern whales, but in addition to the whalelike features already present in *Indohyus*, *Pakicetus* is starting to develop an elongate whalelike head.

- The next fossil whale is *Ambulocetus*, or the walking whale, dated to around 49 million years ago. Although a little larger than *Pakicetus*, about 10 feet long from tip to tail, *Ambulocetus* fossils demonstrate an important change in the cetaceans: They are now starting to inhabit marine in addition to freshwater environments. The move to the oceans had started.

- Another important feature is the presence of a large mandibular foramen. In modern whales, this area of the jaw is filled with fat and helps pass sounds to the inner ear, an important feature for hearing underwater.

- Less than a million years later, evolution has continued to shape and alter the cetaceans. *Remingtonocetus*, a member of the family Remingtonocetidae, is found in more diverse marine environments than its older relatives, with fossils being recovered from nearshore and lagoonal sediments. Its limbs are shorter than previous cetaceans, and although it may have swum in the doggy-paddle style used by *Ambulocetus*, it may also have started to undulate its spine as an aid to swimming.

- Another interesting feature of the Remingtonocetidae is a reduction in the size of the semicircular canals of the inner ear, which regulate balance and are particularly important for active creatures moving around on land. Fred Spoor of University College London, and others, suggested that this change in canal size indicates that these creatures had reached a "point of no return." The cetaceans from now on were destined for an aquatic, or at least semiaquatic, existence.

- At around 48 million years ago, and probably living alongside the Remingtonocetidae, we find fossils belonging to the Protocetidae. Unlike the fossils that we have so far described that were only found in India and Pakistan, the Protocetidae have a much greater global distribution, including Europe and North America, and were probably swimming freely through many of the globe's tropical oceans.

- The Protocetidae are much more whalelike, with the possibility of the development of a fluke, a 2-lobed tail, in some species and the migration of the nasal openings to the top of the skull. For species of the Protocetidae, which may have still been semiaquatic, moving around on land would not have been graceful, probably akin to the way modern seals move today.

- *Basilosaurus* was the first completely aquatic whale. Collectively, these species belonged to a group that are called the Basilosauridae. They were a diverse group, with the basilosaurs themselves at around 60 feet long. These large whales probably occupied the role of top predator in the Eocene oceans between 40 and 34 million years ago, feeding on fish, sharks, and smaller whales.

- It was an odd-looking whale, though—kind of eellike. The vertebrae of *Basilosaurus* were hollow and possibly filled with fluid, which meant that it probably didn't have a deep dive capability and hunted mostly in surface waters. These Eocene monsters also had fairly small brains, meaning that, unlike modern whales, they probably didn't exhibit much complex social behavior.

- It is thought that a group within the Basilosauridae, the dorudontids, would give rise to modern whales. They were much smaller than the elongate basilosaurs, at about 16 feet long, but had overall proportions resembling modern whales. It is from creatures like this that we get the diverse forms of all of today's whales.

- Whales are broadly divided into 2 groups, toothed whales and the baleen whales, who share a common ancestor about 34 million years ago.
  - Toothed whales are characterized by having teeth, and they hunt using echolocation by making a series of clicks at various frequencies. They range in size from the tiny 4.5-foot vaquita to the sperm whale, which can range in size from 33 to 66 feet in length.

  - The baleen whales gulp large volumes of water and sieve out krill, small fish, and other microplankton by squeezing the water back out through their baleen plates, a substance composed of a protein similar to human fingernails. Humpbacks are a popular favorite with whale watchers, but another species of baleen, the blue whale, is possibly the largest animal that has ever lived, at almost 100 feet long.

Cerro Ballena

- By the time we examine the whales at a fossil site in the Atacama Desert of Chile called Cerro Ballena, at around 6 to 9 million years ago, modern-looking whales had already evolved from the dorudontids back in the Eocene. A whole range of marine mammals are found at the site, but the fossil baleen whales are probably the most spectacular.

- Dr. Nicholas Pyenson and collaborating scientists from Chile uncovered a fascinating fossil conundrum: Why are there so many fossil marine mammals present in this location, and why, given that under normal conditions these creatures would be scavenged and their bones scattered, are there so many in such an excellent state of preservation?

- Furthermore, this is not an isolated event. Similar collections of fossils at Cerro Ballena are found in multiple discrete horizons. It would appear that whatever killed and stranded these marine creatures happened around 4 times over a period of 10,000 to 16,000 years.

- Other features of this fossil deposit have helped unravel this mystery. First, these creatures were stranded on a tidal flat, roughly orthogonal to current flow. The whales are also preserved belly up, suggesting that they died at sea and then washed to shore. In addition, high concentrations of iron in the sediments hint at a high algal concentration in the waters in which these animals were swimming.

- It is thought that all of this adds up to a story of a mass stranding caused by a harmful algal bloom. Not all blooms of algae are harmful, but some species of microplankton can cause problems for marine life and have been known to cause whale strandings.

- Our story likely starts with rainfall over the Andes, flushing minerals rich in iron into the Pacific Ocean. This in turn causes a bloom of algae and the death of many marine mammals and fish. A high tide would help wash the animals ashore onto a mudflat, all belly up as decomposing gases start to swell the gut. The ocean hydrodynamically aligned them into neat rows, producing a stranding, just as we find in modern whale strandings today.

- Usually, these whales would be scavenged by creatures living on shore, but back then, as today, a desert existed inland along the coast, restricting the number of creatures and potential scavengers living in the area. Sedimentation rates were also fairly constant, ensuring that the dead were covered rapidly. As such, Cerro Ballena has provided scientists at the Smithsonian's National Museum of Natural History with a wonderful window into marine mammals of the Pacific Ocean more than 6 million years ago.

Questions to consider:

1. If modern whales were removed from the biosphere, which group of mammals might take their place?
2. What was the equivalent of whales during the time of the dinosaurs?

Suggested Reading:

Carwardine, *Smithsonian Handbooks*.

Thewissen, *The Walking Whales*.

Lecture 17 Transcript

Whales: Throwing Away Legs for the Sea

The Earth does not always give up its secrets easily. Paleontologists may have to spend many long hours, days, months, even years hunting for fossils sometimes with no luck at all. As a micropaleontologist, I can't physically hunt down my fossils in the field they are way too small, but I have logged many hours carefully preparing, dissolving, separating and concentrating fossil residues from rocks I collected; excitedly transferring then to a microscope to find nada, zilch, nothing.

Often when the fossils are found they are either poorly preserved or fragmentary. Frequently one is presented with a jigsaw puzzle that takes dedicated workers months, sometimes years to fit together again.

But sometimes, just sometimes, we strike gold. Let's let Dr. Nick Pyenson tell about just one such find.

What I love about paleontology is that it's fundamentally about discovery. Discovery in the world out there. You never know what you're going to find.

For the site in Chile, we just happened to be at the right place at the right time. We heard about this unusual site called Cerro Ballena many times before, and one of our colleagues brought it to our attention because it was exposed by a road construction company.

What we're talking about is a site that's about three football fields in size, covered with the skeletons of fossil whales. The skeletons were exposed right next to the Pan-American Highway, so there are 18-wheelers, large road equipment going by all the time, bringing vibrations that can actually damage and destroy the fossils. Our job is to preserve as much information as possible, and we had precious little time to document anything scientifically relevant about how those skeletons came to rest.

So I remember that before I went to the Atacama I've met two people—Adam and Vince—roaming around the halls here in Paleobiology, and I thought, "I wonder if their grab bag of digital tools would help us.

So we met Nick years ago in the basement of the Natural History museum when we were digitizing dinosaur fossils. He stopped by to see what we were doing; we talked for 20 minutes, and then we didn't hear from him for over a year. Suddenly we got a panicked phone call, and he wanted to bring us down to Chile within a week. And so, within about a week, we found ourselves on an airplane with all of our equipment heading to the desert in Chile.

I think within maybe 30 minutes of our plane landing we were on site and we started 3D scanning. We basically just threw everything we had at the site because we knew we only had five days.

We really developed different workflows and different kinds of ways to document the whale skeletons before us, using a variety of digital techniques because nobody had ever done this before. There is no field guide about how to use 3D in this context.

One of the exciting things about documenting something in 3D on location in another country using new technologies is that you don't know exactly what you've captured until you get it back into the lab and you get to process everything.

Now that we've scanned the site we can actually bring that site back. You have to remember, that site no longer exists.

We can translate it into a variety of outputs, one of which is 3D printing, and what's very exciting is that we're going to be able to print one of the best-preserved fossil whale skeletons. When done, it will be the world's largest 3D print of its kind.

When you document something in 3D as thoroughly as we have, the cool thing is that we might have the information to questions that scientists haven't yet had the opportunity to ask.

So taking Adam and Vince down to Chile was a bit of a leap of faith, but that hunch proofed to be correct, I think, in the long run. Because what they've done is show me what 3D can really do under the right circumstances in the right context, with the right need case. With Cerro Ballena, it was clearly the right thing to do.

As I told you—pure paleontological gold; and isn't it incredible what technology allows scientists to achieve today? An entire paleontological site now preserved electronically even though the original quarry is now paved over.

We will return later and see what this fascinating study has revealed about these incredible fossils. But for the rest of this episode, I would like to consider why would creatures that evolved on land move back into the ocean? Which group of mammals would throw away their legs for the sea? And finally, ask how can we explain the wonderful paleontological treasure at Cerro Ballena?

Since we started to consider animals scientifically comparing them to other creatures in the animal kingdom, it was abundantly obvious that whales or cetaceans to give them their scientific name were mammals that probably had ancestors that lived on land.

We see features such as the bones in flippers that very closely resemble the limbs of land mammals a vestigial hind limb a vertical movement of the spine when swimming that shares more in common with a mammal running than a fish swimming and of course the fact that whales need to breathe air. All of this suggested a land animal link. But why make the move back into the water?

If you have watched my other show *A New History of Life* you will know the difficulties vertebrates had to overcome in leaving the oceans and adopting a life on land. One of the most obvious problems was how to obtain oxygen from air rather than water, but there were other hurdles too.

There is the need for a more robust skeleton that would make up for the buoyancy effects you would no longer enjoy on land. Also, there are adjustments needed to accommodate hearing in a gas environment rather than a liquid, and of course there is the necessity of developing limbs rather than fins for getting around.

In *A New History of Life*, we identified a fish that was pre-adapted to make this break for land—Eusthenopteron. And Tiktaalik—a vertebrate in transition between fish with fins and a tetrapod with limbs.

In that series, we also described an array of early tetrapods like Acanthostega and Ichthyostega and some of the first reptiles like little Westlothiana which with the evolution of amniotic shelled eggs would finally break the ties with the aquatic environment.

So given all the challenges vertebrates faced and adaptations they evolved in making the break for land it might be a surprise that some of them returned to the water; some so completely that they can no longer survive on land. Several groups of vertebrates have made this profound move, but why?

At this point, we must be cautious with the language we use. It is very easy to give the impression that there is some sort of teleological motivation driving evolution. It is important to understand that evolution does not have a goal in mind it is not a matter of design.

Creatures will be selected for in particular environments as environments change based on certain physical and biological shifts through time. The characteristics or mutations that organisms possess are not produced because they choose to evolve things like flippers or echolocation. These features evolve through natural selection over long stretches of time.

To expand on this, a little further let's consider the theories of one Jean Baptiste Lamarck. Lamarck was born in Northern France in 1744 and had an early career as a soldier. His career path would change, though, and he became quite a noted naturalist.

He lived during interesting philosophical and political times and would witness the French Revolution potentially a little worrying as he did come from an aristocratic family, even if it wasn't all that well to do at the time. He was appointed as the keeper of the Royal Botanical Gardens and very tactfully changed its name from Jardin du Roi—Garden of the King—to Jardin des Plantes at the height of the revolution.

Earlier in his career as a naturalist he believed that species were fixed but would later change his views, coming to believe in the transmutation of species over time. Lamarck, although not the first to consider some form of evolution, was one of the first to come up with a reasoned process to explain how it might happen what today we call the concept of Lamarckian inheritance

His hypothesis suggests that creatures pass on characteristics to offspring that they have acquired over a lifetime. Let's consider how Lamarck believed species could change over time with a little help from this Giraffe. In a Lamarckian view of evolution, giraffes got longer necks as they stretched to reach high leaves in high trees. This tendency to generate longer necks would then be passed on from generation to generation.

In contrast, a Darwinian view though would state that giraffes that were born with slightly longer necks would have a selective advantage if they were feeding on tall trees. Over time natural selection would favor those giraffes in this tall tree niche with longer necks, and over time we would see the evolution of the giraffe as we know it today.

It is important to understand this while we consider the evolution of cetaceans from their land-based ancestors. The proto-whales did not consciously move back into the oceans forcing their own evolution in some way. Rather certain forms would have a selective advantage allowing them to inhabit increasingly more aquatic environments in a step-by-step manner over millions of years.

This story is now all the more fascinating as in recent years many wonderful fossils have been found that have turned mere speculation about the evolution of cetaceans into a true, well documented, family history

So, which group of mammals would start the whales along a path to the ocean? Genetically we know that whales fall into a group of mammals that we call the even-toed ungulates or the artiodactyls.

The Artiodactyla include familiar modern animals like pigs, camels, the giraffe, and deer but the closest land-living relatives of the whales today is the hippopotamus. Please note though this is not suggesting that Whales evolved from hippos Rather they share a common ancestor somewhere in the biosphere's deep past.

This concept of a common ancestor is often a source of confusion for some. particularly with regard to our own biological status specifically, we and our closest living relatives the chimpanzee and the bonobo all both share a common primate ancestor back in Earth's history. Humans did not evolve from chimps.

But getting back to the whales: Who is the best paleontological candidate for the common ancestor in this case?

To answer this question, we need to roll the clock back to around 54 million years about 12 million years after the death of the dinosaurs during the early Eocene and in a region of the planet defined by the Tethys Sea.

Back then the Earth was much warmer than it is today and climatically more homogeneous with less difference in temperature between the equator and the poles. Fossils of tropical species of palm trees and crocodiles are found at quite high latitudes. The early Eocene environment had just come off the heels of the Paleocene–Eocene Thermal Maximum which was the warmest period of the Cenozoic just 2 million years earlier.

The Eocene world had started to look a little more like the planet we are familiar with today with continents and oceans in the familiar position and a widening Atlantic Ocean. And India was starting to collide with Asia building the Tibetan plateau and the Himalayas. And in India a small herbivorous deer-like creature about the size of a raccoon could be found living along the edge of rivers and lakes in an area not far above sea level but which is now elevated high in the Himalaya Mountains.

This creature about 2 feet long is Indohyus, a member of a sister group of the cetaceans called the raoellids. Most paleontologists now agree that it was from animals like this that the lineage that would give rise to modern whales would evolve, at about 54 million years ago.

Why do we presume a cetacean relationship to this little beast? It certainly doesn't look like a whale. There are, though, similarities. For example, the auditory bulla the bones that surround the inner ear are very distinctive adapted for hearing underwater and only shared by this group and the cetaceans. In addition, the bones of Indohyus are thickened, not unlike modern hippos; an adaptation that semiaquatic animals possess to help them overcome buoyancy effects and allow them to stay underwater.

You could imagine this little creature probably moving relatively slowly on land but diving into a pond or river to escape predators or nibble on aquatic plants. So if Indohyus or a creature very much like it gave rise to the whales who is the first true member of the whale lineage, the cetaceans?

Allow me to introduce Pakicetus so far the oldest member of the cetaceans for which we have fossil evidence. We are now around 50 million years before present and the little herbivorous deer from which whales evolved has grown to about 6.6 feet long, and this creature has also developed a taste for meat.

It lived on the shores of the Tethys Sea in what is now northern Pakistan. Pakicetus lived in freshwater floodplain environments but was likely a poor swimmer. It might have waded through shallow water, ambushing animals that came to the waters' edge to drink.

This creature still doesn't resemble modern whales, but in addition to the whale-like features present in Indohyus, Pakicetus is starting to develop an elongate whale-like head.

Our next fossil whale is Ambulocetus or the walking whale dated to about 49 million years ago Although a little larger than Pakicetus—about 10 foot from tip to tail—Ambulocetus fossils demonstrate an important change in the cetaceans they are now starting to inhabit marine in addition to freshwater environments the move to the oceans had started.

Another important feature is the presence of a large mandibular foramen. In modern whales, this area of the jaw is filled with fat and helps pass sounds to the inner ear, an important feature to hearing under water.

Less than a million years later evolution has continued to shape and alter the cetaceans. This is Remingtonocetus—a member of the family remingtonocetidae. A creature that is found in more diverse marine environments than its older relatives with fossils being recovered from nearshore and lagoonal sediments. Its limbs are shorter than previous cetaceans and although it may have swum in the doggy paddle style used by Ambulocetus it may also have started to undulate its spine as an aid to swimming.

Let's pause just a minute. Ever considered how you manage to stand up? It's actually a complex interplay of many parts of your physiology and anatomy, but a vital part of the hardware of balance are the semicircular canals in the inner ear. These are 3 tubes filled with a liquid that moves as you move your head in various directions The liquid moves hairs in the canals that trigger nerve impulses to your brain, giving you an idea of the motion you're experiencing. These motion detectors are particularly important if you are an active creature moving around on land.

And this brings us back to our whale story for another interesting feature of the remingtonocetidae is a reduction in the size of their semicircular canals. In a paper authored by Fred Spoor of the University College of London and others, it was stated that this change in canal size indicates that these creatures had reached a point of no return. The cetaceans from now on were destined for an aquatic or at least semiaquatic existence.

At around 48 million years ago and probably living alongside the remingtonocetidae we find fossils belonging to the protocetidae. Unlike the fossils that we have so far described that were only found in India and Pakistan, the Protocetidae have a far greater global distribution including Europe and North America and were probably swimming freely through many of the globe's tropical oceans.

As you can see the protocetidae such as Protocetus here are becoming much more whale-like with the possibility of the development of a fluke a two-lobed tail in some species and the migration of the nasal openings to the top of the skull.

For species of the protocetidae which may have still been semiaquatic moving around on land would not have been graceful probably akin to the way modern seals do today.

But for the next stage in Whale evolution, we need to move to the USA and some interesting fossils recovered from Arkansas and Alabama. In 1843 vertebrae belonging to some fossil animal were sent to the American Philosophical Society in Philadelphia and described by Richard Harlan. Initially, these findings were identified as belonging to a Mesozoic marine reptile. This may in part be due to the fact that marine reptiles like Mosasaurs and Pliosaurs had recently been described in Europe. Accordingly, Harlan gave his find a reptilian suffix, Basilosaurus, meaning "King Lizard."

In 1839, the famous British anatomist Richard Owen who would later go on to found what is today the Natural History Museum in London correctly identified the fossils as belonging to a mammal. He wanted to rename the beast Zeuglodon meaning "yoked tooth," but the rules of taxonomy are strict, and the first name has precedence.

About 10 years after the initial description of Basilosaurus, Albert C. Koch, a fossil hunter and it would appear a man who had the element of the showman about him, traveled to Alabama to find more fossils. He collected a range of material including a Basilosaurus skull that would be assembled into a composite skeleton although he did claim it came from an individual that he called Hydrarchos, "Water King."

In 1845 you could pay 25c to go and see this giant sea serpent that Koch described as being 114 feet long and weighing 7500 lbs. As I said—showman. Despite great skepticism from the scientific community Hydrarchos was a great hit and would eventually tour Europe like a rock star.

The Smithsonian has quite a long history with Basilosaurus too. In 1894, George Brown Goode, then curator of the Smithsonian's National Museum, sent one of his assistant curators of invertebrate Paleontology, Charles Schuchert, to Clarke County in Alabama to collect Basilosaurus specimens. I wonder if they were getting fed up with all the attention Hydrarchos was getting?

Apparently, Basilosaurus fossils were so common in Clarke County that farmers had been using them in the stone walls He collected fossils from various individuals and assembled a complete skeleton that was the first accurately assembled Basilosaurus in the world. You can view it in all its glory today along with other examples of whale evolution in the wonderful Sant Ocean Hall in the National Museum of Nature History.

Basilosaurus was the first completely aquatic whale. Collectively these species belonged to a group that is called the basilosauridae. They were a diverse group with the Basilosaurus themselves at around 60 feet long. These large whales probably occupied the role of top predator in the Eocene oceans between 40–34 million years ago feeding on fish, sharks, and smaller whales.

It was an odd looking whale though kind of eel-like but not quite the sea serpent of Koch's imagination. The vertebrae of Basilosaurus were hollow and possibly filled with fluid which meant it probably didn't have a deep dive capability and hunted mostly in surface waters. These Eocene monsters also had fairly small brains, meaning that unlike modern whales they probably didn't exhibit much complex social behavior.

It is thought that a group within the Basilosauridae, the Durodontids would give rise to modern whales. They were much smaller than the elongate Basilosaurus at about 16 feet long but had overall proportions resembling modern whales. It is from creatures like this that we get the diverse forms of all today's whales.

Whales today are broadly divided into two groups toothed whales and the baleen whales who share a common ancestor around about 34 million years ago. Toothed whales as their name suggests are characterized by having teeth, and they hunt using echolocation by making a series of clicks at various frequencies They range in size from the tiny 4.5 foot Vaquita, a species of porpoise found in the Gulf of California, to the sperm whale that can range in size from 33–66 feet in length. They are also among the great acrobats of the oceans, as anyone who has watched dolphins surfing in the wake of a boat or seen Orcas breaching out of a clear sea can attest.

The other group of whales the baleen whales gulp large volumes of water and sieve out krill, small fish, and other microplankton by squeezing the water back out through their baleen plates a substance composed of a protein similar to your fingernails. Humpbacks are a popular favorite with whale watchers, but it is another species of baleen that is possibly the largest animal that has ever lived—the blue whale. At almost 100 feet long the size of the animal is truly mind blowing especially when you consider that the main source of food is krill tiny crustaceans 0.4-0.8 inches long.

Le'ts just pause here and consider the changes we have described. From the first terrestrially adapted cetaceans like Pakicetus at 50 million years ago to the first obligate aquatic members of the whale family such as Basilosaurus at around 41 million years ago. It took just 9 million years of evolution to radically transform this group of mammals.

It shows how rapidly natural selection can operate when a group of creatures radiates to fill a relatively unoccupied niche; in this case, a niche left vacant by the giant Mesozoic aquatic reptiles following the end-Cretaceous extinction event. It also highlights how paleontology can be used to test the predictions made by evolutionary biologists, the fossils themselves being the acid test of our suppositions regarding our understanding of the history of life.

By the time we examine the whales at Cerro Ballena at around 6 to 9 million years ago modern looking whales had already evolved back in the Eocene. A whole range of marine mammals are found at the site, including two species of phocid earless seals, and Odobenocetops, a bizarre looking walrus, faced whale that had tusks. There was even a semi-aquatic sloth that probably grazed on seaweed and seagrass. But it is the fossil baleen whales that are probably the most spectacular.

Dr. Nicholas Pyenson and collaborating scientists from Chile uncovered a fascinating fossil conundrum. Why are there so many fossil marine mammals present in this location and why—given that under normal circumstances these creatures would be scavenged and their bones scattered—why are there so many in such an excellent state of preservation?

What's more, this is not an isolated event. Similar collections of fossils at Cerro Ballena are found in multiple discrete horizons. It would appear that whatever killed and stranded these marine creatures happened around four times over a period of 10-16 thousand years.

Other features of this fossil deposit have helped unravel this mystery. First, these creatures were stranded on a tidal flat, roughly orthogonal to current flow. The whales are also preserved belly up suggesting they died at sea and then washed to shore. In addition, high concentrations of iron in the sediments hints at a high algal concentration in the waters in which these animals were swimming.

It is thought that all of this adds up to a story of a mass stranding caused by a harmful algal bloom or HAB. Not all blooms of algae are harmful in this image you can see a bloom of coccoliths off the coast of Cornwall, England, during the summer of 1999. But some species of microplankton, including species of dinoflagellates and diatoms, can cause problems for marine life and have been known to cause whale strandings.

Our story likely starts with rainfall over the Andes flushing minerals rich in iron to the Pacific Ocean. This, in turn, causes a bloom of algae and the death of many marine mammals and fish. A high tide would help wash the animals ashore onto a mudflat all belly-up as the decomposing gases start to swell the gut. The ocean hydrodynamically aligned them into neat rows producing a stranding just as we find in modern whale strandings today.

Usually, these whales would be scavenged by creatures living on shore, but back then as today a desert existed inland along the coast restricting the number of creatures and potential scavengers living in the area. Sedimentation rates were also fairly constant ensuring that the dead were covered rapidly.

As such Cerro Ballena has provided scientists at Smithsonian's National Museum of Natural History with a wonderful window into marine mammals of the Pacific Ocean over 6 million years ago. But more than that, it speaks to the fantastic evolutionary journey of this group of mammals and also the incredible diversity that can result from natural selection in an instant of geological time.

It has also shown how traditional excavation techniques hand in hand with modern visualization technology can help preserve and decode complex paleontological sites. Although the fossils are no longer visible, the site is preserved forever and will continue to provide valuable insight into the ecology and evolution of cetaceans for many years to come.

Lecture 18

Insects, Plants, and the Rise of Flower Power

Flowering plants, the angiosperms, have had an important role to play in Earth's transformation beyond an aesthetic one. In providing fruits and cereal crops, they have also helped drive the evolution of human civilization. They are a remarkable part of our biosphere. In this lecture, you will discover what Earth was like before flowers, where and when flowers evolved, how the angiosperms dominated, and how angiosperms and animals were partners in the great floral takeover.

Earth before Flowers

- The first evidence of plants growing on the land comes from the Silurian period, about 433 million years ago, with fossils of simple plants living on water-clogged floodplains. Fossils interpreted as the reproductive spores from these simple plants have been reported from the Ordovician period, about 470 million years ago, but these finds are still controversial.

- By the time we get to the Devonian, plants are invading drier landscapes, and by the Late Devonian, there is an incredible innovation: the development of the seed. This, along with other important developments, such as leaves, wood, and true roots, would finally allow plants to break ties with the water's edge and spread even farther into the centers of the continents.

- Various seed ferns would flourish throughout the rest of the Late Paleozoic, and the vast forests of the Upper Carboniferous (Pennsylvanian) are responsible for the coal in the Northern Hemisphere that would ultimately power the Industrial Revolution.

- Seed ferns would flourish into the Triassic and even continue into the Cenozoic, but it would be the gymnosperms, such as conifers, cycads, and an array of ginkgo-like plants (Bennettitales and gnetophytes), that would really start to advance across the Mesozoic world.

- Gymnosperm means "naked seed," describing the unenclosed condition of their seed, unlike the fruit-encased seeds of the angiosperms. The seeds of gymnosperms form in a variety of ways: on the surface of scales or leaves, on short stalks in ginkgophytes, in cones as found in conifers, in flowerlike structures that occur in the extinct Bennettitales, and in other specialized reproductive structures in groups of gymnosperms that have no equivalents in the modern flora.

- By the Triassic, the first period of the Mesozoic, gymnospermous plants represented about 60% of flora species and around 80% of Jurassic species. Today, there are more than 10,000 species of gymnosperms.

- Today, some gymnosperms, such as the conifers. produce vast amounts of pollen that they shed into the wind with the hope of reaching a female pollen receptor. Not all gymnosperms are wind pollinated, though; in fact, most are insect pollinated.

- Interesting evidence of early plant-insect relationships in the gymnosperms has come from the chemical analyses of the fossil mid-Mesozoic proboscis of kalligrammatid lacewings. Pollen was found associated with the specimens'

head and mouthparts, which has led researchers, such as Dr. Conrad Labandeira at the Smithsonian, to suggest that they were feeding on pollination drops, drops of sugary fluid secreted by gymnosperms to trap pollen grains.

- Dr. Labandeira, and colleagues in France and Spain, have found additional evidence of an early insect-plant pollination relationship from the Lower Cretaceous (110 to 115 million years ago) of the Basque region of Spain. Trapped in amber are small insects called thrips or thunderbugs.

The Evolution of Flowers

- Charles Darwin was particularly perplexed by the fossil record of flowers, which appeared to just suddenly emerge, fully formed, in the Cretaceous period. This was one of 2 such problematic fossil "appearances" that gave Darwin a headache. The other was the apparent sudden appearance of complex creatures, such as trilobites, at the base of the Cambrian—often called Darwin's dilemma. New fossil discoveries would provide context to the evolution of flowers and help clarify Darwin's mystery, but as of yet, not completely.

- An ongoing question about the early evolution of flowering plants regards the environment of their evolution. Some researchers suggest that flowering plants evolved on land, but others have contemplated a possible aquatic, or at least semiaquatic, origin. One way we can consider approaching this question is to examine where some of the most primitive angiosperms, called ANITAs (an acronym that stands for some of the basal angiosperm lineages), are found today.

- Researchers such as Mark Chase at the Royal Botanic Gardens in London and collaborators all around the world have been analyzing the DNA of angiosperms to provide direct genetic evidence from which to base a comparison of plant species. From this evidence, the flowering plants with the greatest similarity are clustered together into what we call clades. The picture that has emerged has required a reassessment of some of the relationships we assumed before this technique was available.

- Modern molecular phylogeny has also revealed a plant that is at the base of the angiosperm family tree. This plant represents the most primitive flowering plant that still exists on the planet today: *Amborella*, the only remaining species of the family Amborellaceae, is found on the island of Grande-Terre of New Caledonia, in the South Pacific.

- Given its basal status, it would seem the tropical upland forest setting of *Amborella* would be a good candidate for the environment in which angiosperms evolved. The problem, despite its primitive status, is a complete lack of fossils of a similar plant early in the evolution of the angiosperms.

- The other possible candidate environment is the aquatic environment. The group of plants that includes water lilies (the Nymphaeales) have an early fossil record going back to the Early Cretaceous (125 million years ago). Although this lineage doesn't occur at the base of the current family tree of flowering plants, it is pretty close.

- Surprisingly, perhaps, support for an aquatic origin also comes from *Amborella*, which, like water lilies, has vestigial gas exchange canals, useful in submerged stems and roots.

- The aquatic hypothesis is further supported by one of the oldest complete flowering plant fossils thus far discovered. The fossil, named *Archaefructus*, was recovered from a rock unit called the Yixian Formation in northeastern China that has become famous for its fossils of feathered dinosaurs, primitive birds, and long-proboscid insects of various kinds.

- Based on other fossils found in the area, it was suggested that *Archaefructus* was growing around 144 million years ago, the Jurassic period, which would easily place it at, or very close to, the origin of flowering plants. But once age-dated, it was found that *Archaefructus* dated to 124.6 million years ago, during the Early Cretaceous—after the appearance of the first angiosperm pollen. Although we are still looking for the first fossil flower, *Archaefructus* is still an example of a very early angiosperm and, as such, might still provide insight into where flowers first evolved.

- There are serious concerns regarding the rapid emergence of a terrestrial flowering plant from an aquatic ancestor. For example, how could plants that had evolved in the water evolve quickly enough to cope with gravity on land? There is no doubt that some flowering plants did evolve in an aquatic environment early in their history, but was this the environment into which flowering plants first appeared? It is still very much a matter of debate.

- Regarding the timing of the evolution of plants, based on molecular evidence, it is thought that angiosperms and gymnosperms last shared a common ancestor sometime that is very broadly called the pre-Cretaceous. Hopefully, new fossil discoveries will provide us with more insight to the early times of flower evolution.

Flower Power

- Around 100 million years ago, during the mid-Cretaceous, there was a great radiation in angiosperm diversity initially noted in the fossil record by angiosperm leaf and pollen remains. By the Late Cretaceous, flowering plants started to take over environments that were formerly dominated by ferns, cycads, Bennettitales, and other gymnosperms.

- There is still some uncertainty as to how this replacement occurred. Was it competitive? In this model, the angiosperms, with their short life cycle and rapid growth, were able to muscle in on the gymnosperms and replace them. Or was it noncompetitive? We know that there was an extinction at the end of the Triassic period that drove many gymnosperm lineages into extinction. Were the angiosperms just occupying ecological empty space?

- To understand the spread of angiosperms, it would be useful to consider the nature of the world into which they were to rise to dominance. If you consider some of the earliest fossil angiosperms, dated at about 125 million years ago, the planet they inhabited was very different than it is today.

- The supercontinent of Pangaea had started to fragment, but there were still large continental blocks with Gondwana to the south and Laurasia to the north. It is likely that the interiors of these continents would be pretty dry and arid—environments that are particularly favorable to angiosperms, providing them with large areas of the landscape where they could spread and diversify. Angiosperms also have an ability to propagate and reproduce very quickly, allowing any colonization to be rapid.

- Expanding into the dry interiors of continents may also explain why their fossil record is so poor at this time, as arid upland areas are erosive regions that are likely susceptible to wildfire events and thus difficult areas for fossils to form.

- There is another factor, though. The mid-Cretaceous was a period of intense warming, probably related to the accelerated fragmentation of Pangaea and associated generation of ocean crust and carbon dioxide production. The global warming that resulted would have added to intense drought conditions in the centers of continents and further selected for the morphologically flexible angiosperms.

- Another hypothesis was put forward in 1986 to explain the explosion of flowers by the noted dinosaur paleontologist Robert Bakker, who suggested that it may have been changing dinosaur communities that allowed for angiosperms to spread. Around about 144 million years ago, there was a change in the style of herbivory in the dinosaurs—from high browsers that were cropping leaves from the branches of high conifers to low browsers.

- Low browsing would mean that gymnosperm seedlings would have a reduced opportunity to reach maturity. This would open up the canopy and allow for angiosperms to spread quickly. Due to their rapid life cycle, they were better adapted to reach maturity and thus produce seeds before they were browsed away.

Angiosperms and Animals

- The sole purpose of a flower is reproduction, and in the vast number of cases, that means reproduction where the male gametes (pollen) are transferred from one plant to another by insects. It is possible that this relationship, and therefore the first flowers, evolved in isolated settings such as islands or an island chain, which might also explain their apparent very-sudden appearance in the fossil record. Such isolated settings may have allowed for the development of a specialized relationship between a plant and an animal—for example, a wasp carrying pollen from one plant to another.

- But pollination via insects was not a new gig, as various insects had been aiding the pollination of gymnosperms before the widespread appearance of flowering plants. It is likely that some insects were preadapted to build this relationship with angiosperms. The angiosperms, though, would take insect—in fact, animal—pollination to a whole new level in a classic example of how 2 major groups of organisms would co-associate over time.

- Angiosperms, and in particular their flowers, would coevolve with animals to produce a whole suite of features—including scent, color, fruit, and

mimicry—that would not have existed if it were not for their intimate relationship.

- Most of the fossil and phylogenetic evidence indicates that the earliest flowers were small and bowl-shaped, not showy. There were some, though, that would have stood out in the Cretaceous landscape. Some of the first true flowers of the Cretaceous may have resembled something like magnolias. These flowers likely only had a pollen reward for their insect pollinators; the pollinators themselves were likely generalist in nature, such as beetles, short-tongued wasps, and flies.

- By the time we get to the explosion of angiosperms in the mid-Cretaceous, new varieties of flowering plants emerge with new structures, such as nectaries (where nectar is produced) and specialized petals designed for a more specialized relationship with insects. By the Late Cretaceous, flowering plants were very diverse. At this time, insects such as wasps and flies were also going through radiations, with species—such as the long-tongued bees and flies—evolving specific structures to exploit flowers.

- The end result of this radiation of the angiosperms during the Cretaceous would be a world with a radically different flora, and companion insect associates to match.

Questions to consider:

1. Why were flowers an “abominable mystery” to Charles Darwin?
2. Are insects the only animals to have a special relationship with flowers?

Suggested Reading:

Benton and Harper, *Introduction to Paleobiology and the Fossil Record*, chap. 18.

Goulson, *A Sting in the Tale*.

Lecture 18 Transcript

Insects, Plants, and the Rise of Flower Power

It's difficult to imagine a world without flowers. They have intrigued and inspired us for generations. The ancient Egyptians had a very special relationship with flowers, including species such as the blue lotus said to be the original container of various deities. Flowers have also been found associated with the afterlife. Human burials over 14,000 years old show that people were bringing flowers to gravesites.

Flowering plants, the angiosperms, have had an important role to play beyond the aesthetic, though. In providing fruits and cereal crops like rice, wheat, and corn—the evolution of which we shall investigate later—they have helped to drive human evolution and civilization too.

But our association with flowering plants is just a tiny part of the impact that the angiosperms have had on the planet. They are a remarkable part of the biosphere especially as—on a geological timescale—they are a relatively recent arrival.

So let's ask in this episode what was garden Earth like before flowers? Where and when did flowers evolve? Let's consider flower power and how did the angiosperms dominate? And also look at angiosperms and animals and see if they were partners together in a great floral takeover?

The first evidence of plants growing on the land comes from the Silurian period about 433 million years ago with fossils of simple plants like cooksonia living on water-clogged floodplains. Fossils interpreted as the reproductive spores from these simple plants have been reported from the Ordovician period about 470 million years ago, but these finds are still somewhat controversial.

By the time we get to the Devonian, plants are invading drier landscapes, and by the Late Devonian, there is an incredible innovation—the development of the seed. This along with other important developments such as leaves, wood, and true roots as opposed to rhizoids—the filamentous hairs you find in plants such as mosses and liverworts—these developments would finally allow plants to break ties with the waters' edge and spread even further into the centers of the continents, with a potential negative impact on the biosphere that we have explored in another episode in this series.

Various seed ferns—the Pteridospermatophyta—would flourish throughout the rest of the late Paleozoic, and the vast forests of the upper Carboniferous—the Pennsylvanian—are responsible for the coal in the Northern Hemisphere that would ultimately power the industrial revolution.

Seed ferns would flourish into the Triassic and even continue into the Cenozoic, but it would be one particular lineage—the gymnosperms like conifers and cycads and an array of ginkgo-like plants, Bennettitales, and gnetophytes—that would really start to advance across the Mesozoic world. Gymnosperm means "naked seed," describing the unenclosed condition of their seed unlike the fruit-encased seeds of the angiosperms.

The seeds of gymnosperms form in a whole variety of ways; on the surface of scales or leaves on short stalks in ginkgophytes, in cones as you find in conifers, in flower-like structures that occur in the extinct Bennettitales, or in other specialized reproductive structures in groups of gymnosperms that no longer have any equivalents in the modern flora

As you can see this is a diverse group of plants. By the Triassic, the first period of the Mesozoic, gymnospermous plants represented about 60% of flora species, and around 80% of Jurassic species. Today, there are more than 10,000 species of gymnosperms.

Today, some gymnosperms like the conifers produce vast amounts of pollen that they shed into the wind with the hope of reaching a female pollen receptor, a strategy, not unlike that used by corals during spawning events when clouds of sperm and eggs are released into the oceans. Not all modern gymnosperms are wind pollinated, though, in fact, many are insect pollinated. Let's have a look as some of the fossil evidence of early plant/insect relationships in the gymnosperms that might have foreshadowed the special relationship between flowering plants and insects that was to follow.

Interesting evidence of this ancient relationship has come from the chemical analyses of the fossil mid-Mesozoic proboscis lacewings. The analysis indicated that specimens did not contain iron in their siphon-like food tubes. If iron was present, they might have been blood suckers. Instead, pollen was found associated with the head and mouthparts which have led researchers like Dr. Conrad La Bandera at the Smithsonian to suggest that they were feeding on pollination drops, a drop of sugary fluid secreted by gymnosperms to trap pollen grains. In this way, these ancient lacewings were acting very much like butterflies well before true butterflies would evolve. You may remember, we met this particular fossil in an earlier lecture, showing that some even possessed large eye spots, probably to scare off predators the way that owl butterflies do today.

Dr. La Bandera and colleagues in France and Spain have found additional evidence of an early insect/plant pollination relationship from the Lower Cretaceous, about 110-115 million years ago, of the Basque region of Spain. Trapped in amber are small about 2mm insects called thrips or another common name for them is thunderbugs. The amber they found contained 6 female thrips with pollen grains stuck to their abdomens and wings. Now this was not a chance association rings of hair, called ring setae, on the abdomen and wing margins appear specifically adapted to maximize pollen adhesion. These pollen collection structures are analogous to the branched hairs found on the mouthparts of modern bees, used for collecting pollen. In addition, the pollen grains, most likely from a ginkgophyte, also appear adapted to be carried by the insects.

How could such a relationship develop? A clue may come from the modern ginkgo biloba trees. Even though these particular ginkgoes are wind-pollinated, ginkgoes are either male or female. Male trees have cones that produce pollen, and the female trees produce ovules on the ends of stalks. Pollen has to travel from the male tree to the female tree to allow for fertilization.

During the early Cretaceous pollen from the ginkgophyte was transported by the thrips between the male and the female plants. It is possible that the bugs were collecting pollen to feed their young nymphs that lived in the ovules of the female tree; ovules that would have provided shelter for a small colony of thrips. The thrips would very selectively transport pollen from the male to the female plant host. It could be that insects like thrips were therefore preadapted to later exploit the flowers of the angiosperm evolution that began to dominate the global flora during the Cretaceous.

So when and where did this revolution occur? Well, I guess we had better introduce the angiosperms before we get ahead of ourselves. Angiosperms are distinct from gymnosperms in producing flowers and fruits that enclose their seeds by two layers of tissue. These seeds also contain endosperm, a genetically distinctive, starchy substance that provides nutrients for the growing seedling. The vast majority of angiosperms rely on insect pollination but as we have seen it appears that insect pollination was ongoing long before the evolution of the angiosperms.

But perhaps the most incredible feature of the angiosperms are their flowers. Flowers are a massive investment for angiosperms, but this is outweighed by the advantages they provide. In utilizing animals to aid in their fertilization, they have employed a very efficient and directed pollen distribution and reception system. The flower contains both male stamens producing pollen and the female pistil containing the ovary and long style, topped by the pollen-receiving stigma. The whole structure is surrounded by petals. Flowers will often have the male and female components mature at different times to avoid self-fertilization.

When an obliging insect visits a flower, it deposits a clump of pollen on the flower's stigma resulting in a pollen tube that grows down from the grain, down through the pistil and into the ovary. Sperm is then transferred into the ovule and fertilization occurs, eventually developing into a seed enclosed by a fruit that may also be dispersed by an animal.

Studies of primitive flowering plants like the yellow water lily—*nuphar advena*—have shown evidence of an ancient genome duplication that likely proceeded the evolution of the ancestral flowering plant. Gene duplication occurs when a redundant second copy of genes is produced. Often this second copy has no effect, but they can mutate over a number of generations developing a new function that might then be expressed in an organism.

This may explain why morphological change, especially of the flower, could have been rapid. This, though, doesn't help us with the particulars of where and when the evolution of this important group of plants occurred.

Charles Darwin was particularly perplexed by the fossil record of flowers. Flowers appeared to just suddenly emerge fully formed in the Cretaceous Period. This was one of two such problematic fossil appearances that gave Darwin a headache. The other was the apparent sudden appearance of complex creatures like trilobites at the base of the Cambrian often called Darwin's Dilemma that we covered in an earlier episode.

In a letter to one of his friends, Joseph Hooker, a renowned botanist and explorer, dated 22 July 1879, he referred to the flower issue as an abominable mystery. As with Darwin's Dilemma new fossil discoveries would provide context to the evolution of flowers and help clarify Darwin's mystery, but as of yet, not completely, as we shall see.

An ongoing question about the early evolution of flowering plants regards the environment of their evolution. Some researchers suggest that flowering plants evolved on land, but others have contemplated a possible aquatic or at least semiaquatic origin. One way we can consider approaching this question is to examine where some of the most primitive angiosperms called ANITA's are found today. ANITA is an acronym that stands for some of the basal angiosperm lineages.

Assessing which plants are the most primitive can be a difficult task involving an analysis of what are regarded as primitive or derived characteristics with plants placed into groups based on this assessment. In doing so, you could build a family tree, which illustrated the evolution from primitive to more complex plant species. Fortunately, though, taxonomists now have a relatively new tool that provides more of a definitive answer—DNA molecular analyses.

Researchers such as Mark Chase at the Royal Botanic Gardens in London and collaborators all around the world have been busy analyzing the DNA of angiosperms to provide direct genetic evidence from which to base a comparison of plant species. From this evidence the flowering plants with the greatest similarity are clustered together into what we call clades.

The picture that has emerged has required a reassessment of some of the relationships we assumed before this technique was available. For example, we now know that a rose is closely related to the squash and consider the beautiful lotus flower obviously closely related to water lilies right? No, in fact, the lotus is more closely related to the American sycamore and London plane tree.

What modern molecular phylogeny has also revealed is a plant that is at the base of the angiosperm family tree. This plant represents the most primitive flowering plant that still exists on the planet today, and it is a plant called amborella.

Amborella is the only remaining species of the family Amborellaceae. To find it, you would have to travel to the Island of Grande Terre of New Caledonia, in the South Pacific.

Amborella is an evergreen, dioecious shrub that means it produces either male or female flowers but not at the same time. This is a characteristic more common in gymnosperms than angiosperms which also hints at its primitive status.

It grows in the mountain forests of Grande Terre in two phases. First as a ground-hugging seedling with a creeping root system and eventually changing into to shrub-like growth when an established deeper root system is present. Given its basal status, it would seem the tropical upland forest setting of amborella would be a good candidate for the environment in which angiosperms evolved.

The problem we face here though despite its primitive status is a complete lack of any fossils of a similar plant early in the evolution of the angiosperms. In truth fossil preservation in such upland environments would be kind of difficult anyway but not impossible, so perhaps confirmation is out there somewhere.

But for now, let's consider the other possible candidate environment—the aquatic environment.

For this, we turn to a group of plants that includes water lilies—the nymphaeales, which do have an early fossil record going back to the early Cretaceous—125 million years ago. Although this lineage doesn't occur at the base of the current family tree of flowering plants, it is pretty close.

Surprisingly perhaps support for an aquatic origin also comes from amborella too. Amborella like water lilies has vestigial gas exchange canals useful in submerged stems and roots.

The aquatic hypothesis is further supported by one of the oldest complete flowering plant fossils thus far discovered. The fossil was recovered from a rock unit in Northeastern China, which has become famous for its fossils of feathered dinosaurs, primitive birds and long proboscid insects of a whole bunch of kinds.

The fossil was discovered by Dr. Ge Sun of Jilin University. After studying the fossil, he contacted a colleague at the University of Florida, Dr. David Dilcher, to get his impression of this unusual plant. The first sample he found consisted of two branches. At the top of the branch were structures that enclosed seeds within carpels, a definitive feature of angiosperms, but unlike modern flowers, the carpels are spread out along the stem with possible stamens, the male parts that would contain the pollen below. This flower though lacks petals.

It was named *Archaefructus*, ancient fruit, and at the time created quite a stir even gaining the attention of the popular media. The reason for all the fuss was in part due to the proposed age of the fossil. Based on other fossils found in the area it was suggested that Archaefructus was growing around 144 million years ago, which would easily place it at or very close to the origin of flowering plants. Could this be the first flower?

Drs. Sun and Dilcher suggested that more recognizable angiosperm flowers would evolve from primitive angiosperms like Archaefructus. Accordingly, the stem that holds the carpels would shrink until they were concentrated close to the male stamens while leaves were simultaneously modified into petals.

The putative Late Jurassic age, however, proved problematic as it conflicted with what we knew about angiosperm fossil pollen. Vast amounts of pollen are produced by plants and they form a common component of sediments if unoxidized deposited in terrestrial settings.

Micropaleontologists and specifically palynologists had found pollen that could be positively identified as angiosperm dating to around 134 million years ago, 10 million years after the first proposed age for Archaefructus. This makes the earlier date for Archaefructus extremely improbable given the sheer volume of pollen produced by plants. There should not be a 10 million-year gap between the first flower and the first pollen.

Many of the fossils in this area were preserved when nearby volcanoes covered them in ash, and it would be this ash that would solve the age problem with Archaefructus. As we have discussed before, volcanic ash is a boon to geochronologists as it often contains zircon crystals that can be dated using isotopic dating techniques. Once age-dated, it was found that Archaefructus dated to 124.6 million years ago, during the early Cretaceous after the appearance of the first angiosperm pollen.

Although we are still looking for the first fossil flower, Archaefructus is still an example of a very early angiosperm and as such might still provide insight into just where flowers first evolved.

It stood about 20 cm and inhabited the margins of a lake probably in very damp or even in shallow-water conditions perhaps accounting for its shallow root system, Is this then, yet further evidence that flowering plants evolved in an aquatic environment?

Well, there are some issues. There are serious concerns regarding the rapid emergence of a terrestrial flowering plant even accepting assistance from gene duplication from an aquatic ancestor. For example, how could plants that had evolved in the water evolve quickly enough to cope with gravity on land?

There is no doubt that some flowering plants did evolve in an aquatic environment early in their history but was this the environment into which flowering plants first appeared? It is still very much a matter of debate.

Regarding the timing of the evolution of plants based on molecular evidence, it is thought that angiosperms and gymnosperms last shared a common ancestor sometimes that is very broadly and nebulously called the pre-Cretaceous. Hopefully, new fossil finds would provide us with more insight into the early times of flower evolution.

Around 100 million years ago during the mid-Cretaceous, there was a great radiation in angiosperm diversity initially noted in the fossil record by angiosperm leaf and pollen remains. By the Late Cretaceous, flowering plants started to take over environments that were formerly dominated by ferns, cycads, Bennettitales and other gymnosperms If you could travel to the Late Cretaceous you would be able to view dinosaurs wandering around plants that you would recognize today such as beech, maple, and magnolia.

There is still some uncertainty as to how this replacement occurred. Was it competitive? In this model the angiosperms with their short life cycle and rapid growth were able to muscle in on the gymnosperms and replace them.

Or was it non-competitive? We know there was an extinction at the end of the Triassic period that drove many gymnosperm lineages into extinction. Were the angiosperms just occupying empty space, ecologically?

To completely understand the spread of angiosperms it would be useful to consider the nature of the world in which they were to rise to dominance. If you consider some of the earliest fossil angiosperms dated at about 125 million years ago the planet they inhabited was very different than today.

The supercontinent of Pangaea had started to fragment, but there were still large continental blocks with Gondwana to the south and Laurasia to the north. It is likely that the interiors of these continents would be pretty dry and arid, environments that are particularly favorable to angiosperms, providing them with large areas of the landscape where they could spread and diversify. The innovations that some of the angiosperms evolved included leathery leaves adapted to prevent water loss, a more efficient water conduction system, and a tough drought resistant seed that would stop an embryo from drying out.

They also have an ability to propagate and reproduce very quickly allowing any colonization to be rapid. Angiosperms are very adaptable and today are found in polar regions, in and along freshwater and seawater, in deserts and numerous habitats in between. Expanding into the dry interiors of continents may also explain why their fossil record is so poor at this time, as arid upland areas are erosive regions likely susceptible to wildfire events and thus difficult areas for fossils to form.

There is another factor though The mid-Cretaceous was a period of intense warming, probably related to the accelerated fragmentation of Pangaea and associated generation of ocean crust and carbon dioxide production Dr. Huber used microfossils to record this event, as was detailed in another episode of this series.

Carbon dioxide may have risen to 4 or 5 times that of today with temperatures even in the usually deep cool water of the oceans warming by 15°C. This global warming would have added to intense drought conditions in the centers of continents and further selected for the morphologically flexible angiosperms.

There is another hypothesis that has been put forward to explain the explosion of flowers, though. A hypothesis put forward by the noted dinosaur paleontologist Bob Bakker in 1986.

Dr. Bakker suggested that it may have been changing dinosaur communities that allowed for angiosperms to spread. Around about 144 million years ago there was a change in the style of herbivory in the dinosaurs, changing from high browsers like brachiosaurs that were cropping leaves from the branches of high conifers to low browsers like ceratopsians. Low browsing would mean that gymnosperm seedlings would have a reduced opportunity to reach maturity. This would open up the canopy and allow for angiosperms to spread quickly. Due to their rapid life cycle, they were better adapted to reach maturity and thus produce seeds before they were browsed away.

Dr. Bakker also proposed that dinosaurs may have been instrumental in spreading flowering plants by eating their fruits and dispersing their seeds. This hypothesis has been largely rejected, however, as no direct relationship has been proven regarding changing dinosaur communities and changes in angiosperms. However, Dr. Karen Chin of the University of Colorado has documented the presence of seeds in the coprolites of some evidently herbivorous dinosaurs during this time interval.

But to close this lecture, let's consider the special relations ship that many angiosperms have with animals. The sole purpose of a flower is reproduction and in the vast number of cases that means reproduction where the male gametes pollen are transferred from one plant to another by insects. It is possible that this relationship, and therefore the first flowers, evolved in isolated settings like islands or on an island chain, which might also explain their apparent very sudden appearance in the fossil record. Such isolated settings may have allowed for the development of a specialized relationship between a plant and an animal, for example, a wasp carrying pollen from one plant to another.

As we have already noted, though, pollination via insects was not a new gig as various insects had been aiding the pollination of gymnosperms before the widespread appearance of flowering plants. It is likely that some insects were pre-adapted to build this relationship with the angiosperms. The angiosperms, though, would take insect in fact animal pollination to a whole new level in a classic example of how two major groups of organisms would co-associate over time.

Angiosperms, and in particular their flowers, would coevolve with animals to produce a whole suite of features that would not have existed if it were not for their intimate relationship. To name just a few.

Scent would evolve as a guidance mechanism which would allow insects to judge how far away a plant was and just what area of the flower they should zero in on to get a reward such as nectar. Scent, of course, would be a particularly useful strategy if you want to attract pollinators like moths in low light conditions.

Scent can also act as a timing mechanism only reaching its maximum potency when the plant is ready to be pollinated. It is thought that scent might have originally evolved from substances that were intended to deter certain animals but ended up attracting useful pollinators. The actual scent of the flowers can also be keyed to particular types of insect preferences. For example, skunk cabbage and *rafflesia*, a flower from Malaysia and Indonesia, are keyed to carrion flies.

Another spectacular example from Indonesia is *Amorphophallus*, the world's largest flower reaching over 10 feet in height. When ready just like in the previous examples it emits a powerful stench of rotting flesh to attract cadaver-associated beetles and carrion flies that it dupes into helping pollinate the species.

There is color again used to attract a specific pollinator to specific plants. White, blue and purple flowers are particularly attractive to insects with reds and oranges being often used to entice birds. In addition, nectar guides on some flowers lead animals to the goodies. Some of these guides are only visible in the ultraviolet area of the spectrum that insects can detect.

And of course, there is fruit, generally used to attract larger animals but also using cues and guides such as color to let animals know when the fruit is ripe, and of course when the seeds contained in the fruit are ready to be distributed. Seed and fruit dispersal, in many ways, are the counterpart to pollination in an angiosperm's reproductive cycle.

But perhaps some of the most fascinating examples of the closer relationship that exists between plants and insects is seen in mimicry. For example, consider the fly orchid a plant native to Europe. The flower resembles a flying insect but more than that it also releases a scent that mimics the female sex pheromones of certain bees and wasps. The males attempting to copulate with the flower get dabbed with pollen. This pollination phenomenon is known as pseudocopulation and may be a source of great frustration amongst the local insect population.

Most of the fossil and phylogenetic evidence indicates that the earliest flowers were small and bowl-shaped, not showy. There were some though that would have stood out in the Cretaceous landscape. Some of the first true flowers of the Cretaceous may resemble something like magnolias. These flowers likely only had a pollen reward for their insect pollinators. The pollinators themselves were likely generalist in nature such as beetles, short-tongued wasps, and flies.

By the time we get to the explosion of angiosperms in the mid-Cretaceous, new varieties of flowering plants emerge with new structures such as nectaries—where nectar is produced—and specialized petals designed for a more specialized relationship with insects. By the Late Cretaceous, flowering plants were very diverse. At this time, insects like wasps and flies were also going through radiations with species evolving specific structures to exploit flowers such as the long-tongued bees and flies.

Of course, flowers have not stopped evolving. Many of the flowers we see today have been influenced by the selective pressure imposed by humans. Some have been selected by humans for the fruit they produce, but it is possible that many domesticated non-food flowers probably started off as weeds in stands of other plants cultivated by humans. It is likely that some of them didn't get weeded out due to their appearance and perhaps were perceived to have beneficial to the crops that were being cultivated in some way.

The end result of this radiation of the angiosperms during the Cretaceous would be a world with a radically different flora and companion insect associates to match. A much more colorful world full of fruit, flowers, and exotic scents. It certainly is a beautiful world even if it is all down to sex.

Lecture 19

The Not-So-Humble Story of Grass

Grasses, which are angiosperms (flowering plants), make up the most economically significant plant family today. They include cereal crops, such as maize, wheat, rice, and barley. Some are used as construction materials (bamboo), while others are fermented to make ethanol biofuels (sugar cane). Altogether, it is estimated that there are probably around 10,000 species of grass. In this lecture, you will consider whether grass is a new plant, when the great grass takeover occurred, what triggered the spread of our grassy planet, and how significant grasses have been on the evolution of animals.

Grasses

- Until recently, the general mantra regarding the evolution of grass was that the first grasses evolved long after the dinosaurs had become extinct at 66 million years ago. The oldest fossil grass came from Tennessee, dated to about 55 million years ago.

- There were hints of earlier grasses from fossil pollen, but grass pollen is very difficult to tell apart from non–grass pollen. As such, images of dinosaurs striding through grass were generally regarded as incorrect renderings of the Mesozoic world.

- A discovery by paleobotanists of Lucknow University and Panjab University in India would turn these ideas around, though. They found coprolites, or fossilized feces, from the Late Cretaceous, toward the end of the dinosaurs' reign on Earth, that appeared to contain phytoliths, which are tiny silica structures found in the leaves of certain plants. They help give grass some of its structural support but may also act as defensive structure, deterring grazing by animals.

- On analyzing the phytoliths, phytolith expert Caroline Strömberg at the University of Washington identified them as coming from various grasses, representing at least 5 species. This is significant, because in addition to showing that they existed at the time of the dinosaurs, the phytoliths demonstrated that grass species in the Late Cretaceous had already diversified. This means that the antiquity of grass was a lot longer than anyone had suspected. Strömberg has suggested that this may have pushed back the origin of grass to about 100 million years ago.

- Dinosaur fossils found in close association with these coprolites come from a rather poorly defined, but still rather exciting, group of sauropod dinosaurs called titanosaurs, which contain some of the largest creatures to have ever walked the Earth. For example, *Argentinosaurus*, weighing in at around 83.2 tons and around 30 meters long, would have dined on a wide variety of vegetation, with other flowering plants, cycads, and conifers forming part of their diets.

- We have recently found evidence of even older grass dated to between 97 and 100 million years ago (early Middle Cretaceous) preserved in amber from Myanmar (Burma). This is research led by Oregon State University's Dr. George Poinar, who is one of the world's leading experts on amber and the fossils found in it.

- Dr. Poinar and his colleagues found a beautifully preserved grass floret preserved in amber, and on its tip is an extinct species of parasitic fungus called *Palaeoclaviceps parasiticus* that is likely closely related to a group of fungi that today we call ergot. Ergot may have a special relationship with grass. It tastes bitter and would deter grazing by herbivores. If an animal ingests enough of it, it can cause serious side effects, such as trembling muscle groups that cause an animal to fall over.

The Great Grass Takeover

- On a planetary scale, probably more significant than the date of evolution of the first grasses is the development and evolution of the ecosystem that they would create—the ecosystem we call grasslands, or, more formally, the grassland biome.

- There are many different types of grassland. There are the savannas and velds of Africa. In North America, there are the prairies. In South America are the pampas and llanos, and in Eurasia are the steppes.

- Today, grasslands cover 40% of all our planet's land surface. Because grasses are mostly annual plants that die every year, they develop large and deep soil profiles.

- Grasslands are generally associated with dry but not desert conditions. Because of their high rate of turnover, grasses and grasslands can support a large animal population, unlike trees and shrubs of forests, where a lot of the useful nutrient-rich material is locked up in the plant.

- Another feature is their tendency to burn, creating grassland fires. Rather than being detrimental to grasslands, fires may be an important part of some grassland ecosystems, removing trees and shrubs and allowing grasslands to spread.

Savanna Grassland

- When do we see the advance of this biologically and economically vital part of today's biosphere? Grasses existed during the Cretaceous, but they were not a particularly significant part of the Earth's flora. Even after the Cretaceous, in the first 2 periods of the Cenozoic era, the Paleogene and the Eocene, the world was mostly tropical—warm and wet and mostly covered in forest. Grasses did not yet dominate.

- By about 40 million years ago, as the Eocene was drawing to a close, the Earth's climate started to change, becoming much drier and cooler. Forests gave way to woodlands and eventually into chaparral and other scrub formations and deserts; still lacking, though, were a major component of grasses.

- But by the time we move into to the Oligocene, 23 to 5.3 million years ago, this desert scrub starts to give way to grasses and their deep, loamy soil profiles. By the Late Oligocene, we have evidence of the grassland biome and an association of grassland-adapted mammals spreading across North America.

The Spread of Our Grassy Planet

- Around 95% of land plants, and some grasses, use what is called the C3 metabolic pathway, a form of photosynthesis that probably evolved in the Paleozoic. Many grasses, and some other plants, use a C4 metabolic pathway. Because of their anatomy, some C4 plants can operate at lower levels of atmospheric carbon dioxide than C3 plants can.

- Because C4 plants don't need as much carbon dioxide, they can afford to keep their stoma (surface pores that allow for gas exchange between the plant and the atmosphere) closed more often than C3 plants can. This means that in dry conditions, plants like C4 grasses have a selective advantage. This makes them perfect plants for arid conditions and for the spread of grasslands.

- Today, C4 plants make up about 5% of Earth's plant biomass and around 3% of known species but account for a massive 20 to 30% of terrestrial plant carbon fixation. It is only in the past 10 million years that C4 plants have become such an important part of the biosphere, with a remarkable C3-to-C4 plant transition occurring globally between 8 and 4 million years ago.

- This transition is highlighted in the geological record isotopically by studying 2 of the stable isotopes of carbon: carbon-12 and carbon-13. Life in general has a preference for carbon using the carbon-12 isotope, but C4 photosynthesis usually will incorporate more of the carbon-13 isotope. This gives us a proxy for the presence of C4 plants recorded isotopically in paleosols (fossil soils) or in the bones and teeth of the animals that were eating the plants.

- The change from C3 to C4 grasslands appears to be associated with the general increase in aridity of global climates, favoring those plants with a C4 metabolic pathway that require less water. This increase appears to occur at different times in different regions, though.

- There were probably other mechanisms that played a role in the spread of the C4 grasses, too. For example, an increase in charcoal in Pacific cores between 12 and 7 million years ago hints at a greater number of fires during this period. This would favor grasses and grasslands due to their rapid recovery rates when compared to trees and shrubs.

- There is also a suggested general reduction in atmospheric carbon dioxide levels during the Cenozoic that may have also favored C4 plants with their more efficient photosynthesis. Even cooler climates may have encouraged the spread of the more efficient C4 grass. This might explain why C4 dominance occurs first in hot equatorial areas such as Kenya, followed by Pakistan, and finally the northern Great Plains.

- The spread of grasslands would encourage more grazers, who, in cropping the tops of plants, would favor the spread of grass over trees and shrubs. It is also likely that there could be local reasons for the spread of grasslands.

- But what would set these changes into motion? A possible cause might be found in India and its dash northward to dock with southern Asia. The Himalayas would rise as India crumpled into Asia over many millions of years, but during the Miocene, the mountains began to rise in a major way.

- This thrust vast amounts of rock up into the atmosphere, which led to the rocks starting to weather. A part of the weathering process involves the transformation of rock silicates into clays by the action of carbon dioxide dissolved in rainwater. In this way, carbon dioxide is effectively washed from the atmosphere, transported down river systems in the form of clay, and deposited as a sediment, causing an overall net drawdown of carbon dioxide.

- Even if there is no direct link between falling carbon dioxide levels, temperature, and the spread of C4 plants, a drawdown of carbon dioxide would certainly impact paleoclimate patterns, such as causing the development of a more arid climate, which might favor the spread of a water-efficient C4 flora across the land surface.

Grasses and the Evolution of Animals

- How has the evolution of a relatively new ecosystem, the grassland biome, impacted the rest of the biosphere? A traditional view, for more than 140 years, was that the evolution of mammals with high-crowned, or hypsodont, teeth was in response to the spread of phytolith grasses. These large teeth would be an adaptation to eating the gritty grasses, and as such, you could use the presence of hypsodont teeth as a morphological proxy for the presence of grasslands.

- Unfortunately, this doesn't quite work everywhere. Additional research by Caroline Strömberg on a section of sediments representing 800,000 years of deposition at Gran Barranca in Patagonia, Argentina, has turned this idea around. Hypsodont cheek teeth were discovered in Patagonia dating to around 38 million years ago, which, if we take them as a proxy for grasslands, indicate that grasslands evolved there 20 million years before anywhere else and could represent the cradle of the grassland ecosystem.

- The deposits are composed of river sediments and windblown volcanic ash, probably from the southern volcanic zone in South America. The plant remains, though, do not indicate grassland but, rather, a tropical forest dominated by palms. It is possible that in this particular area hypsodont teeth evolved as a response to the high amount of abrasive volcanic ash and grit in the environment and phytoliths from other phytolith-bearing plants like palms. This effectively undermines the idea that these teeth can always act as a proxy for grasslands—it's not such a simple relationship.

- In other parts of the world, such as North America, there is more of a link between the spread of grasslands and the presence of hypsodont mammals in the fossil record, although even here there does appear to be a 4-million-year lag between grasslands and the first long-toothed (hypsodont) animals.

- But even if the story of teeth evolution in our grazing mammals is not quite nailed down yet, there is certainly a case for co-association, or perhaps coevolution, between grasses and animals. The snout of many creatures became broader and flatter to allow for effective grazing, and jaws became longer and deeper, permitting more efficient grinding of plant material.

- In addition, unable to hide as effectively in grass as they could in a forest, animals started to show adaptations for running, even if they were not grass browsers. The lower parts of limbs started to elongate, feet became more compact, and muscle mass started to increase at the shoulders and hips to help power a quick getaway.

- Grass may have also had an effect on the evolution of humans. The spread of grasslands was proposed by some as a reason why our ancestors started to walk upright. A very significant part of the human story evolved on the open grasslands; this is probably the environment that shaped us the most. The presence of the grassland biome may have also helped genus *Homo* move out of the cradle in Africa and start to populate the planet.

Questions to consider:

1. How many times has the evolution and spread of flora dramatically impacted the evolution of animals?
2. To what extent has the spread of grasslands impacted our own evolution?

Suggested Reading:

Emling, *The Fossil Hunter*.

Savage, *Prairie*.

Lecture 19 Transcript

The Not-So-Humble Story of Grass

To quote the great Philosopher Homer, "Hmm Beer." I guess I should have just said, Homer Simpson. It has been proposed by some archeologists that it was a beer that built civilizations, well perhaps some wobbly ones anyway, in that our ancestors started to domesticate cereals not to make bread but to brew beer. Priorities, priorities.

The origins of beer likely go back to the early Neolithic period about 9500 B.C. Beer is known to have been consumed in ancient Egypt, and chemical evidence of this wonderful liquid has been found at Godin Tepe, dating to 3500—3100 B.C. in the Zagros Mountains of Iran.

Some of the earliest Sumerian writings reference beer too. It is thought that the prayer to the goddess Ninkasi also served as an oral recipe for brewing. Just imagine if your favorite hymn also contained information on how to mix a really good gin and tonic.

Isn't it wonderful that something so good and refreshing could come from something as humble as grass, well, barley to be specific in the case of beer? Grasses are angiosperms, flowering plants, but unlike a lot of angiosperms, almost all of them rely on wind rather than insect pollination.

True grasses, of the Family Poaceae—graminae of older parlance—are characterized by having very small to almost microscopic flowers arranged in a spike—like structure with long leaves that have parallel veins. And of course, they produce fruit in the form of grain. Specifically, they grow from the base rather than from the tip which means they can keep on growing even if grazed.

The grass family is the most economically significant plant family today. They include cereal crops like maize, wheat, rice and our old boozy friend, barley.

They are also used as construction materials in the case of bamboo and thatch, and some such as sugar cane are fermented to make ethanol biofuels

Altogether it is estimated that there are probably around 10,000 species of grass. This is an important group of plants.

So in this lecture let's consider, grass—is it a new plant? When was it that the grassy carpet started to spread? What triggered the spread of our grassy planet? How significant have grasses been on the evolution of animals?

Before we move onto our investigation of grasses, though, I'd like to introduce you to a paleontologist who had to fight the sensibilities and prejudice of the late 18th and early 19th century to take part in science and scientific discovery.

At this time Science was generally regarded as a pass time for gentile people; people with the correct family background and breeding, and more importantly a ready supply of cash. Oh, and another thing you would most likely be a man too.

The paleontologist in question is one Mary Anning, daughter of a cabinetmaker in the pretty little town of Lyme Regis on the Dorset coast of England. From the cliffs close to her home she would collect fossils that were instrumental in developing our understanding of Mesozoic marine reptiles, and in a broader sense our understanding of how the biosphere has changed dramatically over time. However, because of her position and gender, she never had a forum to directly present her discoveries to her peers. She did become quite famous for her discoveries, though, and it's one of those discoveries I want to consider in this lecture

In the abdominal area of an Icthyosaur, that's a streamlined marine reptile that kind of resembles a modern dolphin; she would discover structures called bezoar stones. Bezoar stones were much sought after, as they were believed to have magical properties, including the ability to cure any case of poisoning. Mary noted that when broken open these stones contained the fossilized bones and scales of fish and sometimes fragmentary remains of other marine reptiles.

In 1829, another famous geologist and friend of Mary—William Buckland, proposed that these were in fact fossilized feces. Buckland named them coprolites, from the classical Greek *kopros*, meaning "dung," and *lithos*, meaning "stone."

He went further in his investigations, though. He found spiral marks on the coprolites that he suggested had been produced by the ridges of the ichthyosaur's intestines. He also thought that very black coprolites might be the result of icthyosaurs eating belemnites, a kind of Mesozoic squid with their ink turning the poop black.

Coprolites are extremely valuable to paleontologists. Although it is not always possible to tie coprolites to specific animals, you can certainly tell what the creature was eating. They also provide a wonderful survey of the possible animals and plants that lived in a particular environment. For example—this is some poop I found while visiting Indonesia produced by the Asian Palm Civet. Some braver coffee addicts than me harvest partly digested coffee beans from this to make kopi luwak coffee. To be honest, I've never really been that tempted to try it.

In coprolites, though, much of the original organic material has been replaced by minerals like calcite or quartz. In 1842 John Stevens Henslow, a professor at Cambridge University developed a technique for extracting phosphate from coprolites, and the mining of fossil poop became big business in Cambridgeshire. The phosphate was used for the British fertilizer and munitions industry.

Incidentally, Henslow is famous for another reason too. He was originally listed as the naturalist on the voyage of the Beagle. A famous ship, of course, that was going to survey South America. His wife persuaded him not to go, and so Henslow suggested a young naturalist of his acquaintance take his place—one Charles Darwin. And that, folks, is how history is made.

But why am I banging on about fossilized poop in a lecture about grass? Until recently, the general mantra regarding the evolution of grass was that the first grasses evolved long after the dinosaurs had become extinct at 66 million years ago. The oldest fossil grass came from Tennessee, dated to about 55 million years ago.

There were hints of earlier grasses from fossil pollen. But grass pollen is very difficult to tell apart from non-grass pollen. As such, images of dinosaurs striding through grass were generally regarded as incorrect renderings of the Mesozoic world. A common exam question for students of paleontology would be, "What's wrong with this picture?"

A discovery by paleobotanists of Lucknow and Panjab University, India, would turn these ideas around, though. They found coprolites from the Late Cretaceous toward the end of the dinosaurs' reign on Earth that appeared to contain phytoliths. Phytoliths are tiny silica structures found in the leaves of certain plants. They help give grass some of its structural support but may also act as a defensive structure, deterring grazing by animals.

They called in phytolith expert Caroline Strömberg, currently at the University of Washington. On analyzing the phytoliths, she identified them as coming from various grasses representing at least 5 species. This is significant because in addition to showing that they existed at the time of the dinosaurs the phytoliths demonstrated that grass species in the Late Cretaceous had already diversified. This means that the antiquity of grass was a lot longer than anyone had suspected. Strömberg has suggested that this may have pushed back the origin of grass to about 100 million years ago we shall see later how right she was.

Dinosaur fossils found in close association with these coprolites come from a rather poorly defined but still rather exciting group of sauropod dinosaurs called titanosaurs. The titanosaurs contain some of the largest creatures to have ever walked the Earth. For example, Argentinosaurus weighing about 83.2 tons and around 30 m long. There is a strong possibility that the smoking poop gun if you'll excuse the analogy, lies with the titanosaurs.

These dinosaurs would have dined on a wide variety of vegetation with other flowering plants, cycads, and conifers forming part of their diets. It would appear that they were browsers, nibbling leaves from trees and bushes, but also grazers mowing grass, which might explain their wide mouths, similar to many modern grazing animals today.

The possibility of grasses at this time may also explain why small mammals called gondwanatherians that were scurrying around the feet of the dinosaurs possessed high-crowned or hypsodont teeth like the teeth of modern horses and cows. These teeth are thought by some to have evolved as a response to the gritty phytoliths in grass, although, as we shall see the grass and teeth relationship has been questioned recently.

As it turns out, we have recently found evidence of even older grass dated to between 97–100 million years ago. That's the mid-Cretaceous preserved in amber from Myanmar Burma This is research led by Dr. George Poinar Oregon State University, who is one of the world's leading experts on amber and the fossils found in it.

As we dealt with in an earlier lecture, amber was originally tree resin. As it runs down trees, it picks up and beautifully preserves insects, small vertebrates, plants and a whole bunch of other things too, associated close to or actually on the tree.

Dr. Poinar and his colleagues found a beautifully preserved grass floret preserved in amber, and on its tip was an interesting dark structure. When investigated, this was found to be an extinct species of parasitic fungus called *Palaeoclaviceps parasiticus* that is likely closely related to a group of fungi that today we call ergot.

Ergot may have a special relationship with grass. It tastes bitter and would deter grazing by herbivores. The problem is if you ingest enough of the stuff it can cause serious side effects. In cattle it can cause a condition called paspalum staggers where the animals' major muscle groups will tremble; they become very clumsy and if they try to run they tend to fall over.

Humans can also suffer from ergot poisoning; it's called ergotism. It can cause hallucinations, irrational behavior, convulsions and vasoconstriction leading to gangrene. It is possible that many reports of sudden deaths of large groups of people in villages and towns during the middle ages may have been caused by people eating rye bread contaminated with ergot.

Indeed, some historians believe that the great fear associated with the French Revolution in 1789 may have been fueled by peasants having to eat wheat that was contaminated by ergot. And in 1976 Linda Caporael suggested that the hysterical behavior of the women during the Salem Witch Trials may have been caused by ergotism, although this hypothesis has been criticized much in recent years.

One of the reasons why ergot has such bizarre effects on animals is that one of the alkaloids extracted from this fungus are lysergic acid derivatives. That is the same stuff from which LSD was synthesized by Swiss chemist Albert Hofmann in 1938. Herbivorous dinosaurs were probably vacuuming up ergot as well as grass during the Cretaceous, not that dinosaurs were not high on LSD, but you could imagine some of the ergot alkaloids having serious effects and no one wants giant dinosaurs stumbling around the landscape.

On a planetary scale probably more significant than the date of evolution of the first grasses, though, is the development and evolution of the ecosystem that they would create the ecosystem we call grasslands, or more formally the Grassland Biome.

There are many different types of grassland. There are savannas and velds of Africa, and here in North America we have the prairies there are the pampas and Llanos in South America and the steppes in Eurasia.

Today, grasslands cover a whopping 40% of all our planet's land surface. As grasses are mostly annual—plants that die every year—they develop large and deep soil profiles.

Grasslands are generally associated with dry but not desert conditions, receiving about 20–36 inches of rain every year. Because of their higher rate of turnover, grasses and grasslands can support a large animal population, unlike trees and shrubs of forests, where a lot of the useful nutrient-rich material is locked up in the plant.

Another feature is their tendency to burn, creating grassland fires. Rather than being detrimental to grasslands, fires may be an important part of some grassland ecosystems, removing trees and shrubs and allowing grasslands to spread.

But when do we see the advance of this biologically and economically vital part of today's biosphere? Let's look at the history of the planet's overall flora, following the extinction that wiped out the Mesozoic world—including the dinosaurs around 66 million years ago.

As we have noted, grasses existed during the Cretaceous, but they were not a particularly significant part of the Earth's flora Even after the Cretaceous, in the first two periods of the Cenozoic Era, the world was mostly tropical; warm and wet and mostly covered with forest. Most of the mammals that existed in those ancient forests were small, a much better strategy for a dense forest environment. Not all creatures were tiny though with relatives of the dinosaur like Gastornis, a large flightless bird standing some 2 m tall, probably terrorizing the local mammal population. A beautiful snapshot of the world at this time, about 48 million years ago, can be found in the fossils of the Messel pits deposit of Germany, and although many modern plant taxa are present there is not a single species of grass that is known from this very well-documented plant community preserved there. Grasses did not dominate, at least not yet.

By about 40 million years ago as the Eocene was drawing to a close, the Earth's climate started to change becoming much drier and cooler. Forests gave way to woodlands and eventually into chaparral and other scrub formations and deserts. Still lacking, though, were a major component of grasses.

By the time we move into to the Oligocene, this desert scrub starts to give way to grasses and their deep soil profiles. By the late Oligocene, we have evidence of the grassland biome and an association of grassland adapted mammals spreading across North America.

But what would be the trigger that would allow the development of our grassy planet?

For this, we need to look at the way that some plants conduct the business of photosynthesis. Around 95% of land plants and some grasses use what is called the C3 metabolic pathway. It is named the C3 pathway for the 3 carbon molecule that are produced as the first product of carbon fixation. This form of photosynthesis probably evolved way back in the Paleozoic. Many grasses and some other plants use a C4 metabolic pathway as the first molecule produced in this case has 4 carbon molecules.

Some C4 plants possess a particular arrangement of cells called the Kranz anatomy where the vascular tissue that transports the products of photosynthesis—that would be sugars—is surrounded by two rings of cells. Without going too much into the biochemistry, the Kranz anatomy provides a site where carbon dioxide can be concentrated in the inner ring of cells called the bundle sheath.

This effectively charges the cells with carbon dioxide for photosynthesis, which means that C4 plants can operate at lower levels of atmospheric CO_2 than a C3 plant can This imparts another advantage, though. Because C4 plants don't need as much CO_2, they can afford to keep their stoma—those are the surface pores that allow gas exchange between the plant and the atmosphere—to be closed more often than can C3 plants since C3 plants need to be gulping in more CO_2. This means that in dry conditions plants like C4 grasses have a selective advantage. For example, at 30°C, C3 plants lose 883 molecules of water per carbon dioxide molecule at a fix, but C4 plants only lose 277. This makes them perfect plants for arid conditions and for the spread of grasslands.

Today, C4 plants make up about 5% of Earth's plant biomass, and around 3% of known species but account for a massive 20–30% of terrestrial plant carbon fixation. C4 plants in grasslands are so efficient at drawing down CO_2 from the atmosphere that some have suggested that they could be a biological solution to reducing our current rising levels of anthropogenic carbon dioxide.

The C4 pathway runs about 6 times faster than the C3 pathway and it is this efficiency that has inspired the C4 rice project. This initiative is attempting to produce rice, which is naturally a C3 plant with a C4 metabolism. It is predicted that this would double the rice yield while using less nutrients and water—one possible weapon in the war against future global famine.

It is only in the past 10 million years though that C4 plants have become such an important part of the biosphere with a remarkable C3 to C4 plant transition occurring globally between 8–4 million years ago.

This transition is highlighted in the geological record isotopically by studying two of the stable isotopes of carbon, carbon-12 and the slightly heavier form with one more extra neutron in it nucleus carbon-13. Life, in general, has a preference for carbon using the carbon-12 isotope, but C4 photosynthesis is slightly less fussy than C3 and will incorporate more of the carbon-13 isotope. This gives us a proxy for the presence of C4 plants recorded isotopically in paleosols fossil soils or in bones and teeth of the animals that were eating the plants.

You can see this here in a diagram adapted from Colin Osborne and David Beerling from the University of Sheffield who explored the increases in carbon-13 values in paleosols from North America, Pakistan and East Africa recording the spread of the C4 grasslands This change appears to be associated with the general increase in aridity of global climates favoring those plants with a C4 metabolic pathway that require less water. This increase appears to occur at different times in different regions, though.

There were probably other mechanisms that played a role in the spread of the C4 grasses too, for example, an increase in charcoal in Pacific cores between 12–7 million years ago hints at a greater number of fires during this period. This would favor grasses and grasslands due to their very rapid recovery rates when compared to trees and shrubs.

There is also a suggested general reduction in atmospheric carbon dioxide levels during the Cenozoic that may have also favored C4 plants with their more efficient photosynthesis.

Thirdly, the spread of grasslands would encourage more grazers who, in cropping the tops of plants would favor the spread of grass over trees and shrubs. And it is also likely that there could be local reasons for the spread of grasslands, too.

But what would set these changes into motion? A possible cause might be found in India and its dash northward to meet up and dock with Southern Asia. The Himalayas would rise as India crumbled into Asia over many millions of years, but during the Miocene, the mountains began to rise in a major way.

This thrust vast amounts of rock up into the atmosphere which started to weather. A process of weathering that involves the transformation of rock silicates into clays by the action of CO_2 dissolved in rainwater. This transformation is a process we call silicate weathering. In this way, carbon dioxide is effectively washed from the atmosphere transported down river systems in the form of clay and deposited as a sediment causing an overall net drawdown of carbon dioxide.

Even if there is no direct link between falling carbon dioxide levels, temperature and the spread of C4 plants a general drawdown of CO_2 would certainly impact paleoclimate patterns such as causing the development of a more arid climate which might favor the spread of water-efficient C4 flora across the land surface.

But how has the evolution of a relatively new ecosystem—the grassland biome—impact the rest of the biosphere? A traditional view for over 140 years was the evolution of mammals with high-crowned hypsodont teeth. This was in response to the spread of phytolith grasses. These large teeth would be an adaptation to eating the gritty grasses, and as such you could use the presence of hypsodont teeth as a morphological proxy for the presence of grasslands.

Unfortunately, this doesn't quite work everywhere. Additional research by Caroline Strömberg on a section of sediments representing 800,000 years of deposition at Gran Barranca in Patagonia, Argentina, has turned this idea around Hypsodont cheek teeth were discovered in Patagonia dating to around. 38 million years ago, which if we take them as a proxy for grasslands, indicate that grasslands evolved there 20 million years before anywhere else and could represent the cradle of the grassland ecosystem.

The deposits are composed of river sediments and windblown volcanic ash, probably from the southern volcanic zone in South America. The plant remains though do not indicate grassland but rather a tropical forest dominated by palms. It is possible that in this particular area hypsodont teeth evolved as a response to the high amount of abrasive volcanic ash and grit in the environment and phytoliths from other phytolith bearing plants like those palms. This effectively undermines the idea that these teeth can always act as a proxy for grasslands. It's not such as simple relationship.

In other parts of the world such as North America, there is a more direct link between the spread of grasslands and the presence of hypsodont mammals in the fossil record although even here there does appear to be a four-million-year lag between grasslands and the first long toothed hypsodont animals.

But even if the story of teeth evolution in our grazing mammals is not quite nailed down yet, there is certainly a case for co-association or perhaps coevolution between grasses and animals. The snout of many creatures becomes broader and flatter to allow for effective grazing and jaws became longer and deeper, permitting more efficient grinding of plant material.

In addition, unable to hide as effectively in grass as you can in a forest animals start to show adaptations for running, even if they were not grass browsers the lower parts of limbs start to elongate; feet become more compact and muscle mass starts to increase at the shoulders and hips to help power a very rapid and quick getaway.

Grass, though, may have had an effect on our evolution too. The spread of grasslands was proposed by some as a reason why our ancestors started to walk upright. Ancestors like the famous australopithecine Lucy, who lived around 3.2 million years ago, may have evolved an upright posture as an adaptation to look for predators in high grass. Walking upright would have also freed their hands to make tools, which may have set a series of events in motion that would ultimately lead to *Homo sapiens* and their big brains.

This hypothesis though was called into question by the discovery of *Ardipithecus ramidus* in 1994. Ardi, as it was nicknamed, was an upright walking ape-like creature with long arms adapted for climbing trees, and feet that had big grasping toes. Ardi though had evolved in a woodland environment, not on an open grassland plain. It would appear our family line was preadapted to stride through the grasslands.

That is not to say that grasses and grasslands have not had a profound effect on our evolution. A very significant part of our own human story evolved on the open grasslands from Lucy right through to modern humans. This is probably the environment which shaped us the most

The presence of the grassland biome may have also helped genus Homo move out of the cradle in Africa and start to populate the planet. Possibly the first adventurer was *Homo erectus* who is first found in the fossil record about 1.9 million years ago. This was a creature who from the neck down would have looked very human, but whose features were still very apelike.

It has been suggested that climatic changes would open up the Levantine corridor—a crescentic strip of land along the eastern margin of the Mediterranean—temporarily changing the environment from desert to grassland savanna probably due to changes in the pattern of the Western African Monsoon.

During green Sahara times, grazing animals would migrate along the corridor followed by bands of *Homo erectus* that would spread into Eurasia. This may have started about 1.8 million years ago with erectus spreading as far as China and the island of Java—facilitated by a carpet of grasses.

So the next time you wander past a humble patch of grass just remember you are looking at a powerhouse of change that in recent geological years has been a major shaper of the world we live in—not so humble after all.

Lecture 20 Australia's Megafauna: Komodo Dragons

In 1926, explorer W. Douglas Burden traveled to Komodo Island, just west of the island of Flores in Indonesia, where he discovered the world's largest predatory lizard: the Komodo dragon. He returned with 12 specimens, 2 of them live and 3 of them stuffed and displayed in the American Museum of Natural History in New York City. Today, the Komodo dragon, an endangered species, is being held from the brink of extinction in part by the efforts of the Smithsonian Institution. In 1992, the Smithsonian's National Zoological Park was one of the first zoos outside of Indonesia to hatch clutches of dragons, with various hatchings distributed to zoos around the world.

Komodo Dragons

- Komodo dragons were unknown to Europeans until 1910, when reports of a "land crocodile" reached the Dutch administrators of what was then the Dutch East Indies. Lieutenant van Steyn van Hensbroek was the first European to capture and kill a specimen, sparking a great interest in these "ancient beasts." Just 17 years later, 2 live specimens were proudly exhibited when the London Zoo's Reptile House opened in 1927.

- These are big animals; an average male is around 2.5 meters long and weighs about 91 kilograms. It is estimated that they have a life span of around 30 years. They have a long muscular tail that is as long as the body and a long flat crocodile-like head with a rounded snout. They have powerful front limbs and long curved claws, which are formidable weapons, and a long forked yellow tongue that flicks in and out of the dragon's mouth. Their skin is armored with osteoderm bone, forming a kind of chain mail pattern.

- Despite their position as top predator of these islands, they will quite happily eat carrion. The dragons will hunt almost anything: invertebrates, eggs, lizards, and mammals, including monkeys, goats, water buffalo, and other items. They are not fussy eaters and will on occasion eat other Komodo dragons. Attacks on humans have also been reported.

- Komodo dragons are not designed to run down prey like wolves do. In common with other reptiles of this genus, they have legs that stick out to the side, giving them a characteristic gait, with their body swaying from side to side.

- They are ambush predator, using short bursts of speed to quickly lunge at a creature and strike. Small prey tend to get bitten in the middle of the neck, sometimes even being knocked over by a swipe from that powerful tail. For larger prey, they adopt a bite-and-retreat strategy, critically wounding a larger creature and then waiting for it to die.

- A common hypothesis is that Komodo dragons kill their prey with the aid of virulent strains of bacteria found in their mouths. It is presumed that the bite from the dragon infects its prey with so much bacteria including that they go into septic shock and die as a result.

- Recently, though, research published in 2013 by a number of scientists, including Dr. Ellie Goldstein of the R. M. Alden Research Laboratory in California, have questioned this model. They suggest that the bacterial flora of the Komodo dragon is really no different from other large predators.

- In a paper published in 2009, Bryan Fry at the University of Queensland, and a number of other researchers, used magnetic resonance imaging (MRI) to analyze the skull of a Komodo dragon and model its bite. They found that the dragon does not have a particularly powerful bite when compared to that of a crocodile. In addition, the Komodo dragon does not have a strong skull or jaw muscles to subdue its prey.

- So, if it doesn't use "dirty mouths" or powerful jaws, how does the Komodo dragon kill its prey? They have strong muscles behind their skull to resist

prey as it attempts to pull away from the dragon. This action involves the slicing and ripping of flesh, aided by very sharp, inch-long, serrated teeth. There is also a venom gland that delivers venom through cavities between the teeth, getting quickly into the bloodstream of the prey. Fry's research showed that Komodo dragon venom prevented blood clotting, lowered blood pressure, and caused the prey to go rapidly into shock.

- Komodo dragons can see objects about 300 meters (985 feet) away but have difficulty distinguishing between nonmoving objects. They also have rather poor night, or dim-light, vision. Their hearing isn't too great, either, and they have a reduced range when compared to humans. They would have difficulty in detecting low- and high- pitched sounds. But they do have a very keen sense of smell—a sense of smell that can detect carrion over 2.5 miles (4 kilometers) away.

- Most of the smell, though, is not detected through the nostrils. Instead, they use their yellow tongue to continually "taste" the air. The Komodo dragon assists this detection by swinging its head from side to side as it walks. Like snakes, these scents are passed to a feature called the Jacobson's organ, which can tell if more "scent molecules" are present on the right or left fork of the tongue, and in this way, the dragon can zero in on dinner.

Big Dragons on Small Islands

- It appears that the presence of the Komodo dragons on a small number of Indonesian islands is part of an older paleontological story that has only relatively recently come to light. Komodo dragons today are found on the islands of Gili Motang, Gili Dasami, Rinca, Komodo, and Flores. This is a pretty isolated distribution and leads to an important question: How do you get a 200-pound lizard to a group of small Indonesian islands?

- One possibility is something called the island effect. A common feature on islands is the way that creatures will either get larger or smaller in response to the conditions they find themselves in.

- For example, consider *Stegodon*, a type of prehistoric elephant with a long geological range, from 11.6 million years ago to relatively recently in the Late Pleistocene. They were large creatures, and like modern elephants, they were probably good swimmers and may have reached the Indonesian islands, such as Flores and others, when sea levels were low. On the islands, they started to shrink, probably as an adaptation to the more limited resources on the island, perhaps becoming about the size of a water buffalo.

- Did the Komodo dragon arrive as a small lizard and grow large as part of the island effect? Or is it possible, as has been suggested by some, that the dragon grew large as a result of a need to prey on the dwarf but still large *Stegodon* elephants that were in the same environment? According to these hypotheses, small monitor lizards get washed onto the islands—perhaps on mats of vegetation, by floods, or even by tsunamis—and then evolve into the large Komodo dragon that we know today.

- An alternative hypothesis is that there has been no change in size in the Komodo dragon. Perhaps they arrived on the islands at their present size, in which case we have the same question: How do you get a 200-pound lizard to a group of small islands?

- We are currently in a relatively warm interval during the current ice age, the last glacial advance of which ended about 12,000 years ago. During the various glacial maxima, sea levels would fall, linking many of the current islands in Indonesia, or at least reducing the amount of ocean between them. At these times, the Malay Peninsula, Borneo, Java, and Sumatra were part of a landmass called Sunda, while Australia and New Guinea were part of a landmass called Sahul.

- The increased landmass and reduced distance between these islands facilitated the migration of certain species into this area. Oceanic channels, such as exist between Bali and Lombok, were still present, which would prevent complete mixing of these islands' biotas, even during times of very low sea level.

- When the ice melted and sea levels rose again, the species that had migrated to these islands were trapped. This led to an isolated evolution of their fauna and flora and explains why this is one of the most biodiverse places on the planet. In a biogeographical sense, this area is called Wallacea, in honor of Alfred Wallace, a coauthor with Darwin on the first paper describing evolution by natural selection.

Australian Megafauna

- Where did the Komodo dragons come from? Recent paleontological evidence suggests that these wonderful beasts are Australians. They are part of an ancient megafauna and, as such, could be thought of as "living fossils."

- Most of the Northern Hemisphere megafauna—giant mammoths and mastodons, wooly rhinoceroses, and saber-toothed tigers—was extinct by about 11,700 years ago. Megafauna are considered to be animals that are more than 100 pounds.

- But another megafauna developed on "island" Australia. By 50 million years ago, Australia had separated from other landmasses in the Southern Hemisphere but was still close enough to allow migration of creatures from South America across an ice-free Antarctica and into a northward-drifting Australia. As time progressed, and as the ocean widened, Australia became isolated, and all those creatures were marooned—and evolution in isolation can do some pretty amazing things.

- That original wave of migration from South America included mammals—principally marsupials, which are mammals that commonly carry their young in a pouch. From what were probably small migrants from South America, a whole array of spectacular animals would evolve, including the marsupial lion *Thylacoleo carnifex* and the giant koala *Diprotodon*.

- But it wasn't all marsupials. *Dromornis*, a large flightless bird, inhabited open woodland and is thought by some to have been carnivorous. There were also giant snakes, such as the Bluff Downs giant python from the Early Pliocene of Queensland.

- There were other reptilian components of this megafauna, and lurking around in the Australian bush was a monitor lizard even larger than the Komodo dragon: *Megalania prisca*, or *Varanus priscus*. No complete skeleton has been found, but estimates of the size of this monster vary from 5.5 to 7 meters. Even on the small end of this range, this is still significantly larger than the Komodo dragon.

- Analysis of the fossils of this beast by a team including venom expert Bryan Fry has suggested that, like the Komodo dragon, this lizard was also venomous. This would have made it quite the formidable predator in Australia and one of the largest venomous creatures to have existed.

- *Megalania* fossils are known from Pleistocene deposits of eastern Australia, aged between 2.6 million and 30,000 years ago, which opens up the possibility that the first humans to range across Australia may have encountered this animal, too.

- Additional research in 2009 from the University of Queensland by Scott Hocknull and colleagues would also put to rest the origin of the Komodo dragon in Indonesia, as fossil evidence shows that the Komodo dragon coexisted with *Megalania*. Their studies show that Komodo dragons had dispersed as far as the island of Flores by 900,000 years ago and to Java between 800,000 and 700,000 years ago.

- The Komodo dragon of today, therefore, represents a relic population of a lineage of giant monitor lizards that was once common in eastern Australia and Wallacea as far as Java. By 2000 years ago, the last remaining member of this group of monster lizards had retracted to Flores and small surrounding islands, such as Komodo.

- In fact, much of the Australian megafauna—with notable exceptions, such as the red kangaroo—were extinct by about 30,000 B.P. (before present), probably due to a combination of factors, including climate change and interaction with humans.

The Future of Komodo Dragons

- What is the status of the last surviving relic of these giant lizards? It is estimated that the population hovers around 3000 individuals in the wild, placing them in a "vulnerable" status.

- The decline in population is due to a number of factors. The Lesser Sunda Islands are highly active volcanically, and such natural disasters can easily tip a delicate island ecosystem into crisis.

- Human interaction, though, has had a serious effect. The dragons have suffered from loss of habitat and poaching of their prey. They have also been deliberately poisoned, and some dragons have been captured, presumably for personal collections or trophies.

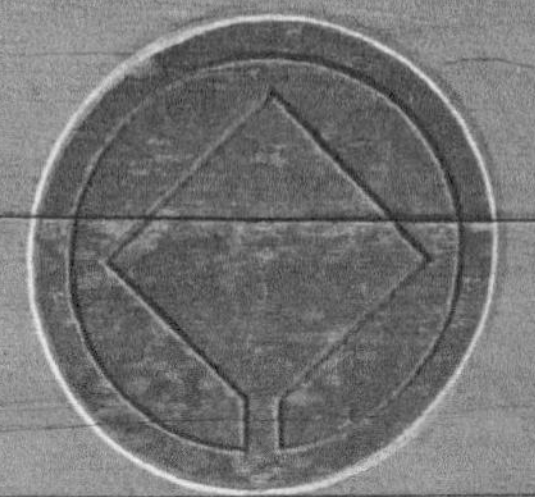

- In 1980, the Komodo National Park was set up to include the islands of Komodo, Rinca, Padar, Wae Wuul, and Wolo Tado as well as reserves on Flores. In total, it covers an area of 1733 square kilometers. In 1991, this area was declared a UNESCO World Heritage Site.

Questions to consider:

1. How many new species have evolved on islands in isolation?
2. How many species of fossil animals may have been venomous?

Suggested Reading:

Fichman, *An Elusive Victorian*.

Long, Archer, Flannery, and Hand, *Prehistoric Mammals of Australia and New Guinea*.

Molnar, *Dragons in the Dust*.

Lecture 20 Transcript

Australia's Megafauna: Komodo Dragons

In 1933, New York City suffered a brutal attack. Records tell of a giant primate terrorizing downtown Manhattan. The beast climbed the Empire State Building and apparently wrought havoc with local air traffic control. The event didn't end well for the belligerent if somewhat misunderstood gorilla, who came to a rather sticky end on the sidewalks of the busy city. No fatalities from falling primate were reported, though.

The specimen was originally captured at great cost by a swashbuckling adventurer from a distant tropical location—Skull Island. The expedition arrived by boat and was almost immediately attacked by fierce cannibals who inhabited that lonely outcrop of land in the Pacific. This was a land where time stood still, a land of steaming jungles full of tall, prehistoric plants reaching for the skies with giant pterosaurs flying overhead. A land full of ancient giants like lumbering brontosaurs and terrifying Tyrannosaurus rex, both of who would provide to be a considerable nuisance in the expeditions attempt to collect a certain primate specimen.

Of course, what I'm describing here is the classic black and white movie King Kong by Merian C Cooper. Cooper though based his 1933 classic on a real expedition in 1926 by the explorer W. Douglass Burden. The island he would travel too was not Skull Island but Komodo Island, just west of the island of Flores, in Indonesia.

It is likely that much of the imagery in the movie was taken from the accounts of Burden's expedition. There are no gorillas on Komodo, but there is a pretty good dinosaur substitute. The world's largest predatory lizard, the Komodo dragon. It has been suggested that Cooper so liked the Kay sound in Komodo dragon that in naming his monster gorilla he used the same K to give us King Kong.

Burden returned from his 1926 expedition with 12 specimens, 2 of them live, and 3 of them stuffed and displayed in another venerable museum, The American Museum of Natural History in New York City.

Today the Komodo dragon, which is an endangered species, is being held from the brink of extinction in part by the efforts of the Smithsonian Institution. In 1992, the Smithsonian National Zoological Park was one of the first zoos outside of Indonesia to hatch clutches of dragons with various hatchings distributed to zoos around the world.

So in this episode, let's ask: Who is the Komodo dragon? Where are the dragons from? What world did the dragons leave behind? What does the future hold for the Komodo dragon?

Let me paint a picture of this fascinating creature for you. There are various names for the dragons. On the Indonesian Islands around Komodo, they are called the *Ora Buaya Darat*, the land crocodile or the *Biawak Rakassa*, meaning giant monitor.

Scientifically, we give them the generic name *Varanus*, from the Arabic word, *waral*, meaning "dragon" or "lizard beast." This is the name given to all monitor lizards, including the Komodo dragon.

The species name *komodoensis* is taken from the island where they are most famously found giving us Monitor Lizard from Komodo, or *Varanus komodoensis*. I'm going to stick with Burden's name as I think it paints a better picture of this wonderful animal—the Komodo dragon.

The dragons were unknown to Europeans until 1910 when reports of a land crocodile reached the Dutch administrators of what was then Dutch East Indies. One Lt. Van Steyn Hensbroek was the first European to capture and kill a specimen that sparked a great interest in these ancient beasts. Just 17 years later, 2 live specimens were proudly exhibited when the London Zoo Reptile House opened in 1927.

I was lucky enough to visit beautiful Komodo Island in February of 2015. We arrived early in the morning, and I can still remember the thrill of peering out of my porthole onto this beautiful island. It really did fit the bill of an exotic skull island. Although the population has expanded in recent years, it still appeared deserted and somehow primitive from the ocean. February is the rainy season in this part of the world, so these deeply dissected volcanic rocks were painted a fantastic lush green color. In front of those green cliffs there was a thin strip of yellow sand, but from the boat, no T-rex, Brontosaurus or Komodo dragons could be seen. At least not yet.

As the island is part of the Komodo National Park, you can only visit with official park guides who escorted us through the beautiful landscape of Komodo. The usually dry Island is a mix of open grass, woodland savanna, and tropical deciduous forests.

This was the first evidence we found of the dragons—poop. It looks pretty much like a large bird dropping—a very large bird dropping. The white material as with birds is urate, that's the pee, and the dark material is the stool, which as reptiles they all pass through the same orifice, the cloaca.

The forest we were moving through was pretty quiet. I saw occasional birds, but there was very little noise apart from the whispers of the others in our party. I was also keeping a wary eye out for snakes as there were a number of poisonous species on the islands, but the snakes were probably more scared of me than I was of them, especially with us crashing through the landscape even though we did think we were being terribly quiet and stealthy.

After a bit of a hike across some hilly terrain we were led to an area where a water hole had been dug to attract the dragons, and after a few minutes to allow our eyes to adjust to the dim light under the trees, there they were—Komodo dragons The dragons were mostly sleeping but still looked formidable You can see me in this picture looking thrilled but slightly nervous as the only protection we had from these big animals were guides with big forked sticks.

These are big animals. An average male is around 2.5 meters long, and about 91 kg The largest recorded specimen was about 3.13 m long and weighed 166 kg that is 365 pounds, although this may have included some undigested food from a recent meal. It is estimated they have a lifespan of around about 30 years.

They have a long muscular tail, as long as the body and a long flat crocodile-like head with a rounded snout, Very much on my mind during my encounter were these powerful front limbs and long curved claws which are formidable weapons And who could not be mesmerized by that long forked yellow tongue that flicks in and out of the dragon's mouth.

They are built a bit like a tank with skin that is armored with osteoderm bone nobbles forming a kind of chain mail pattern.

Despite their position as top predator of these islands, they will quite happily eat carrion. Islanders will often place piles of rocks on graves not to mark the position of a loved one but as a very practical solution to stop Komodo dragons from digging up granny's grave for a quick snack. No one wants to look up from dinner to see a beloved relative being taken for a post mortem dragged by a large lizard.

The dragons will hunt almost anything—invertebrates, eggs, lizards, and mammals including monkeys, goats, water buffalo and other items. Some of their favored prey is the Timor rusa deer, but Komodo dragons are not fussy eaters, and will on occasion eat other Komodo dragons. Attacks on humans have also been reported.

They are not designed run down prey like wolves do. In common with other reptiles of this genus, they have legs that stick out to the side giving them a characteristic gait with their body swaying from side to side as you can see in this specimen that rather unexpectedly decided to cross the path we were on.

The Komodo dragon is an ambush predator using short bursts of speed—around 20 km/h to quickly lunge at a creature and strike. Small prey tends to get bit in the middle of the neck, sometimes even being knocked over by a swipe from that powerful tail. For larger prey, they adopt a bite-and-retreat strategy, critically wounding a larger creature, perhaps like a water buffalo and then waiting for it to die. A large creature like this can be very dangerous when wounded, so it's a much safer option to just let nature and shock take its course.

But is there more to the dragon's bite? A common hypothesis is that the Komodo dragon will kill their prey with the aid of virulent strains of bacteria found in their mouths, often described as being full of rotting flesh. It is presumed that the bite from the Komodo dragon infects its prey with so much bacteria, including *Escherichia coli* and strains of *Staphylococcus* that they go into septic shock and die as a result. This could be termed the sepsis hypothesis.

Recently, though, research published in 2013 by a number of scientists including Dr. Ellie Goldstein of the R. M. Alden Research Laboratory in California, have questioned this model. They suggest that the bacterial flora of the Komodo dragon is really no different from other large predators.

One of the researchers on the team Professor Bryan Fry of the University of Queensland, Australia has suggested why this sepsis hypothesis may have come about in the first place. It centers around water buffalo, a creature only recently introduced to the island, which as adults are really too large to be seriously bothered by the Komodo dragon. When bitten by the Komodo dragon, the buffalo generally have no trouble escaping, but their first instinct is to escape to a wallow hole.

These wallow holes are full of feces and bacteria, and it is probably here that the wounds of the buffalo get infected and develop sepsis. This means they die not as a result of a bacterial infection from the Komodo dragon's mouth, but as result of their own pre-programmed behavior. Of course, the Komodo dragon has no issue with eating a buffalo that has died from an infection in a dirty wallow hole, though, waste not, want not.

Again, though, could there be something else to the dragon's bite? In a paper published in 2009, Fry and a number of other researchers used magnetic resonance imaging—MRI—to analyze the skull of a Komodo dragon and model its bite. What they found was that the dragon does not have a particularly powerful bite when compared to that of a crocodile, for example. In addition, the Komodo dragon does not have a strong skull or jaw muscles to subdue their prey. So if not dirty mouths and powerful jaws just how does the Komodo dragon kill its prey?

First, they have strong muscles behind their skull to resist their prey as it attempts to pull away from the dragon. This action involves the slicing and ripping of flesh aided by very long 2.5 cm serrated teeth.

But there is something else a venom gland that delivers venom through cavities between the teeth, getting quickly into the blood stream of the prey. Fry's research showed that Komodo dragon venom prevented blood clotting, lowered blood pressure and caused the prey to go rapidly into shock.

What else does the Komodo dragon have in its hunting armory? Well, they can see objects about 300 m away, but have difficulty distinguishing between nonmoving objects They also have rather poor night or dim-light vision. So their eyes are OK but certainly not eagle-like.

Their hearing isn't too great either. They have a reduced range when compared to humans. They would have difficulty in detecting low and high pitched sounds, so I guess that rules out Barry White and the Bee Gees on a Komodo dragon's iPod.

But what they do have is a very keen sense of smell, a sense of smell that can detect carrion 2.5miles away. Most of the smell though is not detected through the nostrils. Instead, they use that wonderful yellow tongue that continually tastes the air. The Komodo dragon assists this detection by swinging its head from side to side as it walks. Like snakes, these scents are passed to a feature called the Jacobson's organ. This organ can tell if more scent molecules are present on the right or the left fork of the tongue and in this way the dragon can zero in on dinner.

As you might guess the eating habits of the Komodo dragon are less than elegant, ripping off chunks of flesh or swallowing smaller prey whole. They are very efficient in devouring most of their kill with adults only needing twelve large meals a year. Carrion or a kill will often attract numbers of dragons creating feeding frenzies that have been observed to have some sort of order, with older and larger Komodo dragons getting first dibs at the table.

As with their eating habits, Komodo dragons cannot be said to be sensitive lovers. Mating occurs between May and August. Competing males wrestling sometimes by getting on their hind legs with a tail for support with the looser thrown to the ground.

Females are said to be somewhat antagonistic to mating and may have to be physically held in place by the male. However, male Komodo dragons can coax as well by flicking their tongues over the female's body. I guess it's a Komodo dragon thing.

The female delays laying eggs until September to avoid the heat of the dry season. A clutch of around 20–30 eggs is laid in September and take between 7–8 months to incubate. Komodo dragons will often use the nests of megapode birds like the orange-footed scrubfowl to lay their eggs.

Megapode nests are quite remarkable feats of engineering at around 1m high by 3m wide They consist of a ditch filled with rotting compost that slowly gives off heat onto which the bird's eggs are laid. The ditch is then covered by a layer of insulating sand, a structure which the female Komodo dragons are more than happy to burrow into and co-opt for their own needs.

The female sits on the nest protecting her investment, but as soon as the eggs hatch she abandons her young who are immediately in a precarious state. They only weigh around 100 g and are about 40 cm long, and a potential dinner item for other animals including Komodo dragons who are estimated to eat around 10% of their hatchlings. As a result, the hatchlings tend to live up in trees eating insects, small lizards, and similar fare until they get a little bit larger.

So, why in a series about paleontology am I talking so much about a living species? Well, it appears that the presence of the Komodo dragons on a small number of Indonesian islands is part of an older paleontological story that has only relatively recently come to light.

Komodo dragons today are found on the islands of Gili Motang, Gili Dasami, Rinca, Komodo, and Flores. This is a pretty isolated distribution and leads to an important question, "How do you get a 200 lbs lizard to a group of small Indonesian islands?"

One possibility is something called the island effect. A common feature on islands is the way that creatures will either get larger or smaller in response to the conditions they find themselves in. For example, consider Stegadon, a type of prehistoric elephant, with a long geological range from 11.6 million years to relatively recently in the late Pleistocene. Stegadons were large creatures with some as tall as 3.8 m. Like modern elephants, Stegadons were probably good swimmers and may have reached the Indonesian islands like Flores and others when sea levels were low. On the islands, though the Stegadon started to shrink probably as an adaptation to the more limited resources on the island perhaps becoming about the size of a water buffalo.

The reverse can occur too. For example, the giant *rat Canariomys bravoi* from the island of Tenerife off the coast of northwest Africa clocks in at 1 kg in weight and is 114 cm long. This rat may have got large perhaps due to a reduced number of large predators, but also a larger body and stomach can more efficiently digest poor quality food that you might have to deal with on a small island.

So, did the Komodo dragon arrive as a small lizard and grow large as part of the island effect? Or is it possible as has been suggested by some that the Komodo dragon grew large as a result of a need to prey on the dwarf but still large Stegadon elephants in the same environment? According to these hypotheses, small monitor lizards get washed onto the islands, perhaps on mats of vegetation or alternatively by floods or even by tsunamis and then evolve into the large Komodo dragon that we know of today.

But there is an alternate hypothesis, perhaps there has been no change in size in the Komodo dragon at all. Perhaps they arrived on the islands at their present size, in which case we have the same question How do you get a 200 lbs lizard to a group of small islands? But another one too, if they did come from elsewhere just where did they come from?

To answer the question of how you have to understand that we are currently in a relatively warm interval during the current ice age, the last glacial advance of which ended about 12,000 years ago.

During the various glacial maxima, sea levels would fall linking many of the current islands in Indonesia or at least reducing the amount of ocean between them. At these times, the Malay Peninsula, Borneo, Java, and Sumatra were part of a landmass called Sunda, while Australia and New Guinea were part of a landmass called Sahul.

The increased landmass and reduced distance between these islands facilitated the migration of certain species into this area. Oceanic channels such as exists between Bali and Lombok were still present, which would prevent complete mixing of these island's biotas even during times of very low sea level.

When the ice melted and sea levels rose again, the species that had migrated to these islands were trapped. This led to an isolated evolution of their flora and fauna and explains why this is one of the most biodiverse places on the planet. In a biogeographical sense, this area is called Wallacea, in honor of Alfred Wallace, a co-author with Darwin on the first paper describing evolution by natural selection. Wallace spent many years describing the unique distribution of species in this part of the world, an often forgotten and yet very important figure in science.

But what about the second question? The where did they come from? From the north or the south? Recent paleontological evidence suggests that these wonderful beasts are Australians. Komodo dragons are part of an ancient megafauna, and as such you could think of them as living fossils. Most of us are pretty familiar with the Northern Hemisphere megafauna. Giant mammoths and mastodons, wooly rhinoceroses and fearsome saber-toothed cats like I have behind me here, but most of this fauna was extinct by about 11,700 years ago.

But another megafauna developed on island Australia. By 50 million years ago, Australia had separated from other landmasses in the southern hemisphere but was still close enough to allow migration of creatures from South America across an ice-free Antarctica and into a northward drifting Australia. As time progressed and as the ocean widened, Australia became isolated, and all those creatures were marooned, and evolution in isolation can do some pretty amazing things

That original wave of migration from South America included mammals, but not the placental mammals we are most familiar with in the Northern Hemisphere. Principally marsupials, mammals that carry their young in a pouch. From what were probably small migrants from South America a whole array of spectacular animals would evolve including this the marsupial lion, a top marsupial predator, some of which be as large as 164 kg. Cave paintings from the Kimberly area in northwestern Australia suggest that some of the ancient aboriginal peoples of Australia came face to face with this terrifying creature.

There were also relatives of a familiar Australian icon—the Koala. But not the cute, cuddly Koala we all know and love today, a giant koala—*Diprotodon*—the size of a bear. It has been speculated that Diprotodon may be the beast behind the Aboriginal legends of the Bunyip, a creature or water spirit, said to live in billabongs and creeks and other sources of open water that would prey on unsuspecting people that got too close.

But it's not all marsupials. This is *Dromornis*, a large flightless bird around 10 feet tall that inhabited open woodlands and is thought by some to have been carnivorous.

There were also giant snakes like the Bluff Downs giant python from the Early Pliocene of Queensland. A close relative of this snake is the olive python, but its fossil cousin is estimated to be around 10 m, that's 33 feet, long.

But there were other reptilian components of this megafauna and lurking around in the Australian bush was a monitor lizard, even larger than the Komodo dragon. This is *Megalania* or Varanus priscus to give it it's scientific name. As no complete skeleton has been found, estimates of the size of this monster vary from 5.5 m to about 7 m. Even on the small end of this scale, this is still significantly larger than the Komodo dragon.

Analysis of the fossils of this beast by a team including venom expert Bryan Fry at the University of Queensland has suggested that, like the Komodo dragon, this lizard was also venomous. This would have made it quite the formidable predator in Australia and one of the largest venomous creatures to have existed.

Megalania fossils are known from Pleistocene deposits of Eastern Australia aged between 2.6 million and 30,000 years ago, which opens up the possibility that the first humans to range across Australia may have also encountered this animal too.

Additional research in 2009 from the University of Queensland by Scott Hocknull and colleagues would also put to rest the origin of the Komodo dragon in Indonesia, as fossil evidence shows that the Komodo dragon coexisted with Megalania. Their studies show that Komodo dragons had dispersed as far as the island of Flores by 900,000 years ago, and to Java between 800,000 to 700,000 years ago. The Komodo dragon of today, therefore, represents a relic population of a lineage of giant monitor lizards that was once common in Eastern Australia and Wallacea as far as Java. By 2000 years ago, the last remaining member of this group of monster lizards had retracted to Flores and the small surrounding islands like Komodo.

In fact, much of the Australian megafauna with notable exceptions like the red kangaroo were extinct by about 30,000 years before present. Probably due to a combination of factors including climate change and interaction with humans.

So what is the status of the last surviving relic of these giant lizards? It is estimated that the population hovers around 3000 individuals in the wild, placing them in a very vulnerable status.

The decline in population is due to a number of factors. The lesser Sunda Islands are highly volcanically active, and as such, natural disasters can easily tip a delicate island ecosystem into crisis.

Human interaction though has also had a serious effect. The dragons have been suffering from loss of habitat and poaching of their prey. They have also been deliberately poisoned, and some dragons have been captured, presumably for personal collections or trophies.

In 1980, the Komodo National Park was set up to include the islands of Komodo, Rinca, Padar as well as reserves on Flores. In total, it covers an area of 1733 square km. In 1991, this area was declared a UNESCO World Heritage Site.

I do hope we're not seeing the last days of the Komodo dragon. I hope that the efforts of those involved in protecting the dragons in the wild and breeding programs in various locations around the world will help preserve this remarkable creature. Let's hope they do not go the way of their even scarier Australian ancestors.

Lecture 21

Mammoths, Mastodons, and the Quest to Clone

Some fossils are controversial. Other fossils challenge the way we believed the story of life unfolded on our planet. And a few fossils go beyond that, challenging the very foundations of our perceived reality—our place on Earth. In this lecture, you will learn about one of those fossil species and discover how it is at the center of both a scientific and moral debate that is still raging today.

A Tooth That Shook the World

- In 1705, a Dutch settler in the Hudson River valley near the village of Claverack, New York, found a tooth. The tooth was then traded to a local politician, who subsequently made it a gift for the governor of New York, Lord Cornbury, who was convinced that this was the tooth of a giant—one of the giants thought to have roamed Earth before the flood mentioned in Genesis.

- The tooth was sent to London and became known as *incognitum*, the unknown species. In South Carolina other giant teeth started to turn up. Slaves in that state noted that they looked very much like the teeth of African elephants.

- In addition, tusks and other bones started to be found in the Ohio River valley—fossils that resembled the woolly mammoths recovered from permafrost in Siberia, and as a result, *incognitum* would incorrectly get grouped with them.

- It took a famous French anatomist, Georges Cuvier, to realize that *incognitum* was something different. It all came down to the structure of the teeth. Those found in Ohio and Siberia had ridges, a bit like a running shoe, designed for grazing grasses and the like.

- The teeth of *incognitum* are very different. They have raised cones, which reminded Cuvier of breasts. This creature clearly had a different diet than the mammoths. It is from these breast-like cones that this creature got its most common name: the mastodon. Cuvier soon started to realize that the teeth and bones of these creatures, while resembling modern elephants, were not the same species.

- The general view at that time was that there was an unbroken chain of creatures stretching back to their creation in the Garden of Eden. In addition, when Cuvier was studying these troublesome teeth, the accepted date of the creation of our planet was around 4004 B.C. The discovery of these extinct beasts not only shook ideas of a still-intact creation, but also made scholars start to question the perceived age of the Earth. Around 6000 years didn't appear to be enough time to accommodate all these fantastic animals.

- Cuvier came to believe that most of the fossils he was studying were from "older worlds" destroyed by some sort of catastrophe. He was convinced that many of the fossils and geological formations he examined pointed to the Earth progressing by a series of catastrophes that caused the extinction of many species. These ideas were largely replaced by uniformitarianism, the theory that geological features could be explained by present-day slow processes, such as erosion and deposition of sediments.

- Our modern understanding of extinctions has vindicated Cuvier's views, at least to some extent. The Earth does evolve by slow, almost impercetible changes, as uniformitarianists suggested. But it is also punctuated now and again by extensive catastrophes that today are called mass extinctions.

The Evolutionary History of the Elephant

- Most of the fossils of mammoths and mastodons that were being found in North America during the 1700s date to around 10,000 to 12,000 years B.P.

The origins of the larger group to which they belong, the Proboscidea, has a much older heritage, originating around 9 million years after the death of the dinosaurs along the shores of a vast oceanic body called the Tethys Sea in what is today North Africa, the Middle East, and extending to northern India.

- A very early elephant ancestor is a pig-sized animal called *Phosphatherium*. Today, the closest living relatives of modern elephants include the manatees, dugongs, and hyraxes. It is from unimpressive looking *Phosphatherium*, though, that a wonderful array of elephant-like creatures would evolve.

- By about 37 to 30 million years ago, *Moeritherium* would be wallowing in African swamps like a small hippopotamus. Contemporaneous with *Phosphatherium* was *Phiomia* (about 2.5 meters in length, a more terrestrially adapted creature with 2 sets of short tusks) and *Palaeomastodon* (standing about 2 meters tall and weighing around 2.2 tons, one of the most elephant-looking members of the group at this time, equipped with a trunk and scoop-shaped lower tusks).

- By about 10 million years ago, though, the king of all elephants would evolve: *Deinotherium*, larger than a modern African elephant, some of which weighed about 14.5 tons.

- Rather than upward-curving tusks, those of *Deinotherium* pointed downward, probably to help strip leaves off trees as it fed. This animal probably persisted in Africa until the start of the ice ages; it is likely, therefore, that these creatures interacted with humans.

- In terms of the story of modern elephants, though, *Deinotherium* and its kind were more of a side branch. For the family history of elephants, we need to turn to the *Gomphotheres*, from which we get mammoths and modern elephants.

- From *Palaeomastodon* through the *Gomphotheres*, there would be a radiation of elephant forms adapted to various lifestyles and environments, sadly with just 2 remaining today: the Asian and the African elephants.

The Ice Age Kings of North America

- Mammoths evolved in Africa during the Pliocene and would enter Europe by about 3 million years ago. A European species called the steppe mammoth evolved in eastern Asia and, by around 1.5 million years ago, would cross the Bering Strait across "Beringia" when sea levels were lower than today. The Columbian mammoth would evolve from these pioneering steppe mammoths and populate an area from the northern United States to Costa Rica.

- The Columbian mammoth was about 4 meters (13 feet) at the shoulder and weighed up to 11 tons. Specimens of the Columbian mammoth are quite well known, as many individuals were caught in natural traps.

- There would be another wave of mammoth invasion into North America. From the steppe mammoths that gave rise to the Columbian mammoth, another iconic species would evolve about 400,000 B.P. and cross into North America by 100,000 B.P.: the woolly mammoth.

- Woolly mammoths were smaller than their Columbian cousins, about the same size as an African elephant. They were covered by course hair, probably thicker than that of the Columbian mammoth, and, because they lived in more northerly regions, had small ears, probably an adaptation to conserve heat. The characteristic fatty hump on the mammoths' backs may have been used as a reserve source of nutrients in the more extreme northerly environments.

- We know quite a lot about woolly mammoth anatomy because, unlike the Columbian mammoth, specimens have been found preserved buried in permafrost with soft parts still intact.

- Both the woolly mammoth and Columbian mammoth coexisted in North America, with the woolly mammoth living in the colder, more northerly environments. It is possible that the 2 species interbred where their ranges overlapped. Woolly mammoths even interacted with humans.

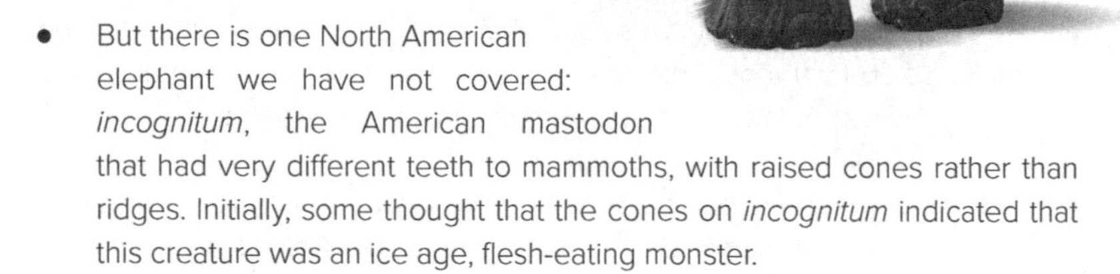

- But there is one North American elephant we have not covered: *incognitum*, the American mastodon that had very different teeth to mammoths, with raised cones rather than ridges. Initially, some thought that the cones on *incognitum* indicated that this creature was an ice age, flesh-eating monster.

- Benjamin Franklin figured out the true nature of the beast. He reasoned, correctly, that mastodons' tusks would have been an impediment for catching prey and suggested, again correctly, that the cone-like teeth would probably be an adaptation to grinding small branches of trees.

- The American mastodon, or *Mammut*, had shorter legs and flatter, longer skulls, and they were more heavily muscled than the mammoths they coexisted with. Large males reached up to 9 feet 2 inches and weighed 5 tons. They also had tusks that were less curved.

- The American mastodon is from a much older root stock than its mammoth cousins, with *Mammut* diverging from the chain of evolution that would lead to the Elephantidae, which includes modern elephants and mammoths, at about 27 million years ago.

- They ranged across North America during the Pleistocene epoch, mostly inhabiting cold spruce woodlands, browsing on trees. Remains of the mastodon have been found frozen in Alaska, and from these the genome of the creature has been sequenced, allowing us to place it fairly accurately within the elephant family tree.

The Disappearance of the Giants

- The mammoths and mastodons of North America did not exist in isolation. Other giant creatures roamed North America, part of a now-extinct Pleistocene megafauna.

- Sharing the Pleistocene landscape were creatures such as the giant ground sloth, *Megatherium*; a North American camel, *Camelops*; giant beavers; the armored *Glyptotherium*, a relative of the armadillo; and the North American bison, who we still have with us today. Giant herbivores like these mean there must have been giant predators, such as the short-faced bear, dire wolves, the bird *Teratornis*, an American lion that was around 25% larger than its African cousin, and the saber-toothed tiger.

- What happened to the megafauna that existed in North America and around the world is still hotly debated, but there are 4 main hypotheses:
  - First, there is hunting. There is a general continent-by-continent extinction of megafauna that follows the migration of humans across the planet. There is archaeological evidence of butchery of some megafauna species by the Clovis people, some of the first human settlers, but some question if small bands of hunter-gathers could have been entirely responsible for the extinction of all these large creatures.

- Some researchers prefer climate change as the vector for extinction. Following a glacial advance during the ice age, there are often associated extinctions of species, possibly related to changes in vegetation patterns as climate warmed during an interglacial.

- A less-favored hypothesis is the possibility of a hyper-disease that killed off many of the large creatures, although it is difficult to imagine how a disease could be fatal to so many different species of animal

- Another less-favored hypothesis is the possibility of an extraterrestrial impact event that caused wildfires, and ultimately a global cooling event, during a period called the Younger Dryas (12,900 to 11,700 B.P.). Evidence for this is somewhat contradictory.

- Perhaps a good explanation could combine aspects of the overhunting and climate change scenarios. Perhaps climate change reduced the population of many of the megafauna genera to such an extent that even low levels of hunting would drive the animals into extinction.

Back from the Grave

- Because megafauna such as mammoths went extinct relatively recently, the chances of finding genetic material is much better than for that of dinosaurs, for which no viable DNA has been found. For megafauna, we even have fleshy material available in frozen remains in the Arctic.

- Given that we have fleshy material available, could we bring back a mammoth from the dead? To answer that, we have to consider the 3 possible methods that we could potentially use to achieve this.
  - We could try somatic cell cloning. For this, we would need a viable cell from all the frozen mammoth meat in the Artic. However, thus far, there has been no living cell recovered from any mammoth carcass, and according to many experts, there never will be.

- We could take whatever DNA fragments we can find and reassemble the genome. Some progress has been made in this regard, and large portions of the mammoth genome has been mapped. But there are even large parts of the human genome that we can't map, never mind the mammoth genome. Even when we extract fragmentary DNA from mammoth material, much of it is contaminated with DNA from other creatures. And even if we could eventually reconstruct the entire genome, we don't know how to wrap it up into chromosomes and insert them into a nucleus.

- Because we know that the Asian elephant is the closest living relative of the mammoth and that it shares about 99% of the mammoth genome, we could recognize the part that is not mammoth and swap it out with parts of the genome that we have positively identified as mammoth. We have the technology to snip off sections of Asian elephant DNA and insert mammoth sections, but we are still a long way from reconstructing a woolly mammoth.

Questions to consider:

1. How much of a role did our species have in the extinction of mammoths?
2. Where should we place the dividing line between what is acceptable and unacceptable in cloning?

Suggested Reading:

Lister and Bahn, *Mammoths*.

Shapiro, *How to Clone a Mammoth*.

Lecture 21 Transcript Mammoths, Mastodons, and the Quest to Clone

Some fossils are controversial. Other fossils challenge the way we believed that the story of life unfolded on our planet And a few fossils go beyond that, challenging the very foundations of our perceived reality, our place on Earth I would like to look at one of those fossil species in this lecture and see how they are at the centre of both a scientific and moral debate that is still raging today.

So in this episode lets ask how could a tooth shake the world? What is the evolutionary history of the elephant? Who were the Ice Age kings of North America? Where did all the giants go? And can we bring them back?

In 1705, a Dutch settler in the Hudson River Valley near the village of Claverack, New York found a tooth. This tooth was then traded apparently for rum to a local politician who subsequently made it a gift for the Governor of New York, one Lord Cornbury. Cornbury was a colorful character who, so his political enemies claimed, who was fond of dressing as his cousin Queen Anne. Cornbury was convinced that this was the tooth of a giant; one of the giants thought to have roamed Earth before the flood mentioned in Genesis.

The tooth was sent to London and became known as incognitum, "the unknown species." This was before dinosaur mania would grip the public imagination. In fact, before the term dinosaur had been created, incognitum was the fossil monster star of its time.

Incognitum would not stand alone for long, though. In South Carolina, other monster teeth started to turn up. Slaves in that state noted they looked very much like the teeth of African elephants.

In addition, tusks and other bones started to be found in the Ohio River Valley. Fossils that resembled the fantastic wooly mammoths were recovered from permafrost in Siberia, and as a result, incognitum would incorrectly get grouped with them.

It took a famous French anatomist George Cuvier to realize that incognitum was something different. It all came down to the structure of the teeth. Those found in Ohio and Siberia had ridges that looked a bit like a running shoe, designed for grazing grasses and the like. The teeth of incognitum are very different. They have raised cones which reminded Cuvier of breasts This creature clearly had a different diet to the mammoths. It was from these breast-like cones that this creature got its most common name—the mastodon.

It is easy today to underestimate the impact these finds would have. Cuvier soon started to realize that the teeth and bones of these creatures, while resembling modern elephants, were not the same species. So what might you ask? So these were extinct species, and in using that word, extinct, you have hit the nail on the head.

As we mentioned in an earlier lecture the general view of all creatures great and small at that time was that there was an unbroken chain of creatures stretching back to their creation in the Garden of Eden. To some, the concept of extinction implied that God's perfect creation was no longer perfect, a heresy.

But there was another problem too. One related to the passage of time and the age of the Earth. When Cuvier was studying these troublesome teeth, the accepted chronology of our planet was one that had been developed by Archbishop James Ussher the Archbishop of Armagh and Primate of all Ireland between 1625 and 1656.

He lived through turbulent times—a time when tensions between Catholics and Protestants were high in England. Also, a time of civil war when England would ultimately execute their king and set up a parliamentarian protectorate under Oliver Cromwell. As a moderate Calvinist, Ussher often had to walk a fine line between two warring parties.

Not too surprising then that he would turn his attention to more academic pursuits. He studied the chronology of the church fathers and the deep history of the world. This required considerable scholarship of not only the Bible, but also other accounts and calendars including those of the Persians, Greeks, and Romans.

He placed the death of Alexander in 323 B.C. and Julius Caesar in 44 B.C. close to those of modern accounts. He would take things much further back, though. Using biblical sources, he placed the creation at the entrance of the night preceding the 23rd day of October, the year before Christ 4004 B.C. In today's calendar that places creation around 6 pm on October 23rd 4004 B.C.

The discovery of these extinct beasts then not only shook ideas of a still intact creation but also made scholars start to question the perceived age of the Earth. Around 6000 years didn't appear to be enough time to accommodate all these fantastic animals. Cuvier himself came to believe that most of the fossils he was studying were from older worlds destroyed by some sort of catastrophe. In his 1766 paper on living and fossil elephants he wrote, "All of these facts, consistent among themselves, and not opposed by any report, seem to me to prove the existence of a world previous to ours, destroyed by some sort of catastrophe."

Cuvier was convinced that many of the fossils and geological formations he examined pointed to the Earth progressing by a series of catastrophes that caused the extinction of many species. These ideas were largely replaced by uniformitarianism, the theory that geological features could be explained by present-day slow processes, such as erosion and deposition of sediments

Our modern understanding of extinctions has vindicated Cuvier's ideas, at least to some extent. The Earth does evolve by slow, almost imperceptible changes as uniformitarianists like Hutton and Lyell suggested. But it also punctuated, just now and again by extensive catastrophes that today we call mass extinctions.

Most of the fossils of mammoths and mastodons that were being found in North America during the 1700s date to around 10,000 to 12,000 years before the present. The origins of the larger group to which they belong, the Proboscidea, has a much older heritage originating around 9 million years after the death of the dinosaurs along the shores of a vast oceanic body called the Tethys Sea in what is today North Africa, the Middle East, and extending to northern India.

A very early elephant ancestor is a pig-sized animal called Phosphatherium. Today the closest living relatives of modern elephants include the manatees, dugongs, and the hyraxes.

It is from the unimpressive Phosphatherium though that a wonderful array of elephant-like creatures would evolve. By about 37–30 million years ago, *Moeritherium* would be wallowing in African swamps like a small hippopotamus. Contemporaneous with Moeritherium was *Phiomia*, about 2.5 meters in length a more terrestrially adapted creature with 2 sets of short tusks and *Paleomastodon*, one of the most elephant-looking members of the group at this time, equipped with a trunk and scoop-shaped lower tusks that it may have been using to dredge up plants from the edges of rivers and lakes This was a large animal too, standing about 2 m and weighing about 2.2 tons.

By about 10 million years ago, though, the king of all elephants would evolve—this is D*einotherium*, larger than a modern African elephant, some of which weighed in at around 14.5 tons.

Rather than upward curving tusks, those of *Deinotherium* pointed downwards probably to help strip leaves off trees as it fed. This animal probably persisted in Africa until the start of the ice ages. It is likely therefore that these wonderful creatures interacted with humans.

In terms of the story of modern elephants though Deinotherium and its kind were kind of a side branch. For the family history of elephants, we need to turn to another group, the *Gomphotheres*. It is from these beasts that we get mammoths and the modern elephants.

So from Paleomastodon, though the Gomphotheres there would be a radiation of elephant forms adapted to various lifestyles and environments sadly with just two remaining today the Asian and the African Elephants. But who were the North American elephants we talked about earlier?

This beautiful skeleton of a mammoth graced the National Museum of National History since 1970. It was recently dismantled so that it could be repaired and returned in a more dynamic pose with the opening of the new fossil hall in 2019.

Actually, this display was constructed from more than one specimen, some of the bones coming from the American Museum of Natural History in New York City. Although described as wooly mammoth parts of its skeleton come from a closely related species the Columbian mammoth.

This is something that you would not find in modern displays but was quite common in the past, some displays being a kind of a hybrid. The museum is currently determining if the specimen represents more than one species and is looking into the possibly of replacing the odd bones to make it a more pure species.

Even so, this wonderful display with its dramatically curving tusks does give us an impression of these fantastic animals that once wandered across North America in a geological sense just yesterday.

Mammoths evolved in Africa during the Pliocene and would enter Europe by about 3 million years ago. A European species called the steppe mammoth evolved in Eastern Asia, and by around 1.5 million years ago would cross the Bering Strait across Beringia when sea levels were lower than today. The Columbian Mammoth would evolve from these pioneering Steppe mammoths and populate an area from the northern U.S.A. to Costa Rica.

The Columbian mammoth was about 4 m at the shoulder, and up to about 11 tons. This must have been a magnificent beast to have seen striding across the open prairies and grasslands. Adults would have likely been immune to attacks from even large predators, but the young and old were probably vulnerable to wolves and some of the big cats of the time.

Specimens of the Columbian mammoth are quite well known as many individuals were caught in natural traps. A good example is the sinkhole at the Hot Springs Mammoth Site in South Dakota. Here a collapsed cavern was filled with warm spring water, creating a steep-sided pond into which many examples of the local megafauna would slip and become trapped over about 700 years at about 26,000 years before the present.

Another really famous trap site is found at the La Brea Tar Pits, near downtown Los Angeles. Here animals became stuck in asphalt pools that still form at the site today from a natural petroleum seep. After struggling to escape, they would exhaust themselves, eventually dying of starvation The corpses would attract large predators like saber-toothed tigers and dire wolves which would also become stuck—a natural predator trap.

There would be another wave of mammoth invasion into North America, though. From the steppe mammoths that gave rise to the Columbian mammoth, another iconic species would evolve about 400,000 before the present and cross into North America by 100,000 before the present—the wooly mammoth.

Woolly Mammoths were smaller than their Columbian cousins, about the same size as an African elephant. They were covered by coarse hair, probably thicker than that of the Columbian mammoth and as they lived in the more northerly regions had small ears probably an adaptation to conserve heat. The characteristic fatty hump on the Mammoth's back was may have been used as a reserve source of nutrients in the more extreme northerly environments.

Incidentally, the closest living relative of mammoths is the Asian elephant, which is more closely related to the extinct mammoths than the African elephant.

We know quite a lot about wooly mammoth anatomy as unlike the Columbian mammoth specimens have been found preserved buried in permafrost with soft parts still intact. For example, this is Lyuba, a frozen mammoth calf found in the Russian Arctic in 2007. Lyuba was so well preserved that hair, trunk, eyes and internal organs were preserved even the contents of his last meal were preserved in the intestine.

Both the wooly mammoth and Columbian mammoth coexisted in North America with the wooly mammoth living in colder, more northerly environments. It is possible that the two species interbred where their ranges overlapped, though.

Wooly mammoths interacted with humans and form an important part of early cave art in Europe. In some areas, wooly mammoths were used by Neanderthals and Cro-Magnon people to make huts. Likewise, the Columbian mammoth was part of the cultural landscape of North America as you can see from this petroglyph from Utah depicting two Columbian mammoths and a bison.

But there is one North American elephant we haven't covered. What about Incognetum? You remember that Incognetum, the American mastodon, had very different teeth to mammoths with raised cones rather than ridges. Initially, some thought that the cones on Incognitum indicated that this creature was an Ice Age, flesh-eating monster.

It would be Benjamin Franklin who would figure out the true nature of the beast while on a diplomatic mission to London. He reasoned, correctly, that mastodon's tusks would have been an impediment for catching prey and suggested that the cone-like teeth would probably be an adaptation for grinding small branches of trees. Right again.

The American mastodon or *Mammut* to give him his correct scientific name had shorter legs, flatter, longer skulls and were more heavily muscled than the mammoths they coexisted with. Large males reached up to 9 feet 2 inches and weighed 5 tons. They also had tusks that were less curved.

The American mastodon is from a much older root stock than their mammoth cousins with Mammut diverging from the chain of evolution that would lead to the Elephantidae that includes modern elephants and mammoths at about 27 million years ago.

They ranged across North America during the Pleistocene Epoch, mostly inhabiting cold spruce woodlands, browsing on trees. Remains of the mastodon have been found frozen in Alaska and from these the genome of the creature sequenced allowing us to place it fairly accurately within the elephant family tree.

As a side note, the discovery of all these giants of North America was important ammunition for Benjamin Franklin when refuting the idea that creatures in the New World were somehow inferior to those from Europe.

This idea was promoted by French Naturalist George-Louis Leclerc, Comte de Buffon, a famous French naturalist. Buffon suggested that climatic conditions and an unprolific land in North America only allowed for degenerate creatures to exist, including humans. He viewed Adam and Eve as being Caucasian, and all other races including Native Americans to be the result of degeneration from the original Caucasian standard due to environmental factors. Don't you just want to give him a slap?

Franklin was having non of this nonsense about American degeneracy and constructed tables demonstrating how many North American species were far larger than those found in Europe. He also pointed to the discovery of the magnificent mammoths and mastodons as clearly refuting Buffon's ideas. You gotta love a man of logic like Franklin.

It's important to remember that the mammoths and mastodons of North America did not exist in isolation. Other giant creatures roamed North America, part of a now-extinct Pleistocene megafauna. Megafauna are considered to be animals that are over 100 lbs. So, yes, we humans are considered megafauna too. Sharing the Pleistocene landscape were creatures like the giant sloth, *Megatherium*, weighing up to 4.4 tons and 20 feet from head to tail. A North American camel, *Camelops* that stood 7 feet at the shoulder; giant beavers up to 7.2 feet long, the fantastic armored *Glyptotherium*, a relative of the armadillo over 6 feet long and weighing over a ton, and the North American Bison who, thankfully, we still have with us today.

Giant herbivores like these mean there must have been giant predators and these are often found in abundance at animal traps like we mentioned earlier. Predators, such as the short-faced bear, some possibly over 12 feet tall; dire wolves the same size as a modern wolf, but more muscular; the bird *Teratornis*, a scavenger and possibly active predator with a wingspan of around 12 feet—it could probably swallow a rabbit with one gulp; and the American lion, around 25% larger than its African cousin, and, of course, the iconic sabre-toothed tiger,

Smilodon. And this is just a small sampling of the wonderful megafauna that existed in North America and around the world So what happened to them?

This is still hotly debated, but there are four main hypotheses. First, there is hunting. There is a general continent by continent extinction of megafauna that follows the migration of humans across the planet. 33 genera of large mammals go extinct in North America with around 15 genera going into extinction around 11,500–10,000 before the present, which is just after the arrival of some of the first human settlers, the Clovis People.

There is archaeological evidence of butchery of some megafauna species by the Clovis culture, but some question if small bands of hunter-gathers could have been entirely responsible for the extinction of all these large creatures. In addition, there is evidence from the Paisley Cave of south-central Oregon that there were pre-Clovis People at 14,000 years ago, perhaps crossing Beringia as early as 15,000 years before the present day and some evidence perhaps of even earlier peoples too.

Some researchers prefer climate change as the vector for extinction. Following the glacial advance during the Ice Age, there are often associated extinctions of species possibly related to changes in vegetation patterns, as the climate warmed during an interglacial. Other less favored hypotheses include the possibility of a hyper-disease that killed off many of the larger creatures, although it is difficult to imagine how a disease could be fatal to so many different species of animal. And the possibility of an extraterrestrial impact event that caused wildfires, and ultimately a global cooling event during a period called the Younger Dryas between 12,900–11,700 before the present day. Evidence for this though is somewhat contradictory.

Perhaps a good explanation could combine aspects of possibilities 1 and 2, the overhunting and climate change scenarios. Perhaps climate change reduced the population of many of the megafauna genera to such an extent, that even low levels of hunting would drive the animals into extinction. But to finish with I'd like to consider another question. Is it possible today to bring these fantastic giants back from the grave?

This, of course, rings of Michael Crichton's *Jurassic Park*, doesn't it? You remember the story—mosquitoes from the time of the dinosaurs trapped in amber are used to collect dinosaur blood upon which they had been feeding. Today, of course, mosquitoes prefer the blood of short, balding Englishmen.

The DNA extracted from the dinosaur blood was used to reconstruct their genome with missing bits substituted with DNA from living animals. Dinosaur chromosomes were then implanted in the egg of another creature such as a bird or crocodilian, creating dinosaur embryos, and ultimately living, breathing dinosaurs. Of course, it all goes horribly wrong, and lots of people with the exception of a hero paleontologist and a few others get eaten in a variety of grizzly ways.

It's a great story and a fun film, but unfortunately, it is very unlikely that we will ever be able to find DNA of that antiquity, even if material was preserved in amber. DNA is a fairly fragile molecule and degrades pretty rapidly, and no viable dinosaur DNA has been found. But what if we set our sight to more recent times? Rather than 10s to 100s of millions ago, how about we look for DNA just 10s to100s of thousands of years old. The time of the megafauna.

Megafauna like mammoths went extinct relatively recently, so the chances of finding genetic material are much better than for that of dinosaurs. We even have fleshy material available in the frozen remains on the arctic. The first extensively documented frozen mammoth was from Russian botanist Michael Friedrich Adams in 1806. Since then, a number of mammoth carcasses have been recovered, including several virtually intact juveniles.

So given that we have fleshy material available, could we bring back a mammoth from the dead? To answer that we have to consider the three possible methods that we could potentially use to achieve this.

Firstly, we could try somatic cell cloning. Cloning is no longer reserved for sci-fi movies; cloning is real. Ever since the first clone—Dolly, the sheep in 1993—cloning has been a reality.

It has even become a commercial enterprise, if you can't stand to be parted from your favorite pooch for the sum of around $100,000, you can have your best friend cloned in South Korea.

How is it done? To understand this, you need to realize that all animals are composed of two basic kinds of cell—somatic cells and germ cells. Germ cells are eggs and sperm while somatic cells are every other cell in your body. Sperm and egg cells unite to form a zygote, the earliest developmental phase of a new organism. At this stage the cells of the zygote are pluripotent stem cells, that is cells that have the potential to become every other specialized cell in the mature organism.

Under certain circumstances, you can convince somatic cells to become stem cells again, and it is from stem cells that you can start the process of cloning. This is what they did with Dolly. Cells were taken from the mammary tissue of a donor animal the one to be cloned and were then stressed, convincing them that they were stem cells.

An egg cell from a different breed of sheep that had had its nucleus removed became the host for the donor material. This was then given an electric shock to start cell division, and the zygote that had formed was placed in a surrogate of yet another variety of sheep. The lamb produced, Dolly, was a clone of the original nuclear donor and shared none of the genetic material of the other sheep involved in the process.

So why not use this process to clone a mammoth? The surrogate, in this case, would be an Asian elephant the closest living relative of the mammoth. All you need to find is a viable cell from all that frozen mammoth meat in the arctic and off you go. Well, that sounds simple enough, but there is a problem. Thus far there has been no living cell recovered from any mammoth carcass, and according to many of those in the know, there never will be either.

The problem is that soon after death bodies start to decay as they are attacked by microbes, oxygen and ultraviolet radiation Without the mechanisms in living organisms that continually repair DNA, this delicate molecule breaks down so that only fragments remain.

OK, so how about a more Jurassic Park method of cloning? How about we take the fragments that we can find and reassemble the genome? Stick that into a nucleus and off we go again. Some progress has been made in this regard, and large portions of the mammoth genome have been mapped.

A problem, though, there are even large parts of the human genome that we just can't map, never mind the mammoth genome. Who knows what role these unknown areas might play in the development of a new individual, even if they don't code for anything important in a living organism.

Let's put that to one side, though, and assume that those difficult areas don't really matter; we still have a problem with mammoth DNA. Even when we extract fragmentary DNA from mammoth material, much of it is contamination with DNA from other creatures.

No problem. We can just compare it with the existing elephant genome using it as a template. But what of all that stuff that is specific to mammoths? The DNA that makes a mammoth a mammoth. How could you tell that from all the contaminating DNA? It is very difficult and even if we could eventually reconstruct the entire genome we don't know how to wrap it up into chromosomes and the insert them in to a nucleus.

Before we get too depressed though we have a third option. We know that the Asian elephant is the closest living relative of the mammoth, In fact, they probably share about 99% of the mammoth genome. All we need to do then is recognize that part that is not mammoth and swap it out with parts of the genome that we have positively identified as mammoth. We could in effect engineer a mammoth.

We have the technology to snip off sections of Asian Elephant DNA and insert mammoth sections. Kevin Campbell of the University of Manitoba has already done this with parts of the genes that code for hemoglobin, the molecule that takes up oxygen in red blood cells. He found that the resurrected mammoth blood that he created was much more efficient at transporting oxygen at low temperatures, just what you would expect from a wooly mammoth.

We are still a long way off from reconstructing a wooly mammoth, though. The problems inherent with all of these techniques have been covered by Dr. Beth Shapiro of the University of California at Santa Cruz, who wrote an excellent book on this subject called *How to Clone a Mammoth*. Beth raises another important question, though. Should we clone mammoths? If we could bring a mammoth back, they would be out of time or anachronisms. The gaps they left now filled as the biosphere moves on. Is there a place for these creatures today?

The research going on into these cloning technologies is still potentially beneficial, though, if not for extinct species then perhaps endangered species. Species where populations have become so small that they lack sufficient diversity to fight off disease. Perhaps in these cases snips of DNA from other vanished populations may help patch them up, increasing their genetic diversity and resistance to disease. But for now, the ethical debate is somewhat academic. I wouldn't expect to see mammoths wandering the prairies anytime soon. No matter how much I might really want to see them.

Lecture 22 The Little People of Flores

Myths and legends are a wonderful part of who we are. Some myths are unique, but many share a common theme that is repeated across many cultures. One of those common myths is the existence of "little people." This lecture will take you to the island of Flores, part of the Lesser Sunda Islands in Indonesia. You will learn about the paleontological secrets that Flores holds as well as the identity of *Homo floresiensis*.

The Island of Flores

- Although little people are common characters in mythology, no one would have suggested that any might actually have existed, at least until a team of Australian and Indonesian archeologists made a fantastic discovery on the island of Flores in 2003.

- Tantalizing stone tools had been discovered previously by Father Theodor Verhoeven in the 1950s. Dutch and Indonesian archeologists in the 1990s dated similar tools to around 700,000 B.P., suggesting that human ancestors, probably *Homo erectus*, had migrated to the islands of Indonesia long before the evolution of modern humans about 200,000 years ago.

- So, Flores was probably a good place to hunt for these earliest migrants making their way across the East Indies.

- In 2001, an Indonesian-Australian team co-led by Dr. Mike Morwood of the University of Wollongong in Australia began excavations at a large limestone cave called Liang Bua. They dug for 2 years, first finding an arm bone and then about a year later a tooth. Then, on September 6, 2003, one of the locally hired excavators, Benyamin Tarus, exposed the top of a skull (LB1) at about 6 meters (around 20 feet) below the cave floor.

- This was a small individual, about 3 feet 6 inches (1.06 meters) and weighing about 66 pounds (30 kilograms). It looked like a female child, until the teeth were examined: The jaw contained wisdom teeth that were fully exposed and also demonstrated signs of wear. This was a tiny adult about 30 years old, dating to geologically very recent times, at about 18,000 years ago.

- Fragments of 12 other individuals were recovered that were associated with this skeleton, as were stone tools. In addition, charcoal was found, suggesting the use of fire. The initial dates obtained for these discoveries overlapped with the time that modern humans are known to have arrived in Indonesia at about 45,000 B.P. Could they have coexisted with modern humans?

- The discovery generated a media storm, and very quickly these little people were named "Flores Hobbits" by the media after the little people in J. R. R. Tolkien's books. Scientifically, they were named *Homo floresiensis*, a member of our own genus and part of the taxonomic tribe called hominins, comprising modern humans and their ancestors.

- The discovery also generated immediate controversy. Part of the problem centered around *floresiensis*' brain, which, at 400 cubic centimeters, places it in the range of chimps. This flies in the face of our understanding of our human evolution, which demonstrates the advance of upright walking apes evolving progressively larger brains. Now here is *floresiensis*, potentially coexisting with modern humans but possessing a tiny brain.

- Many were unhappy with the idea of these fossils being a new species of human. The head Indonesian anthropologist Teuku Jacob, after examining the remains, declared that they resemble modern humans suffering from a condition known as microcephaly. This is a rare neurological disorder generally associated with dwarfism and characterized by people with very small brains and heads, and often a diminished mental capacity, and a characteristic sloping forehead.

- Finding the skull of *Homo floresiensis* allowed the production of an endocast, an internal cast that gives us a picture of the external surface of the brain. When examined, it was found that, unlike many microcephalic brains, *Homo floresiensis*' brain had very well-developed frontal lobes, and Brodmann area 10, the area of the brain involved with higher cognition, was proportionally of similar size to that of modern humans.

- Does this explain the presence of stone tools in the cave and evidence for the use of fire? Could it also explain the butchered remains of extinct pygmy elephants called *Stegodon florensis insularis*? This animal was about the size of a cow but would still be a challenge for an individual shorter than 4 feet. Could this imply group hunting, a complex social structure requiring advanced cognition and perhaps speech?

- Other explanations by pathology—to try and explain these fossils as human with pathologies—also include congenital hypothyroidism, which can be found in the local population of Flores and could also account for small bodies and brains. In 2014, it was suggested that the LB1 skull demonstrates craniofacial asymmetry, a sign of Down syndrome, although Australian researchers counter this and state that this is likely a feature of the skull's preservation and damage during extraction.

Homo floresiensis

- Supposing that this is a real human ancestor, a close relative but not a member of *Homo sapiens*, where does *Homo floresiensis* fit into our story?

- It is thought that *Homo sapiens* likely evolved from a group of *Homo erectus* in Africa about 200,000 years ago and spread out from there, replacing *Homo erectus* and other human species, such as the Neanderthals, and ultimately becoming the only living species of human on Earth.

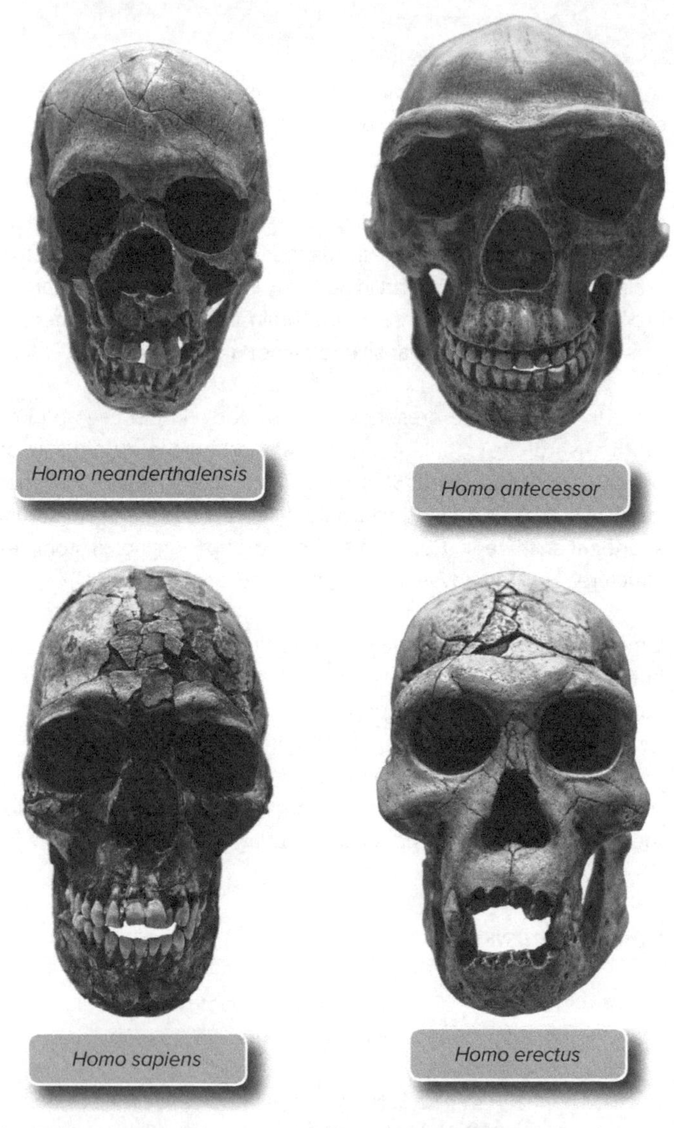

- If *Homo floresiensis* did not evolve from *Homo sapiens*, then from whom in our family tree did they evolve? The most obvious candidate would be *Homo erectus*, our first really human-looking ancestor and the only other human ancestor we have evidence of in Indonesia. But how does a large human ancestor shrink to 3 foot 6 inches? To answer that, we turn to consider some of the strange things that can happen to animals on islands.

- On islands, former mainland species over a number of generations can start to experience dramatic changes in size, effects called insular dwarfism or gigantism. These changes can be due to a number of factors. Growing large on an island may be due to a lack of predators that you no longer have hide from. Shrinking in size may help if resources are limited and the territory available for your seeking of food is much reduced.

- On Flores, there are examples of dwarfism and gigantism. *Homo floresiensis* might have hunted giant rats about twice the size of the average brown rat. There were also giant storks and pigmy elephants. Although probably not an example of the island effect, the intimidating carnivorous Komodo dragons were also ambling around Flores.

- Could the little creature known as *Homo floresiensis* be a dwarf *Homo erectus*? If this is the case, then this is a branch of an extremely successful and widespread human ancestor that survived, hidden, until very recently. Some are unhappy with the island effect hypothesis to explain *Homo floresiensis*, in part because island dwarfism does not usually see a reduction in brain size, like we see in *Homo floresiensis*.

- Could *Homo floresiensis* represent something much more ancient? Is it possible that *Homo floresiensis* has not undergone island shrinking and could have arrived on Flores already small?

- Dr. Matt Tocheri of the Smithsonian's Human Origins Program notes that there are several features in modern human and Neanderthal wrists that are quite different from the wrist structure of great apes and earlier hominins. It would appear, though, that the structure of the wrist of the LB1 fossil shares much more in common with these earlier hominins and the great apes than it does us.

- So, this research confirms the early status of *Homo floresiensis*. In fact, some have wondered if *Homo floresiensis* could be descended from even earlier human ancestors, such as *Homo habilis*, or with the even more ancient australopithecines, such as *Australopithecus afarensis*.

- The problem with this is that unlike *Homo erectus*, the australopithecines and *Homo habilis* are not known out of Africa. It is conceivable, however, that the "Saharan pump"—an opening up of a grassland corridor along the eastern Mediterranean due to changing environmental conditions—could have allowed for their migration just as it did for *Homo erectus*, which possibly reached Java by about 1.5 million years ago. If this is the case, *Homo floresiensis* may represent some of our earliest human wanderers. Perhaps we will find their fossils in the future.

- Perhaps we have another way forward, though. We are already mapping the Neanderthal genome, so what about *Homo floresiensis*? Attempts have been made on the teeth found at Liang Bua, but so far without success. This might be due to the tropical conditions in which the fossils were found. When excavated, they were described as having a "consistency of wet blotting paper." Perhaps future fossil finds will provide better-preserved material and put to rest the origin of these remarkable little humans.

The Fate of Flores's Ecosystem

- What happened to the ecosystem of pygmies and giants on the island of Flores—this ecosystem of tiny humans hunting giant rats and dodging the scary Komodo dragons?

- This is a very volcanically active part of the world. The volcanoes in Indonesia are dangerous, very different from the generally benign volcanic eruptions we see on Hawaii. The volcanoes in Indonesia are explosive and capable of generating pyroclastic flows, hot flows of gas and rock that can reach speeds of 700 kilometers per hour (450 miles per hour) and temperatures of 1000° Celsius (1830° Fahrenheit). They can also be responsible for lahars, a volcanic debris flow composed of a slurry of pyroclastic material and various materials flowing with the consistency of wet concrete.

- The fossils, along with some of the giant storks and pigmy elephants, were found below a thick layer of volcanic ash. Perhaps, then, a volcanic disaster ended the world of the little people of Flores. A large volcanic event in this part of the world could easily disrupt a finely balanced island ecosystem, taking *floresiensis* and the other giants and pigmies of the island into extinction.

- It is possible that they survived only to be wiped out by the modern people of Flores. Unfortunately, the idea that they coexisted with modern humans for an extended period of time may have recently revived a blow.

- Research published in *Nature* in 2016 has provided a much more accurate date for the fossils in the cave at Liang Bua. The oldest fossil remains appear to be about 60,000 years old, with tools dating to about 50,000 years and not the more recent dates that were previously provided. If they ever did coexist with humans, it is likely that as soon as our species started to spread across the Lesser Sunda Islands, they were simply outcompeted and driven into extinction.

- Even so, it would be nice to think that they are still furtively slipping through the forests of Indonesia. Between 2005 and 2009, a camera-trapping project was funded by the National Geographic Society in the hope of snapping a shot of the mysterious creature in the woods—without success.

Questions to consider:

1. How much of our genome is the result of species interbreeding?
2. As far as we know, are we the last remaining member of our genus? Are we an endangered genus?

Suggested Reading:

Stringer, *Lone Survivors*.

Tattersall, *Masters of the Planet*.

Lecture 22 Transcript

The Little People of Flores

Myths and legends are a wonderful part of who we are. Some myths are unique, but many share a common theme repeated again and again across many cultures. One of those common myths is the existence of little people.

In Irish Legend, there are the Leprechaun's; tiny fairies who love gold, making shoes and playing practical jokes.

There is a whole array of the little folk in Celtic folklore, including fairies and elves. They are often associated with mischief and are generally hidden, keeping out of view of the rest of the human population.

Many Native American stories tell of little people who lived in woods, or sometimes hid in rocks near large bodies of water.

Like their European counterparts, they could often be pranksters, but would sometimes also take children away from abusive parents.

But for this lecture, I'd like to take you the island of Flores in Indonesia and accounts of the little folk that mythology talks about there. Flores is part of the Lesser Sunda Islands, which form one of the many island chains of the area. Geologically, Indonesia is complex, sitting at the meeting point of several tectonic plates. In the east, the Pacific plate is in collision with both the Indo-Australian and the Philippine plate, and in the west the Eurasian Plate is colliding with the Indo-Australian plate.

The Island of Flores is part of the Sunda Arc, a volcanic arc that includes the large islands of Sumatra and Java and many smaller islands.

This arc marks where the oceanic part of the Indo-Australian Plate is being subducted below the Eurasian Plate, forming both the Sunda Arc and the deep Sunda or Java Trench. It was the site of the devastating Sumatra–Andaman Earthquake and Tsunami of December 26, 2004.

This grinding tectonic activity is also the reason why this area is one of the most volcanically active areas of the world. It includes Mount Tambora, which erupted in 1815, and is regarded as being the most violent in recently recorded human history; and Krakatoa, which erupted 1883 held to be the loudest noise in recorded history and responsible for the deaths of over 36,000 people.

The complex tectonics of this area is also responsible for the deep ocean trenches, separating the islands which is why even during times of low sea level, such as occurs during ice ages, many of these islands have always been isolated from mainland Asia and Australia.

Flores lies in between two biogeographic boundaries; the Wallace line west of which fauna and flora are Asian in character and the Lydekker's line east of which animals have more of an Australian flavor.

You may remember, we first discussed this area when we were considering the wonderful Komodo dragon in an earlier lecture. The area between these lines is known as Wallacea, named after naturalist and contemporary of Charles Darwin, Alfred Russel Wallace, who noticed these biogeographical divisions.

But getting back to the little people of Flores. In the local Nage language, they are called the Ebu Gogo, which roughly translates into English as Grandmother who eats anything, or perhaps even Granny Glutton. They are described as being fast runners, about 1.5 m tall with broad faces and very hairy bodies The Ebu Gogo are said to have their own language but could repeat parrot-like phrases said to them in the Nage language.

The Nage people hold that the Ebu Gogo still had a presence on the island when Portuguese trading vessels arrived in the 16th century. Some claim that there were occasional sightings of these little folk even into the 20th century.

Nage folklore tells of a somewhat troubled relationship between the two groups of people, though. The Ebu Gogo were often blamed for stealing food and kidnapping children.

The Nage people tell how they finally decided to rid themselves of these troublesome little folk. As a gift, the Nage provided the Ebu Gogo with palm fiber to make clothes.

The Ebu Gogo took this back to their caves, but the villagers threw a firebrand in after them, setting the palm fiber alight and killing the cave dwellers, although legend does tell that a pair escaped into the forest.

Similar stories of little people are found on the island of Sumatra, where they are called the Orang Pendek meaning simply short person, where they are said to live in the forested mountains of the island and stand about 80–150 cm tall. Like the Ebu Gogo, they are blamed for raiding food but are often described as being very strong for their size.

So in this episode, I would like to ask, what paleontological secrets does the island of Flores hold? Where does Homo floresiensis fit into our story? Just who were Homo floresiensis?

Although little folk are common characters in mythology, no one would have suggested that any might actually have existed at least that was what you might have thought until a team of Australian and Indonesian archeologists made a fantastic discovery on the island of Flores back in 2003.

Tantalizing stone tools had been discovered previously by Father Theodor Verhoeven in the 1950s. He was a priest who worked at a seminary on the island. Dutch and Indonesian archeologists in the 1990s dated similar tools to around 700,000 years before the present, suggesting that human ancestors probably *Homo erectus,* had migrated to the islands of Indonesia long before the evolution of modern humans at about 200,000 years ago.

So Flores was probably a good place to go and hunt for these earliest migrants making their way across the East Indies.

In 2001 an Indonesian-Australian team co-led by Dr. Mike Morwood of the University of Wollongong, Australia began excavations at a large limestone cave called Liang Bua meaning "cool cave." They dug for 2 years, first finding an arm bone and then about a year later, a tooth. And then on September 6, 2003, they hit the jackpot. One of the locally hired excavators, Benyamin Tarus, exposed the top of a skull called LB1 at about 6 m around 20 feet below the cave floor.

This was a small individual about 1.06 m weighing about 30 kg and looked like a female child, at least that is what it looked like till the teeth were examined, and then came a surprise. The jaw contained wisdom teeth that were fully exposed and also demonstrated signs of wear. This was no child but a tiny adult about 30 years old dating to very recent geological times; it was suggested about 18,000 years ago.

Fragments of twelve other individuals were recovered associated with this skeleton as were stone tools. In addition, charcoal was found suggesting the use of fire. The initial dates obtained for these discoveries overlapped with the time that humans are known to have arrived in Indonesia at about 45,000 before the present day. Could these little fossils be the remains of the elusive Ebu Gogo and could they have coexisted with modern humans?

The discovery generated a media storm and very quickly these little folk were given the name the Flores Hobbits by the media, after the little people in J.R.R. Tolkien's books. Scientifically, they were named *Homo floresiensis*, a member of our own genus and part of the taxonomic tribe called hominins comprising modern humans and their ancestors.

The discovery also generated immediate controversy. Part of the problem centered around floresiensis' brain which at 400 cc places it in the range of chimps. This flies in the face of our understanding of our human evolution, which demonstrates the advance of upright walking apes evolving progressively larger brains. Now here is floresiensis, potentially co-existing with modern humans, but possessing a tiny brain.

Many were unhappy with the idea of these fossils being a new species of human. The head Indonesian Anthropologist Teuku Jacob, after examining the remains, declared that they resembled modern humans suffering from a condition known as microcephaly. This is a rare neurological disorder generally associated with dwarfism and characterized by people with very small brains and heads and often diminished mental capacity, and a characteristic sloping forehead. This is the condition that the Zika virus can cause.

Finding the skull of *Homo floresiensis*, though, allowed the production of an endocast, an internal cast that gives us a picture of the external surface of the brain, even though the original organ has long since rotted away. When examined, it was found that unlike many microcephalic brains, *Homo floresiensis*' brain had very well developed frontal lobes, and in particular, an area called Brodmann area 10, which proportionally of similar size to that of modern humans.

In modern humans, this is the area of the brain involved with higher cognition and allows for something called cognitive branching. Cognitive branching allows you to come back to something later in your mind so that you can get on with something else that demands your attention right now. As such it may allow you to multi-task. This is a feature of complex analytical thinking.

Does this explain the presence of stone tools in the cave and evidence for the use of fire? Could it also explain the butchered remains of extinct pygmy elephants called *Stegodon florensis insularis*? This animal was about the size of a cow but would still have been a challenge for an individual under 4 feet tall. Could this imply group hunting a complex social structure requiring advanced cognition and perhaps speech?

Other explanations by pathology to try and explain these fossils as human with pathologies also included congenital hypothyroidism, which can be found in the local population of Flores and could also account for small bodies and brains. And more recently in 2014 it was suggested that the LB1 skull demonstrates something called craniofacial asymmetry a sign of Down's Syndrome, although Australian researchers counter this and state that it is likely a feature of the skull's preservation and damage during extraction.

Needless to say, this is a very controversial find. But just for now let's suppose that this is a real human ancestor, a close relative but not a member of *Homo sapiens*. Where does *Homo floresiensis* fit into our story? To understand this, let's review a little of our own family tree.

Molecular evidence suggests that the last common ancestor between humans, gorillas, and chimpanzees occurs sometime as early as 8 million years ago, during the late Miocene. The Miocene is often described as the golden age of the apes with a much higher diversity of apes and possible hominins too.

In truth, fossils from this golden age can be scarce in part due to the acidity of the soils in the tropical forest in which these early apes lived. And so sorting out the direct evolutionary relationship between them can be tricky. But perhaps something like *Ouranopithecus*, an extinct ape recovered from Greece and Bulgaria around 8 million years ago could be close to the first true primate on the human branch.

Closer to us in time at 4.4 million years ago, is Ardipithecus who has been described as a strange mosaic of a creature. It is thought from the position of a feature called the foramen magnum, that Ardipithecus might have walked upright.

In apes, this feature is positioned at the back of the skull, but in upright human walking primates, like humans and Ardipithecus, it is right under the skull reflecting the fact that we are upright and balancing this large melon on the top of our spines

Ardipithecus still possessed many ape-like features, though, with long arms and chimp-like feet with opposable thumbs.

But probably the most famous of our early human ancestors is Lucy or *Australopithecus afarensis*. Lucy was recovered from Hadar in Ethiopia and is dated to 3.2 million years ago. She's a lot more human-looking than Ardipithecus, with fewer tree-climbing adaptations and a much more human looking foot. Even though she walked upright, Lucy still retained some tree climbing characteristics, like strong arms and curved fingers. Although on the path to modern humans, Lucy still had a head and brain similar to that of a chimp with a brain about a third the of the size of the typical human today.

As a species, *Australopithecus afarensis* is found between 3.85 and 2.95 million years ago, making it one of the longest-lived human species.

Lucy was discovered by Don Johanson in 1974, but since then, species that had been found earlier have also been incorporated into this species. With these and new discoveries, we now have around 300 individuals in various collections around the world.

Eventually, Lucy or a similar upright walking ape would give rise to the first member of our own genus, *Homo*. The oldest fossil evidence we have comes from *Homo habilis* or handyman who lived around 2.4–1.4 million years ago. As his name suggests, handyman was demonstrating definite evidence of tool use.

Homo habilis was discovered by the famous Paleoanthropologist Jonathan Leaky in 1960. Its brain was a little larger than Lucy's and slightly less ape-like but still demonstrated protrusive jaws. They were short compared to modern humans about 100–135 cm with an average weight of about 32 kg.

Our first really human-looking ancestor was *Homo erectus*. From a distance you would have probably thought it was a modern human; their bodies looked very much like ours. But if they were to turn and look at you would see a face that was still very ape-like. Their brain although larger than previous hominins at 900 cc was still much smaller than our own—that is, about 1300 cc.

Homo erectus is a considered to be a highly variable species, and many subgroups have been identified. Some of the oldest Homo erectus fossils have been found in Dmanisi, Georgia, *Homo erectus georgicus*. Some have suggested that this opens the possibility that Africa may not necessarily be the cradle of all human ancestors. Perhaps some of our evolutionary history has a bit of an Asian origin too.

Whatever *Homo erectus'* origins, they were certainly travelers. Their travels and likely the migration of other creatures too were aided by changing environmental conditions that would allow for the development of a wet Sahara and the opening up of a grassland corridor along the Eastern Mediterranean, sometimes called Savannahstan.

As grasslands opened up, animals migrate and *Homo erectus* known to be hunters would follow, an effect sometimes called the Saharan Pump It is possible that it would be this pump that would allow *erectus* to reach Java by about 1.5 million years ago.

In fact, it was in Java that the first *Homo erectus* fossil was found by Eugene Dubois. Dubois was born in Limburg in the Netherlands. His father was mayor of the town, but Eugene appeared to be more interested in natural history and in exploring the local mines of Mount Saint Peter, from which a skull of a Mosasaurus was discovered in 1765.

As an adult he had a promising academic career as an anatomist, but became increasingly interested in human origins, perhaps inspired by the Neanderthal finds that were being found in various locations in Europe at that time, It is said that much to the surprise and perhaps dismay of his colleagues, he gave up his university position and joined the Dutch Army to be posted to the Dutch East Indies, now Indonesia, where he was convinced he would find the human missing link,

His logic possibly had something to do with Dubois being intrigued by Wallace's accounts of the area in which he described the presence of numerous caves. At the time caves were the only places human fossils had been found So off he set, to hot, malaria-ridden Sumatra, towing his wife and young daughter with him.

Apparently, his responsibilities in the army were pretty light so he could spend a lot of time looking for fossils. Moving to Java, he would start to amass an enormous collection of animal fossils and eventually struck fossil gold in 1891. He discovered a skullcap, tooth, and a thigh bone of a species he describes as being in between humans and apes. He named it *Pithecanthropus erectus*—ape-human that stands upright—although it was commonly referred to as Java Man.

Today, of course, we know it as *Homo erectus*. During his time in the Dutch East Indies, Dubois would suffer from various tropical illnesses and would tragically lose a daughter to disease too. It took a long time for the scientific community to accept his find as a human ancestor, and this resistance and perhaps the personal price he paid in finding his famous fossil, may explain why he is reported to have died in 1940 a bitter man.

It is thought that *Homo sapiens* likely evolved from a group of *Homo erectus* in Africa about 200,000 years ago and spread out from there replacing *Homo erectus* and other human species like the Neanderthals and ultimately becoming the only living species of human on Earth.

But who was *Homo floresiensis*? This is an important question If not *Homo sapiens*, then from whom in our family tree did *Homo floresiensis* evolve? The most obvious candidate would be *Homo erectus*, the only other human ancestor we have evidence of in Indonesia. But how does a large human ancestor shrink to about 3 foot 6 inches? To answer that we once again turn to consider some of the strange things that can happen to animals on islands.

Like we mentioned in a previous lecture, it is known that on islands, former mainland species over a number of generations can start to experience dramatic changes in size, effects call insular dwarfism or gigantism.

Examples of dwarfism include the 200 kg dwarf hippos on Cyprus or the modern hippo 1500 kg. We find diminutive sauropod dinosaurs from Transylvania, an island in the Cretaceous that were only 6 m long compared to their 30 m long relatives on the mainland. And consider this tiny little Chameleon, *Brookesia micra* discovered on the island of Madagascar in 2007. This is a juvenile, but adults only reach about 29 mm long.

On the other side of the scale, there is the familiar old dodo from Mauritius, which was actually an expanded flightless pigeon about 1m tall; and Haast's Eagle from the South Island of New Zealand, which at 12 kg for a large male probably evolved to prey on the giant Moa. And then there is *Archaeo indris*, a lemur that evolved to the size of a gorilla on Madagascar and may have only gone extinct about 350 years ago.

These changes can be due to a number of factors. Growing large on an island may be due to a lack of predators that you no longer have to hide from. Shrinking in size can help if resources are limited, and the territory available for your seeking food is much reduced.

On Flores, there are examples of dwarfism and gigantism too. *Homo floresiensis* might have hunted giant rats like *Papagomys* with bodies about 45 cm long, and a tail 70 cm long. Twice the size of the average brown rat. I would not want to find any of these living in my basement.

There were giant storks, *Leptoptilos robutus*, and the pigmy elephant, *Stegodon florensis insularis* that we covered earlier We also have to mention, although probably not an example of the island effect, the intimidating carnivorous Komodo dragon, some of which can reach 3m in length that were ambling around Flores and whom we met in a previous lecture.

And of course, there is *Homo floresiensis* herself. Could this little creature be a dwarf *Homo erectus*? If this is the case, then this is a branch of an extremely successful and widespread human ancestor that survived hidden until very recently. Some though are unhappy with the island effect hypothesis to explain *Homo floresiensis* in part because island dwarfism does not usually see a reduction in brain size like we see in *Homo floresiensis*. With all the other evidence presented though it is compelling.

Or could *Homo floresiensis* represent something much more ancient? Is it possible that *Homo floresiensis* has not undergone island-shrinking? Could this little human have arrived on Flores already small?

Dr. Matt Tocheri of the Smithsonian's Human Origins Program was intrigued by the wrist bones of this tiny human.

My Ph.D. research had been looking at the wrist anatomy of not only our species but our close relatives, the great apes, as well as looking at other human fossil wrist bones. And what's really interesting on our wrist, as well as the wrist of Neanderthals, we have several features in our wrist bones that are quite different than what we see in earlier hominins or in African apes. There's been a big change basically in the evolution of our wrist. And we see that change only in Neanderthals and us. And what was so striking to me was the fact that LB1, the specimen of *Homo floresiensis* still has the condition we see in earlier hominins as well as African apes. So, for me, this was very clear evidence that this is not a diseased modern human, this is basically a very close relative to ours, but in different lineage of human.

Could it be that *Homo floresiensis* is descended from earlier ancestors like *Homo habilis* or with the even more ancient *australopithecines*, like Lucy?

The problem with this earlier than erectus migration is this, though. Unlike *Homo erectus* Lucy and her kind the *Australopithecenes* and *Homo habilis* are not known out of Africa, it is conceivable, however, that the Saharan Pump could have allowed for their migration just as it did for Homo erectus. If this is the case, *Homo floresiensis* may represent some of our earliest human wanderers; perhaps we will find their fossils in the future but so far nothing.

Perhaps we have another way forward though We are already mapping the Neanderthal genome, so what about *Homo floresiensis*? Attempts have been made on the teeth found at Liang Bua, but so far without success. This might be due to the tropical conditions in which the fossil were found. When excavated, they were described as having a consistency of wet blotting paper. Perhaps future fossil finds will provide better-preserved material and put to rest the origin of these remarkable little humans.

And what happened to the fantastic ecosystem of pygmies and giants on the island Flores? This ecosystem of tiny humans hunting giant rats and dodging the scary Komodo dragons?

As we already noted, this is a very volcanically active part of the world. The volcanoes in Indonesia are dangerous, very different from the generally benign volcanic eruptions we see on an island say like Hawaii.

These are explosive and capable of generating pyroclastic flows hot flows of gas and rock that can reach speeds of 700 km/h, and temperatures of 1000°C. They can also be responsible for lahars, a volcanic debris flow composed of the slurry of pyroclastic material and various materials flowing with the consistency of wet concrete. You can see one here that developed from Mount Galunggung in West Java in 1982.

The fossils, along with some of the giant storks and pygmy elephants, were found below a thick layer of volcanic ash. Perhaps then a volcanic disaster ended the wonderful world of the little people of Flores. A large volcanic event in this part of the world could easily disrupt a finely balanced island ecosystem, taking *floresiensis* and the other wonderful giants and pigmies of the island into extinction.

It is possible of course that they survived only to be wiped out by the modern people of Flores as described in the legend of the Ebu Gogo. Unfortunately, the idea that they coexisted with modern humans for an extended period of time has recently revived a blow.

Research published in the journal *Nature* in 2016 has provided a much more accurate date for the fossils in the cave at Liang Bua. The oldest fossil remains appear to be about 60,000 years old with tools dating to about 50,000 years and not the previous recent dates that were recorded. If they ever did co-exist with humans, it is likely that as soon as our species started to spread across the Lesser Sunda Islands, they were simply outcompeted and driven into extinction.

Even so, it would be nice to think that they are still furtively slipping through the forests of Indonesia. In addition to the legends of the Orang Pendek on Sumatra, there have been more recent accounts from Dutch colonists in the 1920's of short bipedal creatures in the forest.

Between 2005 and 2009 a camera-trapping project was funded by the National Geographic in the hope of snapping a shot of the mysterious creature in the woods, without success.

Of course, the accounts of little folk in legend or the more recent sightings could just be a case of mistaken identity other primates or perhaps the sun bear which can stand on two legs, but personally, I would love there to be another member of our family out there. We are the only remaining species of the *genus*

Homo a unique circumstance for our family, which has often seen many different species of people sharing this planet. In a way, we are an endangered genus.

Just imagine though what we could learn from our cousins? What stories they could tell us and how they could help us understand our own place in the family tree. Even without living little people from Flores the search to understand human history goes on it's going to be a fascinating ride.

Lecture 23 The Neanderthal among Us

For many years, we had a brutish vision of Neanderthals. The species was regarded as extremely primitive when compared to *Homo sapiens* and not very closely related to us at all. Today, our overall picture of Neanderthals is a rather short, stocky, barrel-chested, and powerfully built people. They had heavier faces than *Homo sapiens* but were a far cry from the brutal shambling ape that was once envisioned. In this lecture, you will learn who the Neanderthals were—specifically, whether they were a species in their own right or just a thick-skulled variety of *Homo sapiens*.

Cognitive Abilities of Neanderthals

- Did Neanderthals just look like humans, or did they actually demonstrate human intelligence and complex social structures? If you consider pure brain size, Neanderthals actually have us beat. The average cranial capacity of *Homo sapiens* is 1400 cubic centimeters, while the capacity of *Homo neanderthalensis* is around 1600 cubic centimeters.

- Brain size is no guarantee of complex social behavior, though. It is possible that a larger part of the Neanderthal brain may have been devoted to vision.

- Perhaps evidence of intelligence can be deciphered from the Neanderthal diet. It would appear that they ate a variety of foodstuffs, including plants and animals, depending on season and location. Neanderthal wooden spears have been found from sites where big game animals had been butchered. To hunt such large animals implies cooperation, social structure, planning, and—perhaps the hallmark of intelligent animals—some sort of complex language.

- This is a difficult thing to prove, but Neanderthal DNA may help us. We know that on chromosome 7 in the human genome there is a gene called FOXP2 that is required for the development of speech and language, controlling the development of various features in the brain, heart, lungs, and gut. Mutations in this gene can cause speech and language disorders.

- There are only 2 differences in the amino acid code of the FOXP2 gene between humans and chimps, and between Neanderthals and humans there is no difference. Does this suggest that they inherited the gene from a common ancestor even further back in time? If this is the case, could we also infer that Neanderthals were equipped with all the language production capacity that we humans have?

- Neanderthals were adept at making sharp flakes of stone for a wide variety purposes, including cutting and scraping. Neanderthal tools have often been thought of as primitive when compared to those produced by *Homo sapiens*, but they are still very versatile and efficient.

- There are examples of social interaction that may also suggest an intelligent species. In Shanidar Cave located in Bradost Mountain in Iraqi Kurdistan, the remains of 10 Neanderthals have been found. A cast of one of those, "Shanidar 1," is held at the Smithsonian's National Museum of Natural History. He was an old man between 40 and 50 years of age who had a withered right arm that was fractured in a number of places, leaving him with very limited or no use of this lower arm and hand. The fractures had healed, though, implying care by his social group.

- For a long time, one of the features of *Homo sapiens* that was regarded as being unique and distinct was the production of art. Art illustrates an ability to conceptualize—to represent the world you see, or perhaps don't see, what is locked in your imagination.

- For a long time, it was thought that Neanderthals did not exhibit artistic abilities, but a geometric crosshatch pattern found at the back of a cave in Gibraltar in 2012 may change that. These symbols are thought to be around 39,000 years old and were discovered below an undisturbed layer of sediment in which Neanderthal tools were found. The image appears to have been made by a point, the artist deliberately deepening the cuts in the hard dolomite rock over many hours.

- There are some who still question whether these markings were made by Neanderthals, but if they were, this is significant. Perhaps the image is a map or perhaps a symbol to mark territory. Whatever its meaning, it shows evidence of abstract thought, and if this is the case, and these are produced by Neanderthals, then they are not unlike those produced by *Homo sapiens*.

Are Neanderthals Related to Us?

- How closely are we related to the Neanderthals? Unlike reading DNA that is many millions of years old, like we would need for the dinosaurs, the prospects of reading DNA that is within the 100,000-year range is much better. Perhaps we can read the genome—the genetic recipe—of Neanderthals.

- There are 2 possible sources of DNA that we can use in genetic studies. Nuclear DNA—that is, the DNA found in the nucleus of our cells—codes for all the structures and functions of our bodies. This is where we store our "blueprints." But there is another source of DNA in small "micromachines," or organelles, that are found in our cells. These are mitochondria, and each of your cells contains around 1700 of them. Mitochondria have their own DNA, independent of the personal blueprint that you have in the nuclei of your cells, that is called mitochondrial DNA (mtDNA).

- Sperm and egg cells also contain mitochondria, but when the nucleus of the sperm fuses with that of the egg to give you the mix of genes from mother and father that we are all composed of, the mitochmdria of the sperm most of the time gets left outside. As a result, it is generally only the maternal mtDNA, that of the egg, that gets passed on down the line through the generations.

- Over time, mtDNA, like nuclear DNA, will undergo random mutations—slight errors when the DNA is copied from one generation to another. We can use these differences between mtDNA in different people and estimate the rate of mutation to determine how far back in time they once shared a common ancestor. We can also find out how long since they shared that ancestor that they have been traveling along their own particular branch of the ancestry road.

- It was using this principle that researchers in the 1980s were able to determine that the common ancestor of all mtDNA—and, therefore, all humans living today—was a woman who lived in Africa around 100,000 to 200,000 years ago: "mitochondrial Eve."

- Tracing the last common ancestor of humans and Neanderthals would be attempted by Svante Pääbo of the Max Plank Institute in Germany. For this study, they would use Neanderthal 1, the type specimen of the species, which acts as a taxonomic reference when a species is first described. Even if DNA is present, most of it will be fragmentary, and the vast majority of the DNA will be contaminants from soil bacteria and from scientists and curators who have studied the specimen over the years.

- Taking these contaminants into account, Pääbo's team finally isolated Neanderthal mtDNA and then compared it to around 1000 mtDNA sequences of modern humans. Among the modern humans, the DNA differed by about 8 mutations but, between the modern humans and the Neanderthals, the average was around 26. This would place the common ancestor of Neanderthals and modern humans around 500,000 years ago.

- Perhaps *Homo heidelbergensis*, who was known to have lived in Europe and Africa at about that time, is close to the split of the Neanderthal and modern lineages. Some reserach, based on tooth structure, has suggested that the split occured even further back in history, perhaps as much as 1 million years ago. This places it in the range of ancestral humans such as *Homo erectus*.

- This ancient divergence was taken by some to suggest that Neanderthals were a completely separate species to humans and that, with around half a miliion years or more of genetic drift between the 2 populations, it would be very unlikely that a human and a Neanderthal could breed to produce living offspring, let alone offspring that may be viable enough to pass on that genetic mixing into the next generation.

- There had been intriguing fossil finds, though, that suggested some interbreeding had occurred. Had there been inbreeding despite the antiquity of the last common ancestor of the 2 species? To answer this, we would have to find some way to read the nuclear DNA of Neanderthals, the blueprint of the organism held in the cell's nucleus. This was the quest for the Neanderthal genome.

- This is something that we have only just recently achieved for our own species in the Human Genome Project, an international effort to read our own human blueprint. The human genome project was initiated in 1990 and completed in April of 2003.

- The Neanderthal genome would be a more challenging endeavor. There would be problems with contamination and degradation of the DNA, plus the added issue of distinguishing Neanderthal DNA from a very close relative: *Homo sapiens*. These were the challenges facing Pääbo's team when the Neanderthal Genome Project was initiated in July of 2006.

- Eventually, a Neanderthal genome was sequenced, with an initial draft published in *Science* in 2010. It was concluded that around 98.7% of the base pairs in the 2 genomes were identical. It was from this that comparisons between the FOXP2 gene in humans and Neanderthals could be made.

- In comparing the Neanderthal genome to people from different racial groupings, they found that around 1 to 4% of Neanderthal DNA is in our genome. The only way it could have gotten there is via interbreeding. Despite the distance in time that separates humans and Neanderthals from their last common ancestor, it would appear that we were still sufficiently similar to produce viable offspring.

- Not everyone has the same part of the genome, but in total it is thought that around 30 to 40% of the Neanderthal genome is floating around in the human population. This indicates that there was not one Neanderthal ancestor, but there must be an entire history of Neanderthal-human interactions.

- A discovery by researchers, including Pääbo, reported in *Nature* in 2015 provides data from an individual much closer to the original interaction. A specimen from a cave in Romania dated to 37,000 to 42,000 B.P. has been found with between 6 and 9% Neanderthal DNA, probably indicating a fully Neanderthal ancestor just 6 generations back.

- This is one possible picture of our messy genome that was published by Chris Stringer of the Natural History Museum in London in *Nature* in 2012. Humans and Neanderthals diverged from a common *heidelbergensis* ancestor, with Neanderthals generating another ancient human group, the Denisovans. Each of these forms included an archaic flow back into the *Homo sapiens* population, contributing in varying degrees some small parts of the modern human genome.

What Happened to the Neanderthals?

- The forerunners of *Homo neanderthalensis* arrived in Europe about 800,000 B.P. *Homo sapiens* arrives at about 40,000 B.P., and just 10,000 years later, the Neanderthals are gone. This could suggest that they were outcompeted for resources or were hunted down and exterminated in acts of interspecies genocide.

- Another possible explanation may relate to rapid climate change. By about 55,000 years ago, climate started to fluctuate rapidly from extremely cold to mild within a few decades, perhaps within the lifetime of an individual. This would have caused rapid changes in the landscape, from woodlands to grasslands, with familiar plants and animals appearing and disappearing.

- It has been suggested that early modern humans during this period would have had more widespread social networks. This would allow them to acquire resources over a greater area. Neanderthals in this model become increasingly isolated and starved of resources. They finally become extinct at about 41,000 to 39,000 B.P., at the start of a very cold snap.

- But perhaps there is another explanation. As modern humans moved into the Neanderthals' territory, perhaps the Neanderthals were just absorbed by interbreeding. Or perhaps there was a combination of all 3 of these ideas: part confrontation, part climate-and-resource related, and part genetic assimilation.

Questions to consider:

1. How do we define consciousness, and is it only a trait that evolved in *Homo sapiens*?
2. Was it inevitable that *Homo sapiens* would become the only member of our genus on Earth? Is it possible that different circumstances could have seen the *Homo neanderthalensis* rise to dominance?

Suggested Reading:

Pääbo, *Neanderthal Man*.

Papagianni and Morse, *The Neanderthals Rediscovered*.

Lecture 23 Transcript

The Neanderthal among Us

Just 8 miles to the east of Düsseldorf, a city renowned for its fashion, world fair, and fine art—and who could possibly forget Germany's premier electronic band Kraftwerk—is a rather unremarkable little kiln that marks the site of a small cave where limestone was quarried. Quarrying was active back in the mid-1850s, but the little cave was lost until it was rediscovered after a determined search back in 1997.

The reason why such a search was mounted was the discovery back then of fossils that were uncovered by two Italian miners in August of 1856. The cave was about 20 m above the valley floor. The Kleine Feldhofer Grotte, as it is now known, had a 1 m opening and was about 3 m wide by 5 m long by 5m high. From it were unearthed a skull cap and several bones.

I say unearthed and not discovered because these precious finds were dumped with debris from the cave down the side of the valley. Fortunately, the fossils came to the attention of the owner of the cave, one Wilhelm Beckershoff, who thinking they belonged to a prehistoric bear took them to a local school teacher, Johann Carl Fuhlrott, who was also an avid fossil collector and naturalist.

In total, around 16 bone fragments, including the impressive skull cap were recovered from the little cave. Fuhlrott recognizing the bones came from something a lot more human-like than a bear.

A report of the find was published in the local newspaper dated September 4, 1856; that would attract the attention of two anatomy professors in Bonn, Professors Hermann Schaaffhausen, and August Franz Josef Karl Mayer.

They recognized the antiquity of the find and called attention to the strong brow ridges above the eyes and the low sloping forehead reminiscent of the great apes. They concluded that it belonged to a native tribe that lived in Germany before modern humans in antediluvian pre-biblical flood times.

It would not be until 1864, though, that the specimen would be given the name we are familiar with today, by Irish geologist William King. King would later decide that they didn't deserve to be included in our own genus, though.

The fossil found in the Neander valley is the type specimen for *Homo neanderthalensis*—Neanderthal 1. A type specimen acts as a reference when species are first described. In truth, though, Neanderthal 1 was not the first specimen to be found, that honor goes to Engis in Belgium. Here in 1829, fragmentary remains of a 2–3-year-old child were found by Dutch naturalist Philippe-Charles Schmerling. At the time of discovery though even though considered to be old they were regarded as being the remains of ancient *Homo sapiens*.

But just who were these Neanderthals? A species in their own right or just a thick-skulled variety of *Homo sapiens*? In this episode, let's have a closer look and ask would you recognize Neanderthals at the supermarket? Were Neanderthals brutish cavemen? Where do Neanderthals fit in our family tree? What happened to *Homo neanderthalensis*?

One of the first reconstructions of a Neanderthal came from a relatively complete skeleton that was discovered in 1908 at the La Chapelle-aux-Saints cave in central France. A specimen a male is regarded to have died about 60,000 years ago at the relatively ancient Neanderthal age of about 40 years old. The old man of La Chapelle was studied by French paleontologist Marcellin Boule, and it is from him that we gain our first reconstruction of the species.

Boule commissioned a reconstruction that portrayed Neanderthals as a brutish and ape-like; he even provided them with a slouching gait drawing a sharp contrast with proud upright walking *Homo sapiens*. The image was printed in the Illustrated London News, who praised the artist, Mr. Kupka, for having given the face the expression it must have worn. As you can guess, it had the face of a wild animal, not a wise man.

It has been suggested that the condition of the old man when he died may have confused Boule. At his death, he was suffering from osteoarthritis, which may have led to an interpretation of a slouching gorilla-like gait, but it is likely that what we are looking at here is more of a case of cultural bias than misinterpretation, Interpretation being guided by the preconceptions of an ape-man.

This brutish vision of Neanderthals would persist for many years. The species was regarded as extremely primitive when compared to *Homo sapiens* and not very closely related to us at all. It would appear that many of the scientific community at this time would like to keep this part of our family strictly at arm's length.

So what is our current view of Neanderthals? If you were to meet one in a supermarket wearing modern clothes and with a tidy haircut would you pay them a second glance?

If you were to meet a Neanderthal, possibly one of the first things you would recognize is their heavy-set face. They had pronounced brow ridges below a low sloping forehead, a rather receding chin, and a large prominent nose

They had large eye sockets with areas of the brain devoted to processing vision. This has led some to suggest that Neanderthals had particularly acute eyesight.

Neanderthals were on average a short species with males around 164–168 cm and females about 152–56 cm tall. At 166 cm, I may have been able to play for a Neanderthal basketball team, but meeting them on a football field or wrestling mat would have been a little scary. Although they were short, they were packed with muscle, perhaps built a bit like Olympic power lifters.

They might have looked a little stubby in the leg having femurs proportionally shorter than we have, but they possessed particularly strong arms and hands They probably hunted large prey at close quarters, using spears that they would lunge with rather than throw from a distance.

Genetic studies, more on this later, have shown that Neanderthals probably had light colored skin some may have even had red hair. The first true Neanderthals were thought to have evolved in Eurasia, a significant distance from the human cradle in Africa. In the North, there was a selective advantage to evolve lighter skins with less melanin, which would allow the synthesis of vitamin D with a less intense sun.

Neanderthals were living in a much cooler world during the last Ice Age. This may in part explain their stocky build—an adaptation to retain heat. It has also been suggested that the large schnoz of the Neanderthal was an adaptation to warming cold air before it entered the lungs, although some have questioned this as a large nose would have been a large area of heat loss, too.

So our overall picture of Neanderthals at the moment is a rather short, stocky, barrel-chested, and powerfully built people. They had heavier faces than *Homo sapiens* but were a far cry from the brutal shambling ape painted by Boule.

So even if Neanderthals didn't look like Boule's brutish ape, did they behave like them? Were Neanderthals just human-looking or did they actually demonstrate human intelligence and complex social structures? If you consider pure brain size, Neanderthals actually have us beat. The average cranial capacity of *Homo sapiens* is 1400 cc, and *Homo neanderthalensis* is around 1600 cc.

Brain size, though, is no guarantee of complex social behavior. As we said, it is possible that a larger part of the Neanderthal brain may have been devoted to vision. So do we have any other evidence of intelligence? Perhaps we can take something from the Neanderthal diet.

It would appear that they ate a variety of foodstuffs including plants and animals depending on season and location. As Neanderthals were not a tropical species, they could not rely on a stable plant diets year-round possibly forcing them to rely more upon meat than their tropical cousins. Neanderthal wooden spears have been found from sites where big game animals had been butchered.

To hunt such large animals implies cooperation, social structure, planning and perhaps the hallmark of intelligent animals some sort of complex language.

Obviously, this is a difficult thing to prove, but once again Neanderthal DNA may help us here. We know that on chromosome 7 in the human genome there is a gene called FOXP2. This is required for the development of speech and language, controlling the development of various features in the brain, heart, lungs and gut. Mutations in this gene can lead to speech and language disorders.

There are only two differences in the amino acid code of the FOXP2 gene between humans and chimps; and between Neanderthals and humans, there is no difference. Does this suggest that they had inherited the gene from a common ancestor even further back in time? If this is the case, could we also infer that Neanderthals were equipped with the language production capacity that we humans have?

Neanderthals were adept at making sharp flakes of stone for a wide variety purposes, including cutting and scraping. Neanderthal tools have often been thought of as primitive when compared to those produced by *Homo sapiens*. However, they are still very versatile and efficient. At a former Neanderthal hunting campsite in eastern Germany where butchered remains of various animals were found, something intriguing was discovered. It would appear that the Neanderthals had developed technology to attach points to spears using an adhesive pitch that they made from birch bark.

In order to do this, they would have to employ a technique known as dry distillation to make a sticky resinous goo, a process that requires temperatures carefully controlled around 343°C with no oxygen. This is not a simple process, and would likely have required experimentation and an ability to imagine possible outcomes to create what may be considered at 40,000 years before the present day, one of the world's first industrial processes.

But there are examples of social interaction that might also suggest an intelligent species too. In Shanidar Cave, located in the Bradost Mountain in Iraqi Kurdistan, the remains of 10 Neanderthals have been found. A cast of one of those, Shanidar 1, is held at the Smithsonian's National Museum of Natural History. He was an old man between 40 and 50 years of age who was certainly showing signs of his age

He was probably blind in one eye from a blow to the head. He also had a withered right arm that was fractured in a number of places. These fractures had healed, but he probably had very limited or no use of this lower arm and hand. He also walked with a noticeable limp.

As these had healed, though, this implies care by his social group. This person, especially in his later years, would have probably required a lot of support from the members of his tribe. He was not abandoned but thought of as important to the rest of the community and worth the effort of helping him live a relatively long life even with his disabilities.

The same was probably true for the old man at La Chapelle-aux-Saints. The individual also had osteoarthritis, had few teeth, and would have required help from others. When he died, this Neanderthal may have even been carefully placed in a shallow grave. This is not an isolated case and shows that thought was given to the dead, a certain reverence and perhaps the concept of an afterlife, revealing the possibility of ability to conceptualize.

For a long time, one of the features of *Homo sapiens* that was regarded as being fairly unique and distinct was the production of art. Like the concept of an afterlife, art illustrates an ability to conceptualize, to represent the world you see or perhaps don't see that is locked in your imagination. The Lascaux and Chauvet Cave paintings in France are beautiful examples of this; the artist laying down beautiful images at 17,300 and around 31,000 years ago, respectively.

The same goes for the many small statuettes and figurines that are associated with the spread of *Homo sapiens* into Europe, such as the Venus of Willendorf.

So do Neanderthals exhibit similar artistic abilities? For a long time, it was thought the answer was no, but a geometric cross-hatch pattern found at the back of a cave in Gibraltar in 2012 may change that. These symbols are thought to be around 39,000 years old and were discovered below an undisturbed layer of sediment in which Neanderthal tools were found The image appears to have been made by a point, the artist deliberately deepening the cuts in the hard dolomite rock over many hours. This was not an accidental artifact of cutting meat or some other substance.

And what does it mean? There are some who still question if these markings were actually made by Neanderthals, but if they were, this is significant Perhaps the image is a map or perhaps a symbol to mark territory. Whatever its meaning it shows evidence of abstract thought—and if this is the case—and these are produced by Neanderthals, then they are not unlike those produced by modern *Homo sapiens*.

Other interesting evidence of the possible cognitive abilities of Neanderthals has come from a 2012 paper by researchers from Spain and Gibraltar. They studied bird bones including jackdaws, vultures, and eagles found at over 1699 Neanderthal sites across Eurasia.

What they discovered were cut marks on the wing bones and evidence of peeling of the bird's feathers. Wings now are not nutritionally useful compared to the rest of the bird, suggesting that the dark feathers of these birds may have been collected perhaps worn in some sort of decoration. If this is the case, this is once again evidence of complex thought.

Add to this the suggestion that Neanderthals may have been using red and black pigments perhaps to paint their bodies, and using shells as ornaments and a much more colorful and thinking human emerges.

So just how similar to *Homo sapiens* were *Homo neanderthalensis*? As we have seen, they looked different but not as brutally different as was originally thought in Boule's days. They appeared to care for members of their kind and likely had significant communication skills and perhaps an appreciation of conceptual art. But what about at the deeper genetic level? Just how closely are we related to the Neanderthals?

Unlike reading DNA that is many millions of years old like we would need for the dinosaurs the prospects of reading DNA that is within the 100s of 1000s of year range is much better. Perhaps we can read the genome, therefore—the genetic recipe—of Neanderthals.

It is important to realize that there are two possible sources of DNA that we can use in genetic studies. Firstly, nuclear DNA that is, the DNA found in the nucleus of our cells, that codes for all the structures and functions of our bodies. This is where we store our blueprints as it were. But there is another source of DNA in small micro-machines or organelles that are found in our cells. These are mitochondria, and each of your cells contains around 1700 of them.

Mitochondria are essentially the power engines of eukaryotic life. That is, life more complex than bacteria prokaryotes. It is thought that they originally were independently living prokaryotes that became engulfed but not digested by its host. The mitochondrial got a home safe from the external environment, and the host cell got free power from the mitochondria. As a result, mitochondria have their own DNA, independent of the personal blueprint that you have in the nuclei of your cells. We call this DNA mitochondrial DNA.

Sperm and egg cells also contain mitochondria, but when the nucleus of the sperm fuses with that of the egg to give you the mix of mom and pop genes that we are all composed of, the mitochondria of the sperm most of the time gets left outside. As a result, it is generally only the maternal mitochondrial DNA, that of the egg, that gets passed on down through the generations.

Over time, mitochondrial DNA, like nuclear DNA, will undergo random mutations, slight errors when the DNA is copied from one generation to another. We can use these differences between mitochondrial DNA in different people and estimate the rate of mutation to determine how far back in time they once shared a common ancestor. We can also find out how long since they shared that ancestor that they have been traveling along their own particular branch of ancestry road.

It was using this principle that researchers in the 1980s were able to determine that the common ancestor of all mitochondrial DNA and therefore all humans living today, was a woman who lived in Africa around 100,000–200,000 years ago—mitochondrial Eve.

Tracing the last common ancestor of humans and Neanderthals would be attempted by Svante Pääbo of the Max Plank Institute in Germany. For this study, they would use an old friend, Neanderthal 1—the type specimen of the species. Even if DNA is present, there are problems, though. Most of the DNA will be fragmentary, and the vast majority of the DNA you would find will be from contaminants from soil bacteria plus healthy doses from scientists and curators who have studied the specimen over the years.

Taking these contaminants into account, though, Svante's team finally isolated Neanderthal mitochondrial DNA and then compared it to around 1000 mitochondrial DNA sequences of modern humans. Among the moderns, the DNA differed by about 8 mutations, but between the moderns and the Neanderthal, the average was around 26. This would place the common ancestor of Neanderthals and modern humans around 500,000 years ago. Perhaps this guy, *Homo heidelbergensis*, who was known to have lived in Europe and Africa at about that time, is close to the split of the Neanderthal and modern lineages.

Some research based on tooth structure has suggested that the split occurred even further back in history, perhaps as much as 1 million years ago. This places it in the range of ancestral humans like *Homo erectus*. This ancient divergence was taken by some to suggest that Neanderthals were a completely separate species to humans and that with around half a million years or more of genetic drift between the two populations, it would be very unlikely that a human and a Neanderthal could breed to produce living offspring, let alone offspring that might be viable enough to pass on that genetic mixing into the next generation.

There had been intriguing fossil finds, though, that suggested some interbreeding had occurred. For example, the Lapedo child found in the limestone Canyon of Lagar Velho in central Portugal. This skeleton of a human child was found in 1998, buried with red ochre and pierced shells dating to about 24,500 before the present day. Although looking like a modern human, the skeleton contained some rather Neanderthal features. Was this a hybrid? Evidence of gene mixing between *sapiens* and *neanderthalensis*. Or was this just a stocky child?

Had there been inbreeding despite the antiquity of the last common ancestor of the two species? To answer this, we would have to find some way to read the nuclear DNA of Neanderthals, the blueprint of the organism held in the cell's nucleus. This was the quest for the Neanderthal genome project.

This is something, which we have only just recently achieved for our own species in the Human genome project, an international effort to read our own human blueprint. But just what is a genome? Basically, a genome is all the genes that themselves are composed of DNA, the famous replicating molecule that code for an organism. The code itself is written in an alphabet of just 4 letters made up of 4 molecules 8–10 atoms wide. These nucleobases are adenine, thymine, cytosine, and guanine.

The nucleobases exist as base pairs with A bonding with T, and C bonding with G. The order in which these letters appear on the DNA strand code for specific amino acids and thereby proteins in the cell. The most common way to determine the sequence of these letters is to use an enzyme called DNA polymerase, an enzyme that synthesizes DNA in our cells.

To sequence a piece of DNA, the enzyme is used to generate a new strand of DNA, but using nucleotides that have been tagged fluorescently This fluorescent signal can be read, the signal varying depending on which nucleotides were used. Short overlapping segments are read and from these a longer sequence is constructed.

The human genome project was initiated in 1990 and completed in April of 2003, an incredible step forward in understanding who we are at our very core. It will hopefully provide new insights into diseases like cancer that have plagued us for so many years

But what of the Neanderthal genome? This would be a more challenging endeavor. There would be all the problems with contamination and degradation of the DNA that we mentioned earlier. There would also be the added issue of distinguishing Neanderthal DNA from a very close relative—us. These were the challenges facing Svante Pääbo's team at the Max Plank Institute when the Neanderthal Genome Project was initiated in July of 2006.

Some of the first specimens to be studied came from the femurs of three female Neanderthals found in the Vindija cave in northern Croatia, dated to about 38,000 years old. The bones were well preserved, suggesting that DNA might be recovered. Using enzymes that eliminated bacterial DNA, the team was able to concentrate the Neanderthal DNA by about 5 times.

Eventually, a Neanderthal genome was sequenced with an initial draft published in the journal *Science* in 2010. This now allowed for comparisons to be made between humans and Neanderthals at the level of their DNA It was concluded that around 98.7% of the base pairs in the two genomes were identical.

It was from this that comparisons between the FOXP2 gene in humans and Neanderthals could be made that we mentioned earlier—but something else too.

In comparing the Neanderthal genome to people from different racial groupings they found around 1–4% of Neanderthal DNA is in our genome. The only way it could have got there is via interbreeding. Despite the distance in time that separates humans from Neanderthals from their last common ancestor, it would appear we were still sufficiently similar to produce viable offspring.

There is more to be gleaned from the data, though. No Neanderthal DNA has been found in Africans, suggesting that as modern humans migrated out of Africa, they met and occasionally bred with Neanderthals that they came across. Not everyone has the same part of the genome, but in total it is thought that around 30–40% of the Neanderthal genome is floating around in the human population. This indicates that was not one Neanderthal ancestor, but there must be an entire history of Neanderthal–human interactions.

As I said, any non-African today has between 1–4% of Neanderthal DNA, but a discovery by researchers, including Svante Pääbo from the Max Planck Institute, reported in the journal *Nature* in 2015 provides data from an individual much closer to the original interaction. A specimen from a cave in Romania dated to 37,000–42,000 before present, has been found with between 6–9% Neanderthal DNA, probably indicating a fully Neanderthal ancestor just 6 generations back.

Today Neanderthal DNA is found to be more common in certain places in our genome than others. For example, around 80% of Eurasians have a Neanderthal keratin gene, possibly advantageous to those members of our African-originating species who were moving into colder environments. In addition to skin toughening, we may have also received a boost to our immune systems, resistance to certain diseases that Neanderthals had battled for generations before they met modern humans.

It's not just good things we get from this interbreeding, though. It's possible that certain undesirable traits have been introduced into the mix including lupus, Crohn's disease, and an increased chance of developing Type 2 diabetes. In the case of diabetes, this could be the result of what was originally a useful trait.

Neanderthals were hunter-gatherers living through times of plenty, but also scarcity There would be a definite advantage for individuals to have an ability to gorge themselves when food was available to make up for time when food was scarce. Useful back then but dangerous today in the west, where food is plentiful and available 24–7. It is important to remember though that the overall negative impact of our Neanderthal DNA is probably very, very small, indeed.

This then is one possible picture of our messy genome that was published by Chris Stringer of the Natural History Museum in London in the journal *Nature* in 2012. Humans and Neanderthals diverged from a common heidelbergensis ancestor with Neanderthals generating another ancient human group, the Denisovans. Each of these forms included an archaic flow back into the *Homo sapiens* population, contributing in varying degrees some small parts of the modern human genome we see today.

So what happened to the Neanderthals? The forerunners of the species arrived in Europe about 800,000 before present. *Homo sapiens* arrives at about 40,000 before present, and just 10,000 years later the Neanderthals are gone. This could suggest that they were outcompeted for resources or a darker view, were hunted down and exterminated in acts of interspecies genocide.

Another possible explanation may relate to rapid climate change. By about 55,000 years ago climate started to fluctuate rapidly from extremely cold to mild within a few decades perhaps within the lifetime of an individual. This would have caused rapid changes in the landscape from woodlands to grasslands with familiar plants and animals appearing and disappearing. It has been suggested that early humans during this period would have had more widespread social networks This would have allowed them to acquire resources over a greater area. Neanderthals in this model became increasingly isolated and starved of resources They finally become extinct at about 41,000–39,000 years before the present, at the start of a very cold snap.

But perhaps there is another explanation. As modern humans moved into the Neanderthal's territory, perhaps the Neanderthals were just absorbed by interbreeding. Or perhaps there was a combination of all three-part confrontation, part climate-and-resource-related and part genetic assimilation. For myself, I am happy to think that at least genetically a part of those ancient peoples are still with us today.

Lecture 24

Paleontology and the Future of Earth

In this final lecture, you will consider what the future holds for our species and what role paleontology plays in this inquiry. The future of paleontology is perhaps universal. With new exoplanets being found around stars every year, who knows where a future fossil hunter may tread. And the Smithsonian's National Museum of Natural History is a hotbed of cutting-edge research, charting the history of the Earth system through time. It is in places like this, and others around the world, that the keys to the past, present, and future will be cast—and, with them, perhaps a more secure future for us all.

Future Changes to the Earth System

- The likely causes of change to the Earth system in the future are difficult to predict. The continents will continue to move about our planet as they have for billions of years. Reconstructions like those of Christopher Scotese from the University of Texas at Arlington predict that a new supercontinent will form in about 1/4 of a billion years from now, part of a supercontinent cycle.

- But printed on the slow movements of the continents are rapid events. Consider Yellowstone, a volcano that is so vast that it is difficult to appreciate that it is actually a volcano from the ground. Yellowstone is a caldera that measures 60 by 32 kilometers that last erupted massively about 640,000 years ago. Yellowstone is "breathing," with the caldera floor rising and falling probably in response to the developing magma chamber below. A major eruption at Yellowstone would likely be catastrophic for human civilization, but even with the park's ups and downs, it is thought it is unlikely to erupt any time soon.

- Other sudden events can strike from the skies rather than rising from the depths. Paleontologists and geologists have been studying the violent end of the Mesozoic biosphere 66 million years ago for decades. The general consensus is that the Earth was hit by a 10-kilometer-diameter object that caused such catastrophic environmental change that it drove the Mesozoic biosphere, including the dinosaurs, into extinction.

- Recently, we have been reminded that impacts from space, like major volcanic events, are still a very reasonable part of our collective future. On February 15, 2013, at around 9:20 am, a bright light was seen streaking above the skies of Russia. The event was captured by closed-circuit television and dashboard cameras all over the southern Ural region.

- It was caused by a 20-meter-diameter object entering the atmosphere at a speed of about 40,000 miles per hour. As a result of its high speed and angle of entry, the object exploded 18.4 miles above the ground, generating a shower of fragments and a significant shock wave. The energy released in this event was equivalent to 500 kilotons of TNT, about 20 to 30 times more energy than the bomb detonated at Hiroshima.

- The event damaged more than 7000 buildings in 6 cities. About 1500 people received injuries, mostly from broken glass, that required medical treatment.

- If this object had hit the ground, it would have generated a substantial crater and probably more damage and potential fatalities—not a mass-extinction-level event, but it certainly shows that the days of impacts from space, just like supervolcanoes, are far from over.

- Another important vector of change in both our short- and long-term future is climate change. The now-famous Keeling curve that shows the yearly changes in carbon dioxide at Mauna Loa, Hawaii, beautifully reflects seasonal variations in the uptake of carbon dioxide by plants over a year. The most startling feature of the data is the increase in carbon dioxide levels over time.

- These results placed in the context of a deeper-time perspective show how current changes in carbon dioxide levels cannot be explained by natural ice age cycles. The increase in global temperature will impact sea level, climatic patterns, species distributions, and a whole range of other potential effects that we cannot yet predict. There is a scientific consensus on climate change that the climate is, in fact, changing and that the changes are in large part caused by human activities.

- We are also witnessing a drop in global biodiversity. As with extinctions, there is often a trigger followed by a cascade of events that cause the actual extinction. Today, as in the past, the cause of these extinctions is varied—including climate change, habitat loss, and pollution—and the trigger this time appears to be us.

- With all these changes to the Earth system, there has been a call by some geologists and paleontologists for the erection of a new geological period, the Anthropocene, to reflect these changing times. In 2008, the International Commission on Stratigraphy received a proposal to make the Anthropocene a unit of the geological timescale.

- There is some debate as to when and how the actual boundary will be drawn. Some favor an "early" boundary around the Neolithic revolution, which saw the transition of many human lifestyles from hunting and gathering to one of agriculture about 12,000 years ago. This, though, kind of places it in conflict with a preexisting period, the Holocene, which begins about 11,700 B.P. and continues to the present day.

- Others favor a more recent definition based on atmospheric evidence of the start of the Industrial Revolution in the late 18th century. It has been suggested that the Thomas Newcomen steam engine, the first practical use of coal and steam power used to drain mines, marks the start of the current changes we are witnessing. That would place the base of the Anthropocene sometime in 1712.

The Relevance of Paleontology

- With all these changes occurring in the present, why should we be dreaming about the distant past? Paleontology is obviously engaging, and it is still important to industry, particularly in the context of biostratigraphy for mineral and hydrocarbon exploration. Paleontology is also vital in these times of rapid change in the Earth system. Paleontology, and the wider disciplines of geology and biology on which it rests, are the only areas of science that provide context to current changes.

- The biosphere has moved through many interesting times. It teaches us that change is perhaps one of the few constants that we can rely on in Earth's history. Change always comes tapping on the door, even after millions of years of stability.

- By understanding the vectors of slow changes in climate caused by the shifting continents—or sudden catastrophic change caused by extreme volcanic activity, or impacts from space—and by recording the reaction of the biosphere to these changes as preserved in rocks and fossils, we can grasp at that vital context we need to fully appreciate current changes in the Earth system.

- For example, it is accepted that we are in a downward trend with regard to biodiversity—a trend of extinctions that is greater than the usual background rate of extinction that we would expect in the biosphere. The vital question is whether the current trend is much less, the same, or in excess of the loss of species that we know occurred in the big 5 past extinction events. This is a fundamentally important question, and one that only paleontology and biology can really answer.

- Our understanding of the rates and extent of the big 5 mass extinctions are continually being refined and updated as new techniques, geological sections, and fossils come to light. A fine example of this is the end-Permian extinction, which, until relatively recently, was a mystery. Now we are gaining insights into the triggers, cascade of ensuing events, timing, and extent of this "mother of all extinctions."

- In 2015, a team of scientists, including biologist Paul Ehrlich of Stanford University in California, suggested that we may be moving toward a mass extinction event. The study considered past rates of extinctions so that they could be compared to current changes in biodiversity. To address criticisms that previous studies about current extinction rates were flawed, they selected what is considered to be a very high estimate of background extinctions between the big 5 mass extinctions.

- The results they generated suggest that current extinction rates may be 100 times higher than the assumed, probably artificially high, background rate. With their calculated extinction rate, we should have seen around 9 vertebrate extinctions since 1900, but 468 more documented vertebrate extinctions have been recorded, including 69 mammals, 80 birds, 24 reptiles, 146 amphibians, and 158 fish. And we may be unaware of other species going extinct because they have vanished before we could find them.

- If these estimates are correct, then these are disturbing trends. The study claims, though, that with conservation and proper environmental management, these trends could be reversed. Paleontology has a vital role to play in these efforts. It speaks to the time and pattern and recovery of ecosystems after extinction and also to the minimum level of biodiversity required to maintain a healthy biosphere. It also provides a warning regarding how rapidly the biosphere can be plunged into crisis.

The Future of Paleontology

- Paleontology has an important role to play in the understanding of the past, present, and future of our planet. We have probably only uncovered a minute fraction of the wealth of information that is held in the Earth's crust. You only have to scan through a paleontology journal to see the rate at which new finds are challenging and changing our views of the history of planet Earth. This, in combination with new techniques for studying fossils, is going to open new perspectives of our planet and its 4.54-billion-year journey through time.

- But is there another direction for the science? Perhaps that step has already been made. The supposed bacteria-like fossils described in 1996 on the Martian meteorite ALH84001 have remained elusive, with no definitive evidence that these structures are the product of a Martian biology. Even so, this did start to take paleontology beyond our home planet; for the first time, paleontology had to seriously consider the possibility of fossil life from another world.

- More recent research regarding the possibility of fossils on Mars was published in 2014 by Nora Noffke of Old Dominion University in Norfolk, Virginia. Noffke is an expert in ancient microbially induced sedimentary structures (MISS), not unlike the stromatolites that are produced by biofilms.

- A particular set of photographs taken by the Curiosity rover caught her attention. They were taken in an area called Yellowknife Bay in a dry lake bed that may have gone through seasonal flooding around 3.7 billion years ago. She noticed particular domes, cracks, and pitted structures that looked like structures that she has seen before in ancient MISS from western Australia. She admits that it is possible that these features could have been produced by erosional processes of salt, water, or wind—but they do strongly resemble ancient MISS from Earth.

- Determining the biogenicity of ancient MISS on Earth is difficult enough, so making determinations from photographs of Mars is speculative, but still intriguing. The only way to sort this out for sure would be to go there. This is why the first mission that takes humans to Mars will most likely include a geologist—at the very least, a geologist with a lot of paleontology and biology under his or her belt. Whether or not these structures turn out to be fossils, paleontology will likely be one of the key tools in determining the possibility of life elsewhere in the universe.

Questions to consider:

1. Has technology made *Homo sapiens* immune to mass extinctions?
2. Are we still evolving?

Suggested Reading:

D'Arcy Wood, *Tambora*.

Keller and DeVecchio, *Natural Hazards*.

Lecture 24 Transcript: Paleontology and the Future of Earth

The Winter of 1815 was fairly standard. It wasn't particularly warm or particularly cold. In England the snow fell, the trees were bare, and no doubt people would stare out of their windows waiting for spring to arrive. The problem with the winter of 1815, though, was it just failed to end.

1816 became known as the year without a summer. Average global temperatures dropped around 0.4–0.7°C. New York State, Maine, and New Hampshire had a record snow fall in June, with Quebec City getting a fall of 30 cm.

There were late frosts and crop failures in the United States, Canada, and Europe. The growing Season of New Hampshire would drop from 120 to 60 days. In Ireland, wheat, oats, and potato harvests failed and with this came famine and an associated outbreak of typhoid that killed around 60,000 people.

The weather would be driven to extremes. High rainfall in some areas and drought in others. In Switzerland, the months of April through August, some 152 days, record 130 days of rain.

In India, the Monsoon was disrupted leading to floods and droughts. Three harvests would fail, and famine was close on the heels of the crop failure. It is thought that these famines would see the development of new strains of typhoid and cholera, causing more misery and death.

The 1816 year without a summer would have other effects too. Consider this famous painting, *Chichester Canal* by J. M. M. Turner—beautiful isn't it? But it is just one example of the many fantastic sunsets when the sun was actually visible that would be painted by artists that year.

This strange summer didn't just impact painting, though. With the exception of sunsets, days were pretty gloomy, and this would act as literary inspiration for some. For example, Lord Byron and his poem "Darkness" in 1816.

One Mary Shelly was traveling with Byron and other friends during June through Switzerland. As we noted, Switzerland was rained out in that summer so with nothing else to do but stare out of the window at the rain; they decided to have a writing contest to see who could come up with the scariest story. Mary came up with a touching little Gothic piece called *Frankenstein*.

In general, the climatic gloom, failure of crops, and the spread of famine and disease were taken by many to be caused by immorality and low church attendance. Many thought that the Apocalypse was just around the corner.

What could have caused such a cascade of events and drive so many people into famine and disease? A clue comes from Sir Stamford Raffles, a British statesman best known for the founding of Singapore, who of course has lent his name to a very famous hotel in that city, where if you have the inclination—I did—and not just once, you can indulge in the original Singapore Sling. He also lends his name to Rafflesia, a giant flower with a rotting flesh stench; that is considerably less appealing that a Singapore Sling.

In his memoir, Raffles reports hearing explosions while he was on Java on April 5, 1815, that were at first mistaken for the sound of a distant cannon.

The source of the detonations was not man-made, though; they came from the island of Sumbawa, and specifically the eruption of Mount Tambora. On April 10, the eruption reached its climax with explosions heard over 1600 miles away on the island of Sumatra.

In total, over 138 cubic miles of material were ejected by the eruption. For comparison, the more famous eruption of Krakatoa in 1883 would only release 6 cubic miles of material. The Tambora eruption would kill around 88,000 in pyroclastic flows and tsunamis. It is also the cause of the year without summer.

It would achieve this through processes we have met before in this series. The ejection of fine ash into the atmosphere would dim sunlight. Large volumes of sulfur dioxide would generate various sulfate aerosols and sulfuric acid that would further dim the planet and contribute to the destruction of ozone.

These effects were compounded by the fact that the eruption occurred during a period of low solar output, a Dalton minimum, which caused shifts in weather patterns. Deep troughs developed from the poles, causing monsoons to fail, and droughts and floods as well. Ash and aerosols would paint lurid sunsets in the sky. Estimates vary, but it is thought that hundreds of thousands of people would probably lose their lives as a consequence of the climatic effects of Tambora.

Tambora was one of the greatest volcanic events in recorded human history and had devastating effects far beyond Southeast Asia. Even so, there was no mass extinction recorded at this time. Geologically, this was a minor event. In this final episode, I would like to ask what does the future hold for our species, and what role does paleontology have to play in this inquiry?

So let's ask. Have *Homo sapiens* already nearly suffered extinction? What are the likely causes of change to the Earth system in the future? Why is paleontology still relevant? And just where is paleontology going?

Have you ever noticed how different we all look? Different facial features, skin colors, statures? A tapestry of humanity all very different—a staggering diversity.

You might be surprised to learn that some claim there may be more genetic variation in a troupe of chimps than there is in all our species.

Some take that even further. If you consider the genetic variation between two individuals in neighboring tribes in East Africa, you will find more genetic variation between them than between, for example, any person from Western Europe and anyone from Southeast Asia—counterintuitive right?

This idea comes from the studies of mitochondrial DNA the little power units within our cells that have their own independent genetic material that can be used to trace human lineages through time It has been suggested that the mitochondrial DNA records may indicate that the population of *Homo sapiens* dwindled to around 2000–5000 individuals living in East Africa around 50,000–100,000 years ago. This is called a human population bottleneck and was claimed to account for the low genetic difference that exists between human beings.

So how do bottlenecks work under this hypothesis? Consider human population before the bottleneck event—as represented by this bottle with colored marbles—consisting of diverse and varied genetic populations

Now take that bottle and shake out some of the marbles—as you will see, not all the marbles will get out—this is our bottleneck event.

If we now place the marbles that escaped the bottleneck into another and allowed them to repopulate, we produce a population that is now less genetically diverse than the pre-bottleneck population. This represents the current East African population.

Now take a sample of that remaining East African diversity. This now represents the fraction of that genetic diversity that would spread through the rest of the world, which is why all other people outside of East Africa would have much less genetic diversity than the population in East Africa.

It has been proposed that this is how we end up with such a restricted human diversity today. But what was the bottleneck—what was it that could have caused such a reduction in the diversity of our species?

For a possible culprit, take a trip to the poles. A sample of any material that ends up in the atmosphere ultimately gets preserved in the snow and ice at the poles. By studying gas bubbles trapped in the ice, we can look back in time through climate history for some 800,000 years. What we find is a spike in sulfuric acid representing 2 billion tons of sulfur dioxide, concentrations that that can only be volcanic in origin.

For comparison's sake consider the eruption of Pinatubo in the Philippines in 1991. This released 22 million tons of sulfur dioxide into the atmosphere, causing a global temperature drop of around about 0.5°C. Although this might sound like a minor change, it is thought to have halted global warming for 1 year. Obviously, if we have 2 billion rather than 22 million tons of sulfur dioxide that suggests something much larger than Pinatubo.

We do have a candidate, and for that, we look to the Indonesian island of Sumatra and the Lake Toba volcano that erupted 75,000 years ago.

If you were to travel to Toba, you would be hard pressed to find evidence of a volcano—you will find the lake, of course, but no familiar volcanic cone This is because Lake Toba is the volcano—or at least the shattered remnants of one.

To get a handle on the sheer scale of the Toba eruption, it is helpful to consider what is known as the Volcanic Explosivity Index. This is a kind of a volcanologists Richter scale that considers total volume of material erupted explosively. The scale is ranked from 0 through to 8 with each point on the scale from 1 upwards representing 10 times more material ejected than the previous point on the scale.

A classic example of a VEI-1 eruption would be Kilauea on Hawaii. Eruptions here are what are called effusive. The basaltic magma that produces these volcanoes is fairly runny, in part due to its low silica content, which permits gases to escape easily and generally emerge nonviolently. It produces features such as lava fountains and sometimes rapidly moving lava flows.

Viscosity is critical to the character of a volcano. Consider these three liquids are demonstrating increasing viscosity or stickiness. Relatively milk would be low viscosity, followed by honey, and then peanut butter.

A basalt would be on the milky end, if you want, of this relative scale, but some magmas are definitely of the peanut butter variety, a result of their higher silica content. Gas dissolved in the high viscosity magma cannot escape as easy as it can in low viscosity magma like we see in Hawaii.

So what happens to all that gas in a volcano with viscous magma? As the magma rises through the volcanic feeder pipe, its pressure drops, which allows the gas to come out of solution—all at once. The gas bubbles grow in an instant and when they touch they fragment the magma in an explosive eruption. This is how volcanic ash is formed.

This is the type of eruption that occurred on Sunday, May 18, 1980, at Mount Saint Helen's. The event generated an eruption column that rose 15 miles into the atmosphere and deadly pyroclastic flows of hot ash and gas traveling at 670 miles an hour along the ground surface for over 23 miles. But this is still only at 5 on the VEI scale.

For a VEI of 6 think Vesuvius during A.D. 79. This was a similar type of eruption to Mount Saint Helen's, but with 10 times more material explosively erupted the destruction of Pompeii and Herculaneum would see around 16,000 fatalities. Krakatoa, the largest noise in recent human history, is also VEI-6.

So what of Toba? Well, have a look at this diagram representing the relative volume of material produced in volcanic events. Here is Saint Helen's at VEI-5. At VEI-6, examples include Pinatubo in 1991, Krakatoa in 1883, and Vesuvius during the early Roman Empire. Then there is Tambora in 1815 the eruption that caused a year without a summer– 10 times greater than Vesuvius at VEI-7 and what of lake Toba? Well, Toba is VEI-8 and would erupt over 2,800 km3 of material. This is why Toba is known as a supervolcano.

Toba produced pyroclastic flows that covered 18,000 km^2 and generated an ash cloud 80 km high. Some have claimed that Toba would have sunk the planet into a global winter over 6 years long. Could it be this that nearly wiped our species from the surface of the planet?

Similar genetic bottlenecks have been suggested for other species such as cheetahs and tigers, the Bornean orangutan, certain gorilla populations, and the central Indian macaque, around 70,000 to 55,000 years ago.

But this remains a controversial hypothesis, And more recent research has cast doubt on the link between Toba and proposed human bottlenecks. For example, there have been problems with the timing of the proposed volcanic eruption and the onset of environmental changes in East Africa. Researchers have identified material from Toba in the sediments of Lake Malawi, indicating that the fallout from Toba traveled a considerable distance. However, there is no indication of significant climate change following the eruption.

Other significant problems in the hypothesis have come from archeological stone tools in India that have been recovered from both below and directly above the Toba ash deposits in that country. Not only do the stone tools suggest a continuity of habitation, but there is also a continuity of technology as well.

It is also important to note that other hominid populations—*Homo floresiensis* and *Homo neanderthalensis*—successfully made it through the Toba catastrophe.

Although popular in the late 90s and early 2000s, many have now moved on from a bottleneck in relation to the Toba eruption, demonstrating how new discoveries are continually shaping our insight into Earth's past, even challenging and dismissing sometimes hypotheses that might appear very neat and appealing at the time. What the Toba eruption does highlight, though, is the fact that catastrophic events are not just a feature of the deep past.

Our planet is a continually evolving dynamic system where things generally happen at a fairly, well, geological pace most of the time. This slow background pace is peppered with rapid events that can cause dramatic and sometimes catastrophic events as we have seen in this series. Even if Toba did not have a significant impact on the numbers of humans 75,000 years ago, if Toba were to occur today with all our vast population reliant on limited food resources, it could be catastrophic for human civilization.

So what are the likely causes of change to the Earth system in the future? Obviously, this is difficult to predict. The continents will continue to move about our planet as they have for billions of years. Reconstructions like those of Christopher Scotese from the University of Texas at Arlington, predict that a new supercontinent will form in about 1/4 of a billion years from now, part of a supercontinent cycle like the beating of a heart.

But as we have seen printed on the slow movements of the continents are rapid events like Toba, a volcano which itself is far from extinct with indications that the magma chamber is currently refilling. Although, as there is currently no surface movement it would be a long time probably before we see another eruption.

Toba is not the only supervolcano candidate, of course. Closer to home we have Yellowstone, another volcano that is so vast that it is difficult to appreciate that it is actually is a volcano from the ground. Yellowstone is a caldera 60 by 32 km across, which last erupted in a VE1-8 about 640,000 years ago. Yellowstone, unlike Toba, is breathing with the caldera floor rising and falling, probably in response to the developing magma chamber below. Like Toba, a major eruption at Yellowstone would likely be catastrophic for human civilization, but even with the parks ups and downs, it is thought it is unlikely to erupt anytime soon.

Other sudden events can strike from the skies rather than rising from the depths. Paleontologists and geologists have been studying the violent end of the Mesozoic biosphere 66 million years ago for decades now. The general consensus is that the Earth was hit by a 10 km diameter object that caused such catastrophic environmental change that it drove the Mesozoic biosphere, including the dinosaurs, into extinction

Just recently we have been reminded that impacts from space, like major volcanic events, are still a very reasonable part of our collective future. On 15th of February, 2013 at around 9:20 am, a bright light was seen streaking above the skies of Russia. The event was captured by CCTV and dashboard cameras all over the southern Ural region.

It was caused by a 20 m diameter object entering the atmosphere at a speed at around 40,000 mph. As a result of its high speed and angle of entry, the object exploded 18.4 miles above the ground, generating a shower of fragments and a significant shock wave. The energy released in this event was equivalent to 500 kilotons of TNT that's around 20–30 times more energy than the bomb detonated at Hiroshima.

The event damaged over 7000 buildings in six cities. About 1500 people received injuries mostly from broken glass that required medical treatment. If this object had hit the ground, it would have generated a substantial crater and probably more damage and potential fatalities; not a mass-extinction-level event, but it certainly shows that the days of impacts from space, just like super volcanos are far from over.

Another important vector of change in both our short and long-term future is climate change. The now famous Keeling curve that shows the yearly changes in carbon dioxide at Mauna Loa in Hawaii beautifully reflects seasonal variations in the uptake of carbon dioxide by plants over a year. Of course, the most startling feature of the data is the increase in carbon dioxide levels over time.

These results placed in the context of a deeper time perspective show how current changes in carbon dioxide levels cannot be explained by natural ice age cycles. The increase in global temperature will impact sea level, climatic patterns, species distributions, and a whole range of other potential effects that we cannot yet predict. There is a scientific consensus on climate change, stating that climate is changing and that these changes are in large part caused by human activities.

We are also witnessing a drop in global biodiversity. As with the extinctions, we have studied in this series there is often a trigger followed by a cascade of events that cause the actual extinction. Today, as in the past, the cause of these extinctions is varied; climate change, habitat loss, and pollution to name but a few—and the trigger—well the trigger, this time, appears to be us.

With all these changes to the Earth system, there has been a call by some geologists and paleontologists for the erection of a new geological period, the Anthropocene, to reflect these changing times. In 2008, the International Commission on Stratigraphy received a proposal to make the Anthropocene a unit of the geological time scale.

There is some debate as to when and how the actual boundary will be drawn. Some favor an early boundary around the Neolithic revolution, which saw the transition of many human lifestyles from hunter-gathering to one of agriculture about 12,000 years ago. This, though, kind of places it in conflict with a pre-existing period—the Holocene—that begins about 11,700 years ago and continues to the present day.

Others favor a more recent definition based on atmospheric evidence of the start of the Industrial Revolution in the late 18th century. It has been suggested that the Thomas Newcomen steam engine, the first practical use of coal and steam power used to drain mines, marks the very start of the current changes we are witnessing. That would place the base of the Anthropocene sometime in 1712.

So with all these changes occurring in the present, why should we be dreaming about the distant past? Paleontology is obviously engaging. It has been called the gateway science by many who see fossils as a hook to intrigue children about the wider world of science. It is also still very important to industry, particularly in the context of biostratigraphy for mineral and hydrocarbon exploration. But is there any other reason why our tax dollars should be spent on paleontology other than it's fascinating.

I personally think "it's fascinating" is a good enough reason on its own but then I would, wouldn't I? But I believe paleontology is also vital in these times of rapid change in the Earth system. Paleontology and the wider disciplines of geology and biology on which it rests are the only areas of science that can provide context to current changes.

James Hutton, the father of modern geology, in his paper concerning the theory of the Earth in 1778 concludes, "The result, therefore, of our present inquiry is, that we find no vestige of a beginning—no prospect of an end."

This idea that the past could be interpreted by understanding processes going on in the present would allow for a revolution in the investigation and interpretation of rocks and fossils. As famous 18th-century geological superstar, Charles Lyell, who was heavily influenced by Hutton, would state, "The present is the key to the past."

But we can turn this around, can't we? For as we have seen in this series, the biosphere has moved through many interesting times. It teaches us that change is perhaps one of the few constants that we can rely on in Earth's history. As we have seen, change always comes tapping on the door even after millions of years of stability.

So perhaps we can paraphrase Lyell, perhaps the past could also be the key to the future. By understanding the vectors of slow changes in climate caused by the shifting continents or sudden catastrophic change caused by extreme volcanic activity or impacts from space, and by recording the reaction of the biosphere to these changes as preserved in rocks and fossils, we can grasp at that vital context we need to fully appreciate current changes in the Earth system.

For example, as we mentioned earlier, it is accepted that we are in a downward trend with regards to biodiversity—a trend of extinctions that is greater than the usual background rate of extinction that we would expect in the biosphere. The vital question is whether the current trend is far less, the same or in excess of the loss of species that we know occurred in the big five past extinction events? This is a fundamentally important question and one which only paleontology and biology can really answer.

Our understanding of the rates and extent of the big five mass extinctions are continually being refined and updated as new geological techniques, sections, and fossils come to light. A fine example of this is the end-Permian extinction, which until relatively recently was a relative mystery. Now we are gaining insights into the triggers, the cascade of ensuing events, the timing and the extent of this mother of all extinctions.

In 2015, a team of scientists, including biologist Paul Ehrlich of Stanford University in California, suggested that we may be moving towards a mass extinction event. The study considered past rates of extinctions so that they could be compared to current changes in biodiversity. To address criticisms that previous studies about current extinction rates were flawed, they selected what is considered to be very a high estimate of background extinctions between the big five mass extinctions.

The results they generated suggest that current extinction rates may be 100 times higher than the assumed, probably artificially high, background rate. With their calculated extinction rate, we should have seen around 9 vertebrate extinctions since 1900, but 468 more documented vertebrate extinctions have been recorded including 69 mammals, 80 birds, 24 reptiles, 146 amphibians, and 158 fish. Of course, we may be unaware of other species going extinct because they have vanished before we could find them.

If these estimates are correct, then these are disturbing trends. The study claims, though, that with conservation and proper environmental management these trends could be reversed. I think paleontology has a vital role to play in these efforts. It speaks to the time and pattern and recovery of ecosystems after extinction, and also to the minimum level of biodiversity required to maintain a healthy biosphere. It also provides a warning as we noted for the end-Permian extinction regarding how rapidly the biosphere can be plunged into crisis.

Ernest Rutherford was claimed to have said, "All science is either physics or stamp collecting." I hope I have demonstrated how untrue that is for paleontology While the classification and taxonomy of fossils is vital and underpins paleontology, the science that we do now upon that bedrock is both amazing and fundamentally important. But what of the future? Where will paleontology go?

As I have said, paleontology has an important role to play in the understanding of the past, present and future of our planet. We have probably only uncovered a minute fraction of the wealth of information that is held in the Earth's crust. You only have to scan through a paleontology journal to see the rate at which new finds are challenging and changing our views of the history of planet Earth. This, in combination with new techniques for studying fossils, is going to open up new perspectives of our planet and its 4.54 billion-year journey through time.

But is there another direction for the science? Perhaps that step has already been made? The supposed bacteria-like fossils described in 1996 on the Martian meteorite ALH8001 have remained elusive with no definitive evidence that these structures are the product of a Martian biology. Even so, this really did start to take paleontology beyond our home planet. For the first time, paleontology had to seriously consider the possibility of fossil life from another world.

More recent research regarding the possibility of fossils on Mars was published in 2014 by Nora Noffke of the Old Dominion University in Norfolk Virginia. Nora is an expert in ancient microbially-induced sedimentary structures or MISS, not unlike the stromatolites produced by biofilms that we mentioned in an earlier episode

A particular set of photographs taken by the Curiosity Rover caught her attention. They were taken in an area called Yellowknife Bay in a dry lake bed that may have gone through seasonal flooding around 3.7 billion years ago.

What she noticed were particular domes, cracks, and pitted structures that looked very familiar. Structures that she had seen before in ancient MISS structures from western Australia. She admits that it is possible that these features could have been produced by erosional processes of salt, water or wind, but they do strongly resemble ancient MISS from Earth.

To be honest, determining the biogenicity of ancient microbially-induced sedimentary structures on Earth is difficult enough, so making determinations from photographs of Mars is speculative, but still intriguing The only way to sort this out for sure would be to go there. This is why the first mission taking humans to Mars will most likely include a geologist, and at the very least a geologist with a lot of paleontology and biology under their belts. Whether or not these structures turn out to be fossils, paleontology will likely be one of the key tools in determining the possibility of life elsewhere in the universe.

So what is the future of paleontology? Perhaps universal. With new exoplanets being found around stars every year who knows where a future fossil hunter may tread. And the future of museums like the Smithsonian's National Museum of Natural History? Far from an institute of paleontological stamp collectors, the museum is a hotbed of cutting-edge research charting the history of the Earth system through time.

It is in places like this and others around the world that the keys to the past, present, and the future will be cast and with them perhaps a more secure future for us all.

Bibliography

Armstrong, Howard, and Martin Brasier. *Microfossils*. Hoboken: Wiley-Blackwell, 2005. A good textbook that covers all of the major microfossil groups.

Beerling, David. *The Emerald Planet: How Plants Changed Earth's History*. Oxford: Oxford University Press, 2008. An interesting account of how plants have influenced the evolution of the Earth system.

Benton, Michael J., and David A. T. Harper. *Introduction to Paleobiology and the Fossil Record*. Hoboken: Wiley-Blackwell, 2009. A standard text in paleontology that includes useful information on the processes of fossilization and the history of paleontology.

Blunt, Wilfrid. *Linnaeus: The Compleat Naturalist*. Princeton, NJ: Princeton University Press, 2002. An interesting review of the life and works of Carl Linnaeus.

Bryson, Bill. *A Short History of Nearly Everything*. London: Penguin Random House, 2003. A very amusing guide to the history of "everything" from the big bang onward, including some of the characters in the development of the science of geology and paleontology.

Carwardine, Mark. *Smithsonian Handbooks: Whales, Dolphins, and Porpoises*. London: DK Penguin Random House, 2002. A useful pocket guide to help you identify modern whales.

Chesterman, Charles W. *The Audubon Society Field Guide to North American Rocks and Minerals*. New York: Knopf, 1994. A useful guide to minerals you might find in North America.

Conway-Morris, Simon. *The Crucible of Creation: The Burgess Shale and the Rise of Animals*. Oxford: Oxford University Press, 1998.An interesting overview of the discovery, research, and fossils found in the Burgess Shale, including the arthropods.

Corfield, Richard. *The Silent Landscape: The Scientific Voyage of* HMS Challenger. Washington, DC: Joseph Henry Press, 2003. A book describing the discoveries of the *HMS Challenger*, the ship responsible for discovering the Challenger Deep.

D'Arcy Wood, Gillen. *Tambora: The Eruption That Changed the World.* Princeton, NJ: Princeton University Press, 2014. A description of the eruption of Mount Tambora in 1815 and the consequences it had for the entire planet.

Emling, Shelley. *The Fossil Hunter: Dinosaurs, Evolution, and the Woman Whose Discoveries Changed the World.* New York: St. Martin's Griffin, 2011. A detailed account of the life, times, and scientific contributions of Mary Anning.

Erwin, Douglas. *Extinction: How Life on Earth Nearly Ended 250 Million Years Ago.* Princeton, NJ: Princeton University Press, 2006. An overview of how Earth suffered the greatest biological crisis in its history at the end of the Permian.

Erwin, Douglas, and James Valentine. *The Cambrian Explosion: The Construction of Animal Biodiversity.* Englewood, CO: Roberts and Company, 2013. Exploring the roots of the Cambrian explosion.

Fedonkin, Mikhail, James G. Gehling, Kathleen Grey, Guy M. Narbonne, and Patricia Vickers-Rich. *The Rise of Animals: Evolution and Diversification of the Kingdom Animalia.* Baltimore: The Johns Hopkins University Press, 2007. An exploration of the earliest animal life on the planet prior to the Cambrian explosion.

Fichman, Martin. *An Elusive Victorian: The Evolution of Alfred Russel Wallace.* Chicago: University of Chicago Press, 2010. A biography of a much forgotten naturalist (and contemporary of Darwin), Alfred Russel Wallace, who formulated many of his ideas in and around Southeast Asia.

Foote, Michael, and Arnold I. Miller. *Principles of Paleontology*. New York: W. H. Freeman and Company, 2007. A more advanced textbook in paleontology that includes details about systematics and taxonomy in paleontology.

Fortey, Richard. *Earth: An Intimate History*. New York: Vintage, 2005. A very readable overview of Earth history.

———. *Trilobite: Eyewitness to Evolution*. New York: Vintage Books, 2001. A great review of this most important extinct group of arthropods.

Goulson, Dave. *A Sting in the Tale: My Adventures with Bumblebees*. New York: Picador, 2013. An interesting read about one of today's common flower visitors, the bumblebee.

Hazen, Robert M. "The Evolution of Minerals." *Scientific American* 3 (2010): 58–65. Robert Hazen's description of the evolution of Earth's rich mineral assemblages.

Keary, Philip, Keith A. Klepeis, and Frederick J. Vine. *Global Tectonics*. Hoboken: Wiley-Blackwell, 2009. A more advanced text on plate tectonic theory.

Keller, Edward A., and Duane E. DeVecchio. *Natural Hazards: Earth's Processes as Hazards, Disasters, and Catastrophes*. Upper Saddle River, NJ: Pearson, 2015. A text covering the various natural disasters facing our planet today and into the future.

Knell, Simon J. *The Great Fossil Enigma: The Search for the Conodont Animal*. Bloomington: Indiana University Press, 2013. This book investigates the discovery of the animal that was ultimately found to be responsible for conodont microfossils.

Knoll, Andrew. *Life on a Young Planet: The First Three Billion Years of Evolution*. Princeton, NJ: Princeton University Press, 2003. An investigation into the biosphere's origins and evolution up to the Cambrian explosion of multicellular life.

Lanham, Url. *The Bone Hunters: The Heroic Age of Paleontology in the American West*. New York: Dover Publications, 2011. A review of the early years of dinosaur hunting in North America.

Levin, Harold L., and David T. King Jr. *The Earth Through Time*. Hoboken: Wiley, 2016. A good introduction to some of the basics of geology and geological time.

Lister, Adrian, and Paul Bahn. *Mammoths: Giants of the Ice Age*. Berkeley: University of California Press, 2007. A review of mammoths and mastodons, their paleobiology, and their extinction.

Lockley, Martin. *Tracking Dinosaurs: A New Look at an Ancient World*. Cambridge: Cambridge University Press, 1991. An interesting look at how trace fossils can be used to help reveal the paleobiology of dinosaurs.

Long, John A., Michael Archer, Timothy Flannery, and Suzanne Hand. *Prehistoric Mammals of Australia and New Guinea: One Hundred Million Years of Evolution*. Baltimore: Johns Hopkins University Press, 2003. A review of the wonderful prehistoric mammals of Australia.

Molnar, Ralph E. *Dragons in the Dust: The Paleobiology of the Giant Monitor Lizard* Megalania. Bloomington: Indiana University Press, 2004. A portrait of the Komodo dragon's larger Australian cousin.

Pääbo, Svante. *Neanderthal Man: In Search of Lost Genomes*. New York: Basic Books, 2014. An account of how the Neanderthal genome was sequenced.

Papagianni, Dimitra, and Michael A. Morse. *The Neanderthals Rediscovered: How Modern Science Is Rewriting Their Story*. London: Thames and Hudson, 2013. A review of hypotheses regarding Neanderthals.

Pim, Keiron. *Dinosaurs: The Grand Tour—Everything Worth Knowing about Dinosaurs from Aardonyx to Zuniceratops*. New York: The Experiment, 2014. A good introduction to the diversity of dinosaurs for both children and adults.

Plummer, Charles, Diane Carlson, and Lisa Hammersley. *Physical Geology*. New York: McGraw-Hill Education, 2016. Good general textbook on geology that covers some of the basics of continental drift and plate tectonics.

Savage, Candace. *Prairie: A Natural History*. Vancouver: Greystone Books, 2011. An interesting guide to the biology and ecology of the prairies of North America.

Seilacher, Adolf. *Trace Fossil Analysis*. New York: Springer, 2007. A detailed review of the interpretation and use of trace fossils.

Shapiro, Beth. *How to Clone a Mammoth: The Science of De-Extinction*. Princeton, NJ: Princeton University Press, 2015. A fascinating look into the possibility of "de-extinction."

Stringer, Chris. *Lone Survivors: How We Came to Be the Only Humans on Earth*. New York: Times Books, 2012. An interesting view on the origins of our species.

Sues, Hans-Dieter, and Nicholas C. Fraser. *Triassic Life on Land: The Great Transition*. New York: Columbia University Press, 2010. A detailed review (aimed at professionals and students of paleontology) of life on land during the Triassic period.

Tattersall, Ian. *Masters of the Planet: The Search for Our Human Origins*. New York: St. Martin's Press, 2012. A more general review of human evolution.

Taylor, Paul. *DK Eyewitness Books: Fossil*. London: Penguin Random House, 2004. A nicely illustrated introductory guide to fossils and fossil collecting.

Thewissen, J. G. M. "Hans." *The Walking Whales: From Land to Water in Eight Million Years*. Oakland: University of California Press, 2014. A nicely illustrated review of the evolution of whales.

Thompson, Ida. *The Audubon Society Field Guide to North American Fossils*. New York: Knopf, 1982. A useful guide to fossils you might find in North America.

Winchester, Simon. *The Map That Changed the World: William Smith and the Birth of Modern Geology*. New York: Harper Collins, 2001. An excellent book that details the many trials and tribulations of William Smith and his use of fossils as timepieces.

Internet Resources

Paleogeographic and Tectonic History of Western North America. http://cpgeosystems.com/wnampalgeog.html. Maps detailing the evolution of western North America, starting 280 million years ago, by Professor Ron Blakey.

PALEOMAP Project. http://www.scotese.com. Reconstructions of the configurations of the continents, covering around 650 million years of Earth's history.

U.S. Department of the Interior: Bureau of Land Management. "Hobby Collection." http://www.blm.gov/wo/st/en/prog/more/CRM/paleontology/fossil_collecting.html. A useful resource for those wishing to start their own fossil collections.

Image Credits

Page 9: © metha1819/Shutterstock.

Page 25: © JuliusKielaitis/Shutterstock.

Page 48: ©LuFeeTheBear/Shutterstock.

Page 67: © netsuthep/Shutterstock.

Page 87: © Georgios Kollidas/Shutterstock.

Page 93: © leonello/iStock/Thinkstock.

Page 109: © Nuttapong/Shutterstock.

Page 129: © BarryTuck/Shutterstock.

Page 148: © Billion Photos/Shutterstock.

Page 169: © Comstock Images/Stockbyte/Thinkstock.

Page 195: © larslentz/iStock/Thinkstock.

Page 230: © Merlin74/Shutterstock.

Page 235: © alslutsky/Shutterstock.

Page 314: © Fresnel/Shutterstock.

Page 338: © Christian Musat/Shutterstock.

Page 356: © WathanyuSowong/Shutterstock.

Page 361: © BrianPIrwin/Shutterstock.

Page 378: © JaySi/Shutterstock.

Page 398: © Mario Susilo/Shutterstock

Page 403: © Ethan Daniels/Shutterstock.

Page 418: © Jakub Krechowicz/Shutterstock.

Page 420: © StanOd/Shutterstock.

Page 440: © creativemarc/Shutterstock.

Page 460–461: © Sebastian Kaulitzki/Shutterstock.

Page 483: ©Microgen/Shutterstock.